LSE Mathematical Series

Calculus

This book is about the calculus of several variables. It also contains an
introduction to elementary differential and difference equations. The
emphasis is on practical problem-solving rather than the proof of formal
theorems. Many worked examples are supplied as well as problems for the
student to solve, together with their solutions. The techniques are illustrated
with applications drawn chiefly from economics, statistics and operational
research. Some elementary knowledge of the calculus of one variable is
assumed but revision material is supplied throughout the text.

A confident approach to problem-solving is not possible without some
understanding of the background theory. In this book the theory is presented systematically but informally. Wherever possible, geometric arguments
are used and the text is illustrated with numerous diagrams. Particular care
has been taken to make the main body of the text suitable for students
who are studying independently of a taught course.

The book will interest students at universities and other higher education
institutions. At the London School of Economics, the course on which
this book is based is attended by students reading for a variety of different
degrees and with a wide disparity in their previous levels of mathematical
training. Some are graduates and some are first-year undergraduates. It is
hoped that this book will attract a similar audience: not only of economists,
statisticians and other social scientists but also physical scientists, engineers
and mathematicians.

This series is to consist of a number of interlocking and overlapping books based on the London School of Economics' teaching programme in mathematical methods. The aim is to provide an integrated course in mathematical methods: from very elementary material for those who know only a little mathematics to advanced material suitable for research workers. The London School of Economics is a school of the social sciences and the books have been written with the needs of economists, statisticians, management scientists and accountants in mind but the books will also be found useful by engineers, physical scientists and mathematicians.

The emphasis throughout is on 'how to do it'. Many practical examples are provided as well as problems for the student to solve, together with their solutions. The necessary theory is described in an informal manner with explanations presented in geometric terms wherever possible.

It is initially planned to include four books: one providing basic mathematics for economists; two higher-level books suitable for a wider audience on calculus and linear algebra respectively; and a book on more advanced mathematical methods.

Calculus

K. G. BINMORE

Professor of Mathematics
London School of Economics

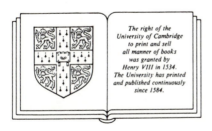

The right of the
University of Cambridge
to print and sell
all manner of books
was granted by
Henry VIII in 1534.
The University has printed
and published continuously
since 1584.

CAMBRIDGE UNIVERSITY PRESS

Cambridge
New York   Port Chester
Melbourne   Sydney

Published by the Press Syndicate of the University of Cambridge
The Pitt Building, Trumpington Street, Cambridge CB2 1RP
32 East 57th Street, New York, NY 10022, USA
10 Stamford Road, Oakleigh, Melbourne 3166, Australia

First published 1983
Reprinted 1986 1989

Printed in Great Britain at the
University Press, Cambridge

Library of Congress catalogue card number: 82–19728

British Library cataloguing in publication data

Binmore, K.G.
Calculus.–(LSE mathematics series)
1. Calculus
I. Title   II. Series
515   QA303

ISBN 0 521 24771 3 hard covers
ISBN 0 521 28952 1 paperback

AL

Contents

Preface

At the London School of Economics, a whole spectrum of courses on mathematical methods is given ranging from very elementary courses for those who know virtually no mathematics to courses for those researching in mathematical economics and the like. It is our intention to produce a number of books which cover the material in these courses. Each of the books will be an independent entity but it is our hope that the whole will prove greater than the sum of its parts.

The current book is a 'second-level' work on calculus. The main topic is the calculus of several variables but elementary differential and difference equations are also treated at some length. The emphasis is very strongly on 'how to do it' aspects of these topics rather than their theoretical basis. However, there is little point in learning formulae by rote (except in so far as this helps in passing examinations set by rote). To use a technique in practice it is necessary to have some understanding of why it works. We therefore supplement the description of the various techniques with brief explanations of their theoretical background. But formal proofs are never attempted and, wherever possible, geometrical arguments are used.

At the end of each chapter, a variety of applications are given. These are drawn from economics, statistics and operational research reflecting our interests at LSE. However, the mathematical techniques described in the book are, of course, far more widely applicable and we hope that the book will be found useful not only by those studying mathematics with a view to applications in the social sciences but also by physical scientists and engineers. Returning to the applications given at the end of each chapter, these are quite advanced compared with the general level of the text. The aim has been to generate some excitement about the potentialities of the mathematical techniques rather than to usurp the role of those teaching applied courses. We have therefore included material on such topics as the duality theorem of linear programming, the Kuhn–Tucker theorem, the Slutsky equations, the cobweb model and a wide range of statistical topics including the central limit theorem and the gambler's ruin problem. Students who find the mathematical techniques described in the main body of the text troublesome to grasp would be advised not to try and puzzle out the details of these applications but to come back to them when they are encountered again at a later stage. Scientists and engineers, incidentally, might well benefit from observing the scope of these applications in the social sciences.

It is assumed that readers of this book will have some previous knowledge of the calculus of functions of one real variable. It is quite unsuitable for those with no knowledge whatsoever of this subject. On the other hand, the material treated in this book is taught at LSE to an exceedingly disparate group of students from all over the world. Some of these know very little. Others are graduate students brushing up their knowledge. We have therefore found it necessary to provide a substantial amount of revision material on topics which it would be better for readers to know properly before starting on this book. It is surprising how large the holes can be in the knowledge even of those whose previous mathematical education is entirely orthodox. Our experience has been that it is unprofitable to place this revision material on the calculus of one variable in a block at the beginning of the book. The temptation to neglect it altogether is then almost irresistible. It has therefore been somewhat slyly interwoven with the main body of the text in the hope that all readers will at least skim through it before moving on to new topics.

Some knowledge of linear algebra is also assumed. However, the level of understanding required is not a high one and sections explaining the basic ideas are included where appropriate. But it should be appreciated that these brief passages are not intended as a substitute for a course in linear algebra. (At LSE, students take a concurrent course in linear algebra while studying the material covered in this book.) As in the case of the calculus of one variable, those who are totally ignorant of the subject should begin with a more elementary book than this.

Sections which are intended as revision material and hence survey the ideas covered rather than explain them are marked with the symbol ¤. These sections should at the very least be scanned quickly to make sure the notation and techniques are all familiar. Certain other sections are marked with the symbol †. These should be omitted altogether by those who find the text difficult to cope with.

Finally, attention should be drawn to the examples and problems. When studying a 'how to do it' book, the criterion of success is whether or not one has learned 'how to do it'. Thus, a reader should count himself successful if and only if he is able to solve a substantial percentage of the problems which are given. This remains the case even for those who are not too sure whether they understand the foundations of the subject. When a formal subject like mathematics is presented informally as in this book, it is inevitable that all but those who are unusually gifted will have doubts about their grasp of the theory. Those who wish to dispel their doubts should consult a book in which the emphasis is on theoretical matters and proofs are given in a formal and precise way (e.g. the author's book *Mathematical Analysis: a straightforward approach*, also published by CUP). But, as far as the current book is concerned, a reader would be wise to accept that his understanding of the basic theory must be reasonably good if he can solve most of the problems since someone with little grasp of the theory would make no headway with the problems at all.

In any case, it is quite pointless to attempt to read this book without making a commitment to tackle the problems. Certainly far more time should be spent on attempting problems than on reading the text. To assist in this task, solutions are given at the end of the book to every other problem. Those for which no solution is given are marked with the symbol *. The usual (but not inevitable) pattern is that a starred question follows a rather similar unstarred question. In attempting a starred question it will therefore often be helpful to begin by first trying the preceding unstarred question and then consulting the solution given for this should this prove necessary. Obviously, however, little will be gained if the solution is consulted prematurely.

K. G. Binmore

1

Vectors and matrices

This book takes for granted that readers have some previous knowledge
of the calculus of real functions of one real variable and also some knowledge of linear algebra. However, for those whose knowledge may be rusty
from long disuse or raw with recent acquisition, sections on the necessary
material from these subjects have been included where appropriate. Although
these revision sections (marked with the symbol ¤) are as self-contained
as possible, they are *not* suitable for those who are entirely ignorant of
the topics covered. The material in the revision sections is surveyed rather
than explained. It is suggested that readers who feel fairly confident of
their mastery of this surveyed material scan through the revision sections
quickly to check that the notation and techniques are all familiar before
going on. Probably, however, there will be few readers who do not find
something here and there in the revision sections which merits their close
attention.

The current chapter is concerned with the fundamental techniques from
linear algebra which we shall be using. This will be particularly useful for
those who may be studying linear algebra concurrently with the present
text.

Algebraists are sometimes neglectful of the geometric implications of their
results. Since we shall be making much use of geometrical arguments, particular attention should therefore be paid to §1.16–§1.21, inclusive, in
which the geometric relevance of various vector notions is explained. This
material will be required almost immediately in chapter 2. The remaining
material will not be needed until chapter 4. Those who are not very confident
of their linear algebra may prefer leaving §1.31 until then.

1.1 Matrices¤

A matrix is a rectangular array of numbers. We usually enclose the array in
brackets as in the examples below:

$$A = \begin{pmatrix} 4 & 1 \\ 0 & -1 \\ 3 & 2 \end{pmatrix}; \qquad B = \begin{pmatrix} 1 & 0 & -1 \\ 2 & 1 & 0 \end{pmatrix}.$$

A matrix with m rows and n columns is called an $m \times n$ matrix. Thus A is a 3×2 matrix and B is a 2×3 matrix.

The numbers which appear in a matrix are called *scalars*. Sometimes it is useful to allow the scalars to be complex numbers but our scalars will always be *real numbers*.

1.2 Scalar multiplication[¤]

One can do a certain amount of algebra with matrices and in this and the next few sections we shall describe the mechanics of some of the operations which are possible.

The first operation we shall consider is called *scalar multiplication*. If A is an $m \times n$ matrix and s is a scalar, then sA is the $m \times n$ matrix obtained by multiplying each entry of A by s. For example,

$$2A = 2\begin{pmatrix} 4 & 1 \\ 0 & -1 \\ 3 & 2 \end{pmatrix} = \begin{pmatrix} 2 \times 4 & 2 \times 1 \\ 2 \times 0 & 2 \times -1 \\ 2 \times 3 & 2 \times 2 \end{pmatrix} = \begin{pmatrix} 8 & 2 \\ 0 & -2 \\ 6 & 4 \end{pmatrix}.$$

Similarly,

$$5B = 5\begin{pmatrix} 1 & 0 & -1 \\ 2 & 1 & 0 \end{pmatrix} = \begin{pmatrix} 5 & 0 & -5 \\ 10 & 5 & 0 \end{pmatrix}.$$

1.3 Matrix addition and subtraction[¤]

If C and D are two $m \times n$ matrices, then $C + D$ is the $m \times n$ matrix obtained by adding corresponding entries of C and D. Similarly, $C - D$ is the $m \times n$ matrix obtained by subtracting corresponding entries. For example, if

$$C = \begin{pmatrix} 1 & -1 & 0 \\ -2 & 3 & 1 \\ 4 & 1 & 0 \end{pmatrix} \text{ and } D = \begin{pmatrix} 2 & 1 & 5 \\ -1 & -3 & 4 \\ -3 & 2 & 1 \end{pmatrix}$$

then

$$C + D = \begin{pmatrix} 1+2 & -1+1 & 0+5 \\ -2-1 & 3-3 & 1+4 \\ 4-3 & 1+2 & 0+1 \end{pmatrix} = \begin{pmatrix} 3 & 0 & 5 \\ -3 & 0 & 5 \\ 1 & 3 & 1 \end{pmatrix}$$

and

$$C - D = \begin{pmatrix} 1-2 & -1-1 & 0-5 \\ -2+1 & 3+3 & 1-4 \\ 4+3 & 1-2 & 0-1 \end{pmatrix} = \begin{pmatrix} -1 & -2 & -5 \\ -1 & 6 & -3 \\ 7 & -1 & -1 \end{pmatrix}.$$

Note that

$$C + C = \begin{pmatrix} 1+1 & -1-1 & 0+0 \\ -2-2 & 3+3 & 1+1 \\ 4+4 & 1+1 & 0+0 \end{pmatrix} = \begin{pmatrix} 2 & -2 & 0 \\ -4 & 6 & 2 \\ 8 & 2 & 0 \end{pmatrix} = 2C$$

and that

$$C - C = \begin{pmatrix} 1-1 & -1+1 & 0-0 \\ -2+2 & 3-3 & 1-1 \\ 4-4 & 1-1 & 0-0 \end{pmatrix} = \begin{pmatrix} 0 & 0 & 0 \\ 0 & 0 & 0 \\ 0 & 0 & 0 \end{pmatrix}.$$

The final matrix is called the 3 × 3 *zero matrix*. We usually denote any zero matrix by 0. This is a little naughty because of the possibility of confusion with other zero matrices or with the scalar 0. However, it has the advantage that we can then write

$$C - C = 0$$

for any matrix C.

Note:
It makes *no* sense to try and add or subtract two matrices which are not of the same shape. Thus, for example,

$$A + B = \begin{pmatrix} 4 & 1 \\ 0 & -1 \\ 3 & 2 \end{pmatrix} + \begin{pmatrix} 1 & 0 & -1 \\ 2 & 1 & 0 \end{pmatrix}$$

is an entirely *meaningless* expression.

1.4   Matrix multiplication[¤]

If A is an $m \times n$ matrix and B is an $n \times p$ matrix, then A and B can be multiplied to give an $m \times p$ matrix AB.

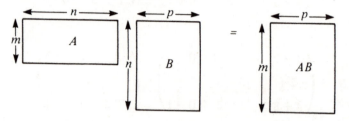

To work out the entry c of AB which appears in its jth row and kth column, we require the jth row of A and the kth column of B as illustrated below.

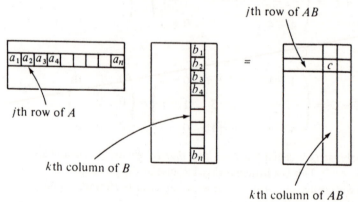

The entry c is then given by

$$c = a_1 b_1 + a_2 b_2 + a_3 b_3 + \ldots + a_n b_n.$$

Example 1.5.[¤]
We compute the product AB of the matrices

$$A = \begin{pmatrix} 0 & 1 & 2 \\ 2 & 0 & 1 \end{pmatrix}; \qquad B = \begin{pmatrix} 1 & 0 \\ 2 & 1 \\ 0 & 2 \end{pmatrix}.$$

Since A is a 2 × 3 matrix and B is a 3 × 2 matrix, their product AB is a 2 × 2 matrix.

$$AB = \begin{pmatrix} 0 & 1 & 2 \\ 2 & 0 & 1 \end{pmatrix} \begin{pmatrix} 1 & 0 \\ 2 & 1 \\ 0 & 2 \end{pmatrix} = \begin{pmatrix} a & b \\ c & d \end{pmatrix}.$$

To calculate c, we require the second row of A and the first column of B. These are indicated in the diagram below:

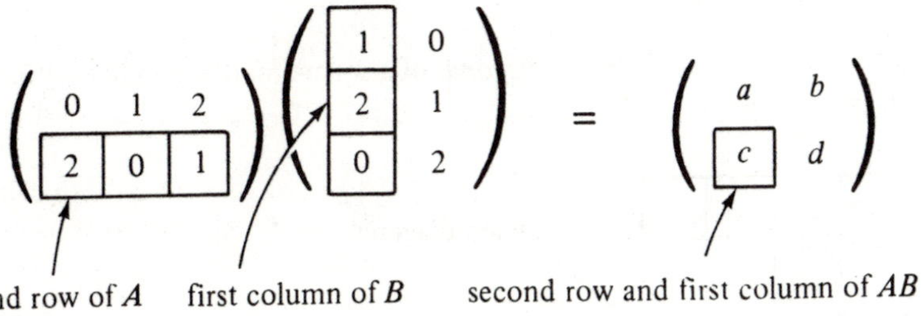

second row of A first column of B second row and first column of AB

We obtain that

$$c = 2 \times 1 + 0 \times 2 + 1 \times 0 = 2 + 0 + 0 = 2.$$

Similarly,

$$a = 0 \times 1 + 1 \times 2 + 2 \times 0 = 0 + 2 + 0 = 2$$
$$b = 0 \times 0 + 1 \times 1 + 2 \times 2 = 0 + 1 + 4 = 5$$
$$d = 2 \times 0 + 0 \times 1 + 1 \times 2 = 0 + 0 + 2 = 2.$$

Thus

$$AB = \begin{pmatrix} 0 & 1 & 2 \\ 2 & 0 & 1 \end{pmatrix} \begin{pmatrix} 1 & 0 \\ 2 & 1 \\ 0 & 2 \end{pmatrix} = \begin{pmatrix} 2 & 5 \\ 2 & 2 \end{pmatrix}.$$

Note:
It makes *no* sense to try and calculate AB unless the number of columns in A is the same as the number of rows in B. Thus, for example,

$$\begin{pmatrix} 1 & 0 \\ 2 & 1 \\ 0 & 2 \end{pmatrix} \begin{pmatrix} 0 & 1 & 2 \\ 2 & 0 & 1 \\ 2 & 1 & 3 \end{pmatrix}$$

is an entirely *meaningless* expression.

1.6   Identity matrices ⊓

An $n \times n$ matrix is called a *square matrix* for obvious reasons. Thus, for example,

$$A = \begin{pmatrix} 1 & 2 & 3 \\ 3 & 1 & 2 \\ 2 & 3 & 1 \end{pmatrix}$$

is a square matrix. The main *diagonal* of a square matrix is indicated in the diagram below:

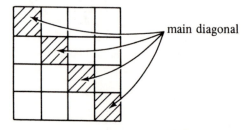 main diagonal

   The $n \times n$ *identity matrix* is the $n \times n$ matrix whose main diagonal entries are all 1 and whose other entries are all 0. We usually denote an identity matrix by I. The 3×3 identity matrix is

$$I = \begin{pmatrix} 1 & 0 & 0 \\ 0 & 1 & 0 \\ 0 & 0 & 1 \end{pmatrix}.$$

Note that an identity matrix *must* be square. Just as a zero matrix behaves like the number 0, so an identity matrix behaves like the number 1. Specifically, we have that, if A is an $m \times n$ matrix, B is an $n \times p$ matrix and I is the $n \times n$ identity matrix, then

$$AI = A \quad \text{and} \quad IB = B.$$

   Examples 1.7⊓

$$\text{(i)} \begin{pmatrix} 0 & 1 & 2 \\ 2 & 0 & 1 \end{pmatrix} \begin{pmatrix} 1 & 0 & 0 \\ 0 & 1 & 0 \\ 0 & 0 & 1 \end{pmatrix} = \begin{pmatrix} 0 & 1 & 2 \\ 2 & 0 & 1 \end{pmatrix}$$

$$\text{(ii)} \begin{pmatrix} 1 & 0 & 0 \\ 0 & 1 & 0 \\ 0 & 0 & 1 \end{pmatrix} \begin{pmatrix} 1 & 0 \\ 2 & 1 \\ 0 & 2 \end{pmatrix} = \begin{pmatrix} 1 & 0 \\ 2 & 1 \\ 0 & 2 \end{pmatrix}.$$

1.8 Determinants[¤]

With each *square* matrix there is associated a scalar called the *determinant* of the matrix. We shall denote the determinant of the square matrix A by $\det(A)$ or by $|A|$. (There is some risk of confusing the latter notation with the modulus or absolute value of a real number. Note that the determinant of a square matrix may be plus *or* minus.)

The general definition of a determinant is rather complicated and we therefore shall only explain how to calculate the determinants of 1×1, 2×2 and 3×3 matrices.

(i) $\underline{1 \times 1 \text{ matrices.}}$ A 1×1 matrix $A = (a)$ is just a scalar and $\det(A) = a$.

(ii) $\underline{2 \times 2 \text{ matrices.}}$ The determinant of the 2×2 matrix

$$A = \begin{pmatrix} a & b \\ c & d \end{pmatrix}$$

is given by

$$\det(A) = \begin{vmatrix} a & b \\ c & d \end{vmatrix} = ad - bc.$$

(iii) $\underline{3 \times 3 \text{ matrices.}}$ The determinant of the 3×3 matrix

$$A = \begin{pmatrix} a & b & c \\ d & e & f \\ g & h & i \end{pmatrix}$$

is given by

$$\det(A) = \begin{vmatrix} a & b & c \\ d & e & f \\ g & h & i \end{vmatrix} = (aei + bfg + cdh) - (ceg + afh + bdi).$$

This is most easily remembered by drawing the diagram below:

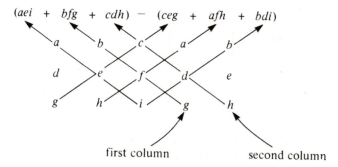

first column   second column

Examples 1.9[H]

(i) The determinant of the 1×1 matrix $A = (3)$ is simply $\det(A) = 3$.

(ii) The determinant of the 2×2 matrix

$$A = \begin{pmatrix} 1 & 2 \\ 3 & 4 \end{pmatrix}$$

is

$$\det(A) = \begin{vmatrix} 1 & 2 \\ 3 & 4 \end{vmatrix} = 1 \times 4 - 2 \times 3 = 4 - 6 = -2.$$

(iii) We find the determinant of the 3×3 matrix

$$A = \begin{pmatrix} 1 & 2 & 3 \\ 3 & 1 & 2 \\ 2 & 3 & 1 \end{pmatrix}$$

using the diagram below:

$$(1 \quad + \quad 8 \quad + \quad 27) \quad - \quad (6 \quad + \quad 6 \quad + \quad 6)$$

Thus

$$\det(A) = \begin{vmatrix} 1 & 2 & 3 \\ 3 & 1 & 2 \\ 2 & 3 & 1 \end{vmatrix} = (1 + 8 + 27) - (6 + 6 + 6) = 36 - 18 = 18$$

1.10   Inverse matrices[□]

We have dealt with matrix addition, subtraction and multiplication and found that these operations only make sense in certain restricted circumstances. The circumstances under which it is possible to *divide* by a matrix are even more restricted.

A *non-singular* matrix is a *square* matrix whose determinant is *non-zero*. Each of the matrices of example 1.9 is therefore non-singular.

> Suppose that A is a matrix. Then there is another matrix B such that
>
> $$AB = BA = I$$
>
> if and only if A is non-singular.

In fact, if A is non-singular there is precisely *one* matrix B such that $AB = BA = I$. We call this matrix B the *inverse matrix* to A and write $B = A^{-1}$.

Thus a non-singular matrix A has an inverse matrix A^{-1} which satisfies

$$AA^{-1} = A^{-1}A = I.$$

If A is an $n \times n$ matrix, then A^{-1} is an $n \times n$ matrix as well (otherwise the equation above would make no sense).

If A is *not* square or if A is square but its determinant is *zero* (i.e. A is singular), then A does *not* have an inverse in the above sense.

If B is an $m \times n$ matrix and A is an $n \times n$ *non-singular* matrix, then one can define B/A by BA^{-1}. Note, however, that such a definition of division is severely restricted in its range of application.

1.11   Transpose matrices[□]

In describing how to compute the inverse of a non-singular matrix, we shall need the idea of a transpose matrix. This is also useful in other connexions.

If A is an $m \times n$ matrix, then its *transpose* A^T is the $n \times m$ matrix whose first row is the first column of A, whose second row is the second column of A, whose third row is the third column of A and so on.

$$A = \begin{array}{|c|c|c|c|c|} \hline a_1 & b_1 & c_1 & d_1 & e_1 \\ \hline a_2 & b_2 & c_2 & d_2 & e_2 \\ \hline a_3 & b_3 & c_3 & d_3 & e_3 \\ \hline \end{array} m \qquad A^T = \begin{array}{|c|c|c|} \hline a_1 & a_2 & a_3 \\ \hline b_1 & b_2 & b_3 \\ \hline c_1 & c_2 & c_3 \\ \hline d_1 & d_2 & d_3 \\ \hline e_1 & e_2 & e_3 \\ \hline \end{array} n$$

Alternative notations for the transpose are A' or A^t.

An important special case is that when A is a square matrix for which $A = A^T$. Such a matrix is called *symmetric*.

Examples 1.12

(i)   If $A = \begin{pmatrix} 4 & 1 \\ 0 & -1 \\ 3 & 2 \end{pmatrix}$, then $A^T = \begin{pmatrix} 4 & 0 & 3 \\ 1 & -1 & 2 \end{pmatrix}$.

Note that

$$(A^T)^T = \begin{pmatrix} 4 & 0 & 3 \\ 1 & -1 & 2 \end{pmatrix}^T = \begin{pmatrix} 4 & 1 \\ 0 & -1 \\ 3 & 2 \end{pmatrix} = A.$$

(ii)   If $A = \begin{pmatrix} 1 & 3 & 5 \\ 3 & 2 & 0 \\ 5 & 0 & 4 \end{pmatrix}$, then $A^T = \begin{pmatrix} 1 & 3 & 5 \\ 3 & 2 & 0 \\ 5 & 0 & 4 \end{pmatrix}$.

Thus $A = A^T$ and so A is *symmetric*.

1.13   Cramer's rule

Cramer's rule is a method for working out the inverse of a non-singular matrix. Other methods exist but Cramer's rule is usually easiest for 1×1, 2×2 and 3×3 matrices.

In the case of 1×1 and 2×2 matrices, one might as well learn the result of using Cramer's rule by heart.

(i) 1 × 1 matrices. A 1×1 non-singular matrix $A = (a)$ is just a non-zero scalar and

$$A^{-1} = \left(\frac{1}{a}\right)$$

provided $a \neq 0$.

(ii) <u>2×2 matrices</u>. A 2×2 non-singular matrix

$$A = \begin{pmatrix} a & b \\ c & d \end{pmatrix}$$

is one for which $\det(A) = ad - bc \neq 0$. Its inverse is given by

$$A^{-1} = \frac{1}{ad - bc} \begin{pmatrix} d & -b \\ -c & a \end{pmatrix}$$

provided $ad - bc \neq 0$.

The formulae can easily be confirmed by checking that AA^{-1} and $A^{-1}A$ actually are equal to I. For example, in the 2×2 case

$$A^{-1}A = \frac{1}{ad - bc} \begin{pmatrix} d & -b \\ -c & a \end{pmatrix} \begin{pmatrix} a & b \\ c & d \end{pmatrix}$$

$$= \frac{1}{ad - bc} \begin{pmatrix} da - bc & 0 \\ 0 & -bc + ad \end{pmatrix} = \begin{pmatrix} 1 & 0 \\ 0 & 1 \end{pmatrix} = I.$$

To describe Cramer's rule in general, we need the idea of a minor.

The *minor M* corresponding to an entry m in a square matrix A is the determinant of the matrix obtained from A by deleting the row and the column containing m.

Suppose, for example, that

$$A = \begin{pmatrix} b & c & d \\ e & f & g \\ h & i & j \end{pmatrix}.$$

Then we compute the minor G corresponding to the entry g by deleting the second row and third column of A as below:

$$\begin{pmatrix} b & c & d \\ e & f & g \\ h & i & j \end{pmatrix}.$$

The minor G is then the determinant of what remains – i.e.

$$G = \begin{vmatrix} b & c \\ h & i \end{vmatrix} = bi - hc.$$

Cramer's rule gives an expression for the inverse A^{-1} of a non-singular matrix A in terms of its minors. In the case of the 3 × 3 matrix above,

$$A^{-1} = \frac{1}{\det(A)} \begin{pmatrix} +B & -C & +D \\ -E & +F & -G \\ +H & -I & +J \end{pmatrix}^T$$

provided $\det(A) \neq 0$. If

$$A = \begin{pmatrix} b & c & d & e \\ f & g & h & i \\ j & k & l & m \\ n & p & q & r \end{pmatrix}$$

then

$$A^{-1} = \frac{1}{\det(A)} \begin{pmatrix} +B & -C & +D & -E \\ -F & +G & -H & +I \\ +J & -K & +L & -M \\ -N & +P & -Q & +R \end{pmatrix}^T$$

provided $\det(A) \neq 0$.

Thus, each entry of A is replaced by the corresponding minor and the sign is then altered according to the checker-board pattern illustrated below:

$$\begin{pmatrix} + & - & + & - & + & \cdots \\ - & + & - & + & - & \cdots \\ + & - & + & - & + & \cdots \\ \vdots & & & & & \end{pmatrix}$$

The inverse matrix A^{-1} is then obtained by multiplying the *transpose* of the result by the *scalar* $\{\det(A)\}^{-1}$.

Examples 1.14□

We compute the inverses of the non-singular matrices of examples 1.9

(i) $A = (3);$ $\det(A) = 3 \neq 0;$

      $A^{-1} = (\tfrac{1}{3})$

(ii) $A = \begin{pmatrix} 1 & 2 \\ 3 & 4 \end{pmatrix};$ $\det(A) = -2 \neq 0;$

$$A^{-1} = \frac{1}{-2}\begin{pmatrix} 4 & -2 \\ -3 & 1 \end{pmatrix} = \begin{pmatrix} -2 & 1 \\ \tfrac{3}{2} & -\tfrac{1}{2} \end{pmatrix}$$

(iii) $A = \begin{pmatrix} 1 & 2 & 3 \\ 3 & 1 & 2 \\ 2 & 3 & 1 \end{pmatrix};$ $\det(A) = 18 \neq 0.$

We begin by replacing each entry of A by the corresponding minor and obtain

$$\begin{pmatrix} -5 & -1 & 7 \\ -7 & -5 & -1 \\ 1 & -7 & -5 \end{pmatrix}.$$

Next, the signs are altered according to the checker-board pattern

$$\begin{pmatrix} + & - & + \\ - & + & - \\ + & - & + \end{pmatrix}$$

to yield the result

$$\begin{pmatrix} -5 & +1 & +7 \\ +7 & -5 & +1 \\ +1 & +7 & -5 \end{pmatrix}$$

We now take the transpose of this matrix – i.e.

$$\begin{pmatrix} -5 & 1 & 7 \\ 7 & -5 & 1 \\ 1 & 7 & -5 \end{pmatrix}^{T} = \begin{pmatrix} -5 & 7 & 1 \\ 1 & -5 & 7 \\ 7 & 1 & -5 \end{pmatrix}.$$

Finally, the inverse is obtained by multiplying by the scalar $\{\det(A)\}^{-1} = \tfrac{1}{18}$.

Thus

$$A^{-1} = \frac{1}{18}\begin{pmatrix} -5 & 7 & 1 \\ 1 & -5 & 7 \\ 7 & 1 & -5 \end{pmatrix}.$$

Exercises 1.15[¤]

1.[¤] The matrices A, B, C and D are given by

$$A = \begin{pmatrix} 6 & 1 & 0 \\ 1 & 0 & 3 \end{pmatrix}; \quad B = \begin{pmatrix} 3 & 1 \\ 1 & 0 \\ 0 & 2 \end{pmatrix};$$

$$C = \begin{pmatrix} 1 & 2 & 3 \\ 4 & 1 & 2 \\ 3 & 5 & 1 \end{pmatrix}; \quad D = \begin{pmatrix} 1 & 2 & 1 \\ 2 & 1 & 2 \\ 3 & 3 & 3 \end{pmatrix}.$$

Which of the following expressions make sense? Where the expressions make sense, calculate the matrix which they represent.

(i) $2D$	(ii) $A + B$	(iii) $B - C$
(iv) $C + D$	(v) $2C - 3D$	(vi) $2A - 3D$
(vii) AB	(viii) BA	(ix) AC
(x) CA	(xi) BC	(xii) CB
(xiii) CD	(xiv) DC	(xv) $\det(A)$
(xvi) $\det(B)$	(xvii) $\det(C)$	(xviii) $\det(D)$
(xix) $\det(AB)$	(xx) $\det(CD)$	(xxi) A^T
(xxii) B^T	(xxiii) C^T	(xxiv) D^T.

2.[¤] The matrices A, B, C and D are given by

$$A = \begin{pmatrix} 2 & 1 \\ 4 & 3 \end{pmatrix}; \quad B = \begin{pmatrix} 1 & 2 \\ 2 & 4 \end{pmatrix};$$

$$C = \begin{pmatrix} 1 & 2 & 3 \\ 4 & 1 & 2 \\ 3 & 5 & 1 \end{pmatrix}; \quad D = \begin{pmatrix} 1 & 2 & 1 \\ 2 & 1 & 2 \\ 3 & 3 & 3 \end{pmatrix}.$$

Which of these square matrices are non-singular? Find the inverse matrices of those which are non-singular and check your answers by multiplying the inverses and the matrices from which they were obtained.

3.[¤] Even if the expressions AB and BA both make sense, it is *not* necessarily true that $AB = BA$. Check this fact using the matrices A and B of question 2.

4.[⌐] It is *always* true that $A(BC) = (AB)C$ provided that one side or the other makes sense. Check this result in the case when

$$A = (2, 4); \quad B = \begin{pmatrix} 1 & 2 \\ 3 & 4 \end{pmatrix}; \quad C = \begin{pmatrix} 3 \\ 5 \end{pmatrix}.$$

5.[⌐] It is *always* true that $(AB)^{-1} = B^{-1}A^{-1}$ provided that A and B are $n \times n$ non-singular matrices. It is also *always* true that $(AB)^T = B^T A^T$ provided that AB makes sense. Check these results in the case when

$$A = \begin{pmatrix} 2 & 1 \\ 4 & 3 \end{pmatrix}; \quad B = \begin{pmatrix} 1 & 2 \\ 3 & 4 \end{pmatrix}.$$

6.[⌐] Let A, B and C be $n \times n$ matrices with the property that $AB = I$ and $CA = I$. Prove that $B = C$.

1.16   Vectors[⌐]

An n-dimensional *row vector* is a $1 \times n$ matrix – i.e. an object of the form

$$(x_1, x_2, x_3, \ldots, x_n)$$

in which $x_1, x_2, \ldots, x_n$ are real numbers. An n-dimensional *column vector* is an $n \times 1$ matrix – i.e. an object of the form

$$\begin{pmatrix} x_1 \\ x_2 \\ \vdots \\ x_n \end{pmatrix}$$

We shall use the notation $\mathbf{x}$ to indicate a column vector. If

$$\mathbf{x} = \begin{pmatrix} x_1 \\ x_2 \\ \vdots \\ x_n \end{pmatrix}$$

then the real numbers $x_1, x_2, \ldots, x_n$ are called the *co-ordinates* or *components* of the vector $\mathbf{x}$. Note that it is often more economical of space to use the transpose notation of §1.11 and write $\mathbf{x} = (x_1, x_2, \ldots, x_n)^T$.

The definitions of matrix addition and scalar multiplication given in §1.2 and §1.3 apply to vectors. Thus

$$\mathbf{x} + \mathbf{y} = \begin{pmatrix} x_1 + y_1 \\ x_2 + y_2 \\ \cdot \\ \cdot \\ \cdot \\ x_n + y_n \end{pmatrix} ; \quad \alpha\mathbf{x} = \begin{pmatrix} \alpha x_1 \\ \alpha x_2 \\ \cdot \\ \cdot \\ \cdot \\ \alpha x_n \end{pmatrix} .$$

These definitions have a simple geometric interpretation which is easiest to illustrate in the case $n = 2$.

A two-dimensional vector $\mathbf{x} = (x_1, x_2)^T$ may be thought of as a point in the plane referred to rectangular Cartesian axes. Alternatively, one can think of $\mathbf{x}$ as an arrow with its blunt end at the origin and its sharp end at the point $(x_1, x_2)^T$.

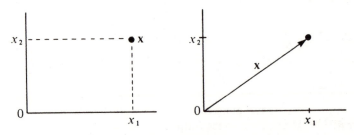

Vector addition and scalar multiplication are illustrated in the diagrams below. For obvious reasons, the rule for adding two vectors is called the *parallelogram law*.

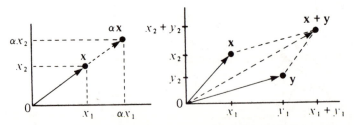

When thinking of a vector as an arrow, it is not always convenient to draw the blunt end at the origin. There is no harm in placing the blunt end elsewhere provided that one is willing to agree that *all* arrows with the same length and direction represent the *same* vector.

It is, for example, often quite useful to represent the vector $\mathbf{y} - \mathbf{x}$ as an arrow with its blunt end at $\mathbf{x}$.

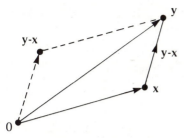

Examples 1.17

Let $\mathbf{x} = (1, 2, 3)^T$ and $\mathbf{y} = (2, 0, 5)^T$. Then

$$\text{(i)} \quad \mathbf{x} + \mathbf{y} = \begin{pmatrix} 1 \\ 2 \\ 3 \end{pmatrix} + \begin{pmatrix} 2 \\ 0 \\ 5 \end{pmatrix} = \begin{pmatrix} 3 \\ 2 \\ 8 \end{pmatrix}$$

$$\text{(ii)} \quad 2\mathbf{x} = 2\begin{pmatrix} 1 \\ 2 \\ 3 \end{pmatrix} = \begin{pmatrix} 2 \\ 4 \\ 6 \end{pmatrix}.$$

1.18 Length and distance

The *length* or *norm* of an n-dimensional vector $\mathbf{x}$ is defined by

$$\|\mathbf{x}\| = \{x_1^2 + x_2^2 + \ldots + x_n^2\}^{1/2}.$$

When $n = 2$, this definition is motivated by Pythagoras's theorem.

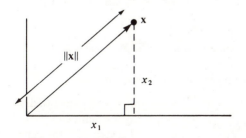

The distance between two points $\mathbf{x}$ and $\mathbf{y}$ is defined by

$$\|\mathbf{y} - \mathbf{x}\| = \{(y_1 - x_1)^2 + (y_2 - x_2)^2 + \ldots + (y_n - x_n)^2\}^{1/2}.$$

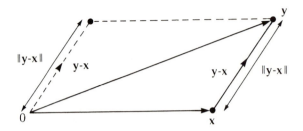

Example 1.19[H]

Let $x = (1, 2, 3)^T$ and $y = (2, 0, 5)^T$. Then the length of x is

$$\|x\| = \{1^2 + 2^2 + 3^2\}^{1/2} = \sqrt{(14)}$$

and the distance between x and y is

$$\|y - x\| = \{(2-1)^2 + (0-2)^2 + (5-3)^2\}^{1/2}$$
$$= \{1 + 4 + 4\}^{1/2} = 3.$$

1.20   Inner product

It makes no sense to talk about the matrix product of two n-dimensional column vectors x and y (unless $n = 1$). However, the product of the row vector x^T and the column vector y does make sense. We have that

$$x^T y = (x_1, x_2, \ldots, x_n) \begin{pmatrix} y_1 \\ y_2 \\ \vdots \\ y_n \end{pmatrix} = x_1 y_1 + x_2 y_2 + \ldots + x_n y_n.$$

We call $x^T y$ the *inner product* of x and y and use the notation $\langle x, y \rangle$. Thus

$$\langle x, y \rangle = x_1 y_1 + x_2 y_2 + \ldots + x_n y_n.$$

Since $\langle x, y \rangle$ is always a scalar, it is sometimes also called the scalar product of x and y.

It is easy to check the following properties of the inner product.

(i) $\langle x, x \rangle = \|x\|^2$

(ii) $\langle x, y \rangle = \langle y, x \rangle$

(iii) $\langle \alpha x + \beta y, z \rangle = \alpha \langle x, z \rangle + \beta \langle y, z \rangle$.

The geometric interpretation is important. The cosine rule will be familiar to those readers who have studied trigonometry. It states that

$$c^2 = a^2 + b^2 - 2ab \cos C.$$

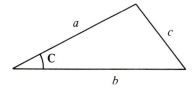

Rewriting this in terms of vectors, we obtain that

$$\|x - y\|^2 = \|x\|^2 + \|y\|^2 - 2\|x\|\,\|y\| \cos \theta$$

where θ is the angle between x and y. But also,

$$\|x - y\|^2 = \langle x - y, x - y \rangle$$
$$= \langle x, x - y \rangle - \langle y, x - y \rangle$$
$$= \langle x, x \rangle - 2 \langle x, y \rangle + \langle y, y \rangle$$
$$= \|x\|^2 + \|y\|^2 - 2 \langle x, y \rangle.$$

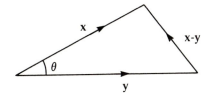

From this it follows that

$$\langle x, y \rangle = \|x\|\,\|y\| \cos \theta$$

This formula provides a means of calculating the angle between two vectors from a knowledge of their co-ordinates.

Example 1.21.
Find the cosine of the angle θ between the vectors $x = (1, 2, 3)^T$ and $y = (2, 0, 5)^T$. We have that

$$\|x\| = \{1^2 + 2^2 + 3^2\}^{1/2} = \sqrt{(14)}$$
$$\|y\| = \{2^2 + 0^2 + 5^2\}^{1/2} = \sqrt{(29)}$$
$$\langle x, y \rangle = 1 \times 2 + 2 \times 0 + 3 \times 5 = 17.$$

Hence

$$\cos \theta = \frac{\langle \mathbf{x}, \mathbf{y} \rangle}{\|\mathbf{x}\| \, \|\mathbf{y}\|} = \frac{17}{\sqrt{(14)}\sqrt{(29)}}.$$

1.22   Directions

With the exception of the zero vector

$$\mathbf{0} = (0, 0, 0, \ldots, 0)^T,$$

all vectors determine a direction. For example, the two-dimensional vector $(2, -2)^T$ points in a 'south-easterly' direction.

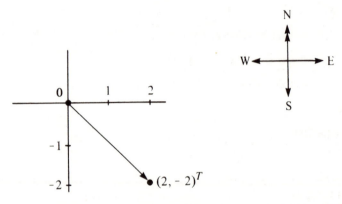

A *unit vector* $\mathbf{u}$ is a vector of length 1 — i.e. with $\|\mathbf{u}\| = 1$. When using a vector $\mathbf{v}$ to specify a direction, it is often convenient to replace $\mathbf{v}$ by the unit vector which points in the same direction.

For example, the vector $\mathbf{v} = (1, 1, 2)^T$ is *not* a unit vector because

$$\|\mathbf{v}\| = \{1^2 + 1^2 + 2^2\}^{1/2} = \sqrt{6}.$$

But the vector

$$\mathbf{u} = \|\mathbf{v}\|^{-1}\mathbf{v} = \frac{1}{\sqrt{6}}\mathbf{v} = \left(\frac{1}{\sqrt{6}}, \frac{1}{\sqrt{6}}, \frac{2}{\sqrt{6}}\right)^T$$

is a unit vector which points in the same direction as $\mathbf{v}$.

The co-ordinates of a unit vector $\mathbf{u}$ are called the directional cosines of the direction in which $\mathbf{u}$ points. For example, if $\mathbf{j} = (0, 1, 0)^T$ and $\mathbf{u} = (u_1, u_2, u_3)^T$, then

$$u_2 = \langle \mathbf{j}, \mathbf{u} \rangle = \|\mathbf{j}\| \, \|\mathbf{u}\| \cos \theta_2.$$

But $\|\mathbf{j}\| = \|\mathbf{u}\| = 1$ and so

$$u_2 = \cos \theta_2.$$

Evaluating u_1 and u_3 in a similar way, we obtain that

$$\mathbf{u} = (\cos\theta_1, \cos\theta_2, \cos\theta_3)^T.$$

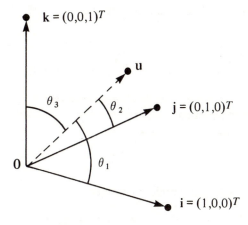

1.23 Lines

Consider the equation

$$\mathbf{x} = \boldsymbol{\xi} + t\mathbf{v}$$

in which $\boldsymbol{\xi}$ and $\mathbf{v}$ are *n*-dimensional vectors with $\mathbf{v} \neq \mathbf{0}$, and t is a scalar.

Think of $\mathbf{x}$ and $\boldsymbol{\xi}$ as determining points and $\mathbf{v}$ as determining a direction. The diagram below illustrates the situation:

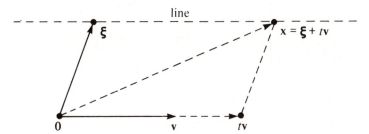

As t varies, $t\mathbf{v}$ moves along the line through $\mathbf{0}$ in the direction $\mathbf{v}$. It follows that $\mathbf{x}$ moves along the parallel line through $\boldsymbol{\xi}$ – i.e. as t varies, $\mathbf{x}$ describes the straight line through the point $\boldsymbol{\xi}$ in the direction $\mathbf{v}$.

We say that

$$\mathbf{x} = \boldsymbol{\xi} + t\mathbf{v}$$

is the parametric equation of the line through $\boldsymbol{\xi}$ in the direction $\mathbf{v}$. Each value of the parameter t yields a point $\mathbf{x}$ on this line.

If **v** is a *unit vector*, then

$$\|\mathbf{x} - \boldsymbol{\xi}\| = \|t\mathbf{v}\| = |t| \, \|\mathbf{v}\| = |t|$$

and so the distance between **x** and $\boldsymbol{\xi}$ is t if **x** lies on one side of $\boldsymbol{\xi}$ and $-t$ if **x** lies on the other side.

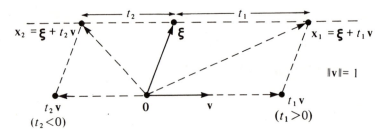

If the equation $\mathbf{x} = \boldsymbol{\xi} + t\mathbf{u}$ is written out in full, it becomes

$$\begin{pmatrix} x_1 \\ x_2 \\ \cdot \\ \cdot \\ \cdot \\ x_n \end{pmatrix} = \begin{pmatrix} \xi_1 \\ \xi_2 \\ \cdot \\ \cdot \\ \cdot \\ \xi_n \end{pmatrix} + t \begin{pmatrix} v_1 \\ v_2 \\ \cdot \\ \cdot \\ \cdot \\ v_n \end{pmatrix} = \begin{pmatrix} \xi_1 + tv_1 \\ \xi_2 + tv_2 \\ \cdot \\ \cdot \\ \cdot \\ \xi_n + tv_n \end{pmatrix}$$

which is the same thing as the system of equations

$$\left. \begin{aligned} x_1 &= \xi_1 + tv_1 \\ x_2 &= \xi_2 + tv_2 \\ & \cdot \\ & \cdot \\ x_n &= \xi_n + tv_n. \end{aligned} \right\}$$

We eliminate t from these equations and obtain

$$\frac{x_1 - \xi_1}{v_1} = \frac{x_2 - \xi_2}{v_2} = \ldots = \frac{x_n - \xi_n}{v_n}.$$

The latter system of equations is therefore an alternative form for the equations of a line through $(\xi_1, \xi_2, \ldots, \xi_n)^T$ in the direction $(v_1, v_2, \ldots, v_n)^T$ when none of the co-ordinates of **v** are zero.

A line passing through two distinct points $\boldsymbol{\xi}$ and $\boldsymbol{\eta}$ has the parametric form

$$\mathbf{x} = (1 - t)\boldsymbol{\xi} + t\boldsymbol{\eta}.$$

This is because $\boldsymbol{\eta} - \boldsymbol{\xi}$ points in the direction of the line.

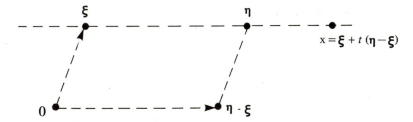

$$x = \xi + t\,(\eta - \xi)$$

Example 1.24.
The line passing through $(1, 2, 3)^T$ and $(3, 1, 2)^T$ has the form

$$\begin{pmatrix} x_1 \\ x_2 \\ x_3 \end{pmatrix} = (1-t)\begin{pmatrix} 1 \\ 2 \\ 3 \end{pmatrix} + t\begin{pmatrix} 3 \\ 1 \\ 2 \end{pmatrix}.$$

This can be expressed in the alternative form

$$\frac{x_1 - 1}{3 - 1} = \frac{x_2 - 2}{1 - 2} = \frac{x_3 - 3}{2 - 3}.$$

1.25 Orthogonal vectors

Two vectors $\mathbf{x}$ and $\mathbf{y}$ are *orthogonal* (or *perpendicular* or *normal*) if and only if

$$\langle \mathbf{x}, \mathbf{y} \rangle = 0.$$

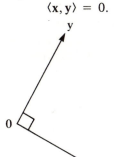

The reason for this definition is that $\langle \mathbf{x}, \mathbf{y} \rangle = \|\mathbf{x}\| \, \|\mathbf{y}\| \cos \theta$ and the cosine of a right angle is zero.

Example 1.26.
The vectors $\mathbf{x} = (1, 2, 3)^T$ and $\mathbf{y} = (-3, 3, -1)^T$ are orthogonal because

$$\langle \mathbf{x}, \mathbf{y} \rangle = 1 \times (-3) + 2 \times 3 + 3 \times (-1) = -3 + 6 - 3 = 0.$$

1.27 Hyperplanes

Consider the equation

$$\langle x - \xi, v \rangle = 0$$

in which ξ and v are *n*-dimensional vectors and $v \neq 0$.

Think of x and ξ as determining points and v as determining a direction. The diagram below illustrates the situation in three dimensions:

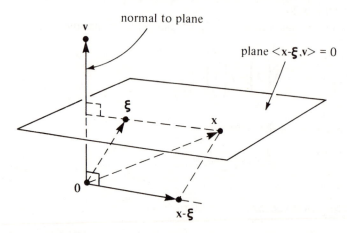

Three dimensions

Thus, in three dimensions, $\langle x - \xi, v \rangle = 0$ is the equation of a *plane* through the point ξ orthogonal to the vector v. We say that v is a *normal* to the plane.

In two dimensions, $\langle x - \xi, v \rangle = 0$ is the equation of a *line* through ξ orthogonal to the vector v.

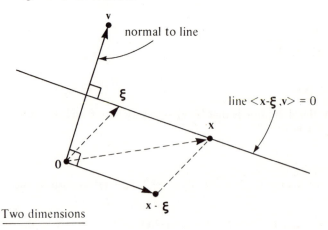

Two dimensions

In n dimensions, we say that $\langle \mathbf{x} - \boldsymbol{\xi}, \mathbf{v} \rangle = 0$ is the equation of a *hyperplane* through the point $\boldsymbol{\xi}$ with *normal* $\mathbf{v}$. If we write $c = \langle \boldsymbol{\xi}, \mathbf{v} \rangle$, then the equation of a hyperplane takes the form

$$\langle \mathbf{x}, \mathbf{v} \rangle = c.$$

When $\mathbf{v}$ is a *unit vector*, we have that

$$c = \langle \boldsymbol{\xi}, \mathbf{v} \rangle = \| \boldsymbol{\xi} \| \, \| \mathbf{v} \| \cos \theta = \| \boldsymbol{\xi} \| \cos \theta$$

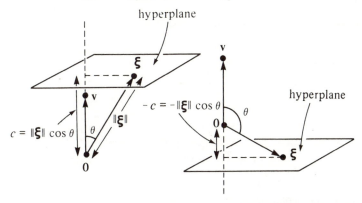

Thus c is the shortest distance from $\mathbf{0}$ to the hyperplane when $\mathbf{0}$ is 'underneath' the hyperplane (i.e. when $\mathbf{v}$ points from $\mathbf{0}$ to the hyperplane) and $-c$ is the shortest distance when $\mathbf{0}$ is 'above' the hyperplane.

If the equation $\langle \mathbf{x}, \mathbf{v} \rangle = c$ is written out in full, it becomes

$$x_1 v_1 + x_2 v_2 + x_3 v_3 + \ldots + x_n v_n = c.$$

Thus a hyperplane is defined by one 'linear' equation. In this equation $\mathbf{v} = (v_1, v_2, \ldots, v_n)^T$ is the normal to the hyperplane and, if $\mathbf{v}$ is a unit vector, $\pm c$ is the distance from $\mathbf{0}$ to the hyperplane.

Example 1.28.
The equation

$$3x + 4y = 5$$

is the equation of a hyperplane in two dimensions. In two dimensions a hyperplane is a line. The vector $(3, 4)^T$ is *normal* to this line. A *unit* normal to the line is the vector $(\frac{3}{5}, \frac{4}{5})^T$. If the equation is rewritten in the form

$$\tfrac{3}{5}x + \tfrac{4}{5}y = 1,$$

we therefore obtain that the distance of the line from $(0, 0)^T$ is 1.

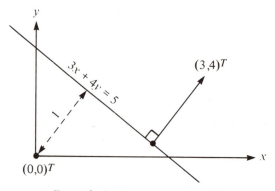

Example 1.29.
The equation

$$x + 2y + 2z = 5$$

is the equation of a hyperplane in three dimensions. In three dimensions a hyperplane is just a plane. A normal to this plane is the vector $(1, 2, 2)^T$. A *unit* normal is the vector $(\frac{1}{3}, \frac{2}{3}, \frac{2}{3})^T$. If the equation is rewritten in the form

$$\tfrac{1}{3}x + \tfrac{2}{3}y + \tfrac{2}{3}z = \tfrac{5}{3},$$

we obtain that the distance from $(0, 0, 0)$ to the plane is $\frac{5}{3}$.

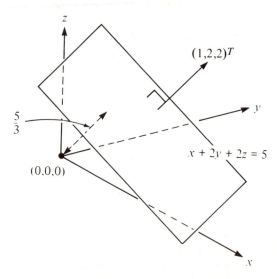

Example 1.30.
A plane passes through the points $(1, 2, 3)^T, (3, 1, 2)^T$ and $(2, 3, 1)^T$. What is its distance from the origin?

Let the plane have equation $ux + vy + wz = c$. Then

$$u1 + v2 + w3 = c$$

$$u3 + v1 + w2 = c$$

$$u2 + v3 + w1 = c,$$

i.e.

$$\begin{pmatrix} 1 & 2 & 3 \\ 3 & 1 & 2 \\ 2 & 3 & 1 \end{pmatrix} \begin{pmatrix} u \\ v \\ w \end{pmatrix} = c \begin{pmatrix} 1 \\ 1 \\ 1 \end{pmatrix}.$$

Hence

$$\begin{pmatrix} u \\ v \\ w \end{pmatrix} = \begin{pmatrix} 1 & 2 & 3 \\ 3 & 1 & 2 \\ 2 & 3 & 1 \end{pmatrix}^{-1} c \begin{pmatrix} 1 \\ 1 \\ 1 \end{pmatrix} = \frac{c}{18} \begin{pmatrix} -5 & 7 & 1 \\ 1 & -5 & 7 \\ 7 & 1 & -5 \end{pmatrix} \begin{pmatrix} 1 \\ 1 \\ 1 \end{pmatrix}$$

$$\begin{pmatrix} u \\ v \\ w \end{pmatrix} = \frac{c}{18} \begin{pmatrix} 3 \\ 3 \\ 3 \end{pmatrix} = \frac{c}{6} \begin{pmatrix} 1 \\ 1 \\ 1 \end{pmatrix}.$$

We next impose the further condition that $(u, v, w)^T$ be a *unit* vector — i.e. $u^2 + v^2 + w^2 = 1$. Then

$$1 = u^2 + v^2 + w^2 = \frac{c^2}{36} + \frac{c^2}{36} + \frac{c^2}{36} = \frac{c^2}{12}.$$

Hence the required distance is

$$c = \sqrt{(12)}.$$

1.31 Flats

Let A denote a $p \times q$ matrix and let $\mathbf{c}$ be a p-dimensional vector. Then the set of q-dimensional vectors $\mathbf{x}$ which satisfy the equation

$$A\mathbf{x} = \mathbf{c}$$

is called a *flat* (or an affine set).

If the equation $A\mathbf{x} = \mathbf{c}$ is written out in full, it becomes

$$\begin{pmatrix} a_{11} & a_{12} \ldots a_{1q} \\ a_{21} & a_{22} \ldots a_{2q} \\ \vdots \\ a_{p1} & a_{p2} \ldots a_{pq} \end{pmatrix} \begin{pmatrix} x_1 \\ x_2 \\ \vdots \\ x_q \end{pmatrix} = \begin{pmatrix} c_1 \\ c_2 \\ \vdots \\ c_p \end{pmatrix}$$

which is the same thing as the system of equations

$$
\left.\begin{array}{l}
a_{11}x_1 + a_{12}x_2 + \ldots + a_{1q}x_q = c_1 \\
a_{21}x_1 + a_{22}x_2 + \ldots + a_{2q}x_q = c_2 \\
\quad \cdot \\
\quad \cdot \\
\quad \cdot \\
a_{p1}x_1 + a_{p2}x_2 + \ldots + a_{pq}x_q = c_p.
\end{array}\right\}
$$

A flat is therefore determined by m 'linear' equations.

Let $\mathbf{a}_k = (a_{k1}, a_{k2}, \ldots, a_{kn})^T$. If $\mathbf{a}_k \neq \mathbf{0}$, the kth linear equation above is then the equation of the hyperplane $\langle \mathbf{x}, \mathbf{a}_k \rangle = c_k$. Expressing our system of linear equations in this form — i.e.

$$
\left.\begin{array}{l}
\langle \mathbf{x}, \mathbf{a}_1 \rangle = c_1 \\
\langle \mathbf{x}, \mathbf{a}_2 \rangle = c_2 \\
\quad \cdot \\
\quad \cdot \\
\langle \mathbf{x}, \mathbf{a}_p \rangle = c_p,
\end{array}\right\}
$$

we see that a flat consists of the points $\mathbf{x}$ which are common to p hyperplanes.

We illustrate some possibilities below when $p = 1, 2$ and 3 and $q = 3$.

(i) $p = 1, q = 3$. In this case A is a 1×3 matrix and the flat $A\mathbf{x} = \mathbf{c}$ reduces to the single equation

$$
\langle \mathbf{x}, \mathbf{a}_1 \rangle = c_1.
$$

The flat is therefore, in general, a *plane*.

(ii) $p = 2, q = 3$. In this case A is a 2×3 matrix and the flat $A\mathbf{x} = \mathbf{c}$ reduces to the pair of equations

$$
\left.\begin{array}{l}
\langle \mathbf{x}, \mathbf{a}_1 \rangle = c_1 \\
\langle \mathbf{x}, \mathbf{a}_2 \rangle = c_2.
\end{array}\right\}
$$

The flat is therefore, in general, the intersection of two planes and hence is a *line*.

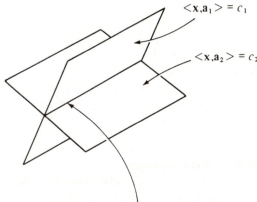

$\langle \mathbf{x}, \mathbf{a}_1 \rangle = c_1$

$\langle \mathbf{x}, \mathbf{a}_2 \rangle = c_2$

flat consisting of those **x** which satisfy

$$\begin{pmatrix} a_{11} & a_{12} & a_{13} \\ a_{21} & a_{22} & a_{23} \end{pmatrix} \begin{pmatrix} x_1 \\ x_2 \\ x_3 \end{pmatrix} = \begin{pmatrix} c_1 \\ c_2 \end{pmatrix}$$

(iii) $p = 3, q = 3$. In this case A is a 3×3 matrix and the flat $A\mathbf{x} = \mathbf{c}$ reduces to the system of equations

$$\left.\begin{aligned} \langle \mathbf{x}, \mathbf{a}_1 \rangle &= c_1 \\ \langle \mathbf{x}, \mathbf{a}_2 \rangle &= c_2 \\ \langle \mathbf{x}, \mathbf{a}_3 \rangle &= c_3. \end{aligned}\right\}$$

The flat is therefore, in general, the intersection of three planes and hence is a *point*.

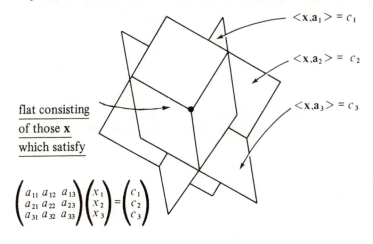

$\langle \mathbf{x}, \mathbf{a}_1 \rangle = c_1$

$\langle \mathbf{x}, \mathbf{a}_2 \rangle = c_2$

$\langle \mathbf{x}, \mathbf{a}_3 \rangle = c_3$

flat consisting
of those **x**
which satisfy

$$\begin{pmatrix} a_{11} & a_{12} & a_{13} \\ a_{21} & a_{22} & a_{23} \\ a_{31} & a_{32} & a_{33} \end{pmatrix} \begin{pmatrix} x_1 \\ x_2 \\ x_3 \end{pmatrix} = \begin{pmatrix} c_1 \\ c_2 \\ c_3 \end{pmatrix}$$

Note:

In the above examples we have been careful to insert the words 'in general' in (i), (ii) and (iii). These words mean that the result is as stated except in degenerate cases. In (ii), for example, things degenerate if $a_1 = 0$ or $a_2 = 0$ or if a_1 and a_2 point in the same (or opposite) directions. In the latter case, the two planes $\langle x, a_1 \rangle = c_1$ and $\langle x, a_2 \rangle = c_2$ are parallel and hence the flat will contain no points at all unless the two planes happen to be identical in which case the flat will be this plane.

The degenerate cases occur when the vectors $a_1, a_2, \ldots, a_p$ are *linearly dependent*. We shall not attempt to define this notion here. It is worth noting, however, that, when $p = q$, the vectors $a_1, a_2, \ldots, a_p$ are linearly independent if and only if the matrix A is non-singular and, in this case, the flat $Ax = c$ consists of the single point $A^{-1}c$.

It is a familiar fact that the set of all $(x, y)^T$ which satisfy an equation of the form

$$y = mx + c$$

is a straight line. We call this straight line the graph of the equation. In §4.3 we shall be concerned with graphs of equations

$$y = Mx + c,$$

in which M is an $m \times n$ matrix and c is an $m \times 1$ column vector. What does such a graph look like?

Observe to begin with that the graph is a set of objects of the form $(x, y)^T = (x_1, x_2, \ldots, x_n, y_1, y_2, \ldots, y_m)^T$. But $y = Mx + c$ may be written as m 'linear equations' in the variables $x_1, x_2, \ldots, x_n$, $y_1, y_2, \ldots, y_m$. From our preceding discussion, it follows that the graph of $y = Mx + c$ is simply a flat in $n + m$ dimensional space.

Exercises 1.32

1. Which of the following are unit vectors? Are any pair of these vectors orthogonal?

(i) $(\frac{1}{6}, \frac{1}{3}, \frac{1}{2})^T$ (ii) $(\frac{1}{3}, \frac{1}{3}, -\frac{1}{3})^T$ (iii) $(\frac{1}{3}, \frac{2}{3}, -\frac{2}{3})^T$.

2. A line in three-dimensional space is defined by the equations

$$\frac{x_1 - 3}{1} = \frac{x_2 - 1}{2} = \frac{x_3 - 2}{1}.$$

Find a unit vector parallel to this line.

3. Write down equations for a line through the points $(1, 2, 1)^T$ and $(2, 1, 2)^T$.

4. Describe the dets defined by
   (i) $x = \xi + tv \quad (t \geqslant 0)$
   (ii) $x = (1 - t)\xi + t\eta \quad (0 \leqslant t \leqslant 1)$.

5. Write down the equation of a plane which passes through $(1, 2, 1)^T$ and which is normal to $(2, 1, 2)^T$. What is the distance of this plane from (i) the origin, (ii) the point $(1, 2, 3)^T$?

6. Find the equations of all lines in two-dimensional space which are orthogonal to the vector $(1, 2)^T$.

7. A plane passes through $(1, 1, 2)^T, (1, 2, 1)^T, (2, 1, 1)^T$. What is its distance from $(0, 0, 0)^T$?

8. Write down a vector which is normal to the plane $x_1 + 2x_2 + 3x_3 = 6$. Find a point which lies on this plane.

9. Which of the planes $x_1 + 2x_2 - 3x_5 = 5$ and $x_1 + 2x_2 - 3x_3 = 6$ lies further from the origin?

10. Let $(p_1, p_2, p_3)^T$ and $(q_1, q_2, q_3)^T$ be two non-zero vectors. Explain why (r_1, r_2, r_3) is orthogonal to both if and only if

and
$$
\left.
\begin{aligned}
p_1 r_1 + p_2 r_2 + p_3 r_3 &= 0 \\
q_1 r_1 + q_2 r_2 + q_3 r_3 &= 0.
\end{aligned}
\right\}
$$

   Deduce that the vector $(p_2 q_3 - p_3 q_2, -p_1 q_3 + p_3 q_1, p_1 q_2 - p_2 q_1)^T$ is orthogonal to both $(p_1, p_2, p_3)^T$ and $(q_1, q_2, q_3)^T$.

11. Find the equations of the planes which
   (i) pass through $(3, 1, 2)^T$ and are parallel to the vectors $(1, 1, 1)^T$ and $(1, -1, 1)^T$,
   (ii) pass through $(3, 1, 2)^T$ and $(1, 2, 3)^T$ and are parallel to the vector $(1, 1, 1)^T$.
[Hint: Use question 10.]

12. The flat defined by

$$
\begin{pmatrix} 1 & 2 & 3 \\ 2 & 1 & 0 \end{pmatrix}
\begin{pmatrix} x_1 \\ x_2 \\ x_3 \end{pmatrix}
= \begin{pmatrix} 1 \\ 1 \end{pmatrix}
$$

is a line. Write down the parametric equation of this line.
[Hint: Use question 10.]

13. The flat defined by

$$
\begin{pmatrix} 1 & 2 & 3 \\ 3 & 1 & 2 \\ 2 & 3 & 1 \end{pmatrix}
\begin{pmatrix} x_1 \\ x_2 \\ x_3 \end{pmatrix}
= \begin{pmatrix} 1 \\ 2 \\ 3 \end{pmatrix}
$$

is a point. Write down the co-ordinates of this point.

14. Explain why the vectors $(1, 1, 1)^T$ and $(-1, -1, -1)^T$ point in opposite directions. Show that the flats

$$(i)\begin{pmatrix} 1 & 1 & 1 \\ -1 & -1 & -1 \end{pmatrix}\begin{pmatrix} x_1 \\ x_2 \\ x_3 \end{pmatrix} = \begin{pmatrix} 2 \\ -2 \end{pmatrix} \quad (ii)\begin{pmatrix} 1 & 1 & 1 \\ -1 & -1 & -1 \end{pmatrix}\begin{pmatrix} x_1 \\ x_2 \\ x_3 \end{pmatrix} = \begin{pmatrix} 2 \\ 2 \end{pmatrix}$$

are, respectively, a plane and the empty set.

15. A flat is defined by

$$\begin{pmatrix} 1 & 2 & 3 \\ 2 & 1 & 0 \\ 3 & 3 & 3 \end{pmatrix}\begin{pmatrix} x_1 \\ x_2 \\ x_3 \end{pmatrix} = \begin{pmatrix} 1 \\ 1 \\ 2 \end{pmatrix}.$$

Prove that the matrix is singular and the flat is a line.

SOME APPLICATIONS (OPTIONAL)

1.33   Commodity bundles

An important use for vectors is to represent *commodity bundles*. A housewife who visits a supermarket and purchases the items on the shopping list,

$$\left.\begin{array}{lll} \text{flour} & \text{4 kilograms} \\ \text{sugar} & \text{1 kilogram} \\ \text{milk} & \text{3 litres} \\ \text{yeast} & \text{0.1 kilogram,} \end{array}\right\}$$

can be described by saying that she has acquired the commodity bundle

$$\mathbf{y} = \begin{pmatrix} 4 \\ 1 \\ 3 \\ .1 \end{pmatrix}.$$

Of course, one needs to know in considering $\mathbf{y}$ what kind of good each co-ordinate represents and in what units this good is measured.

1.34 Linear production models

Suppose that $\mathbf{x}$ is an $n \times 1$ column vector, $\mathbf{y}$ is an $m \times 1$ column vector and A is an $m \times n$ matrix. We can then use the equation

$$\mathbf{y} = A\mathbf{x}$$

as a model for a simple production process. In this process, $\mathbf{y}$ represents the commodity bundle of raw materials (the *input*) required to produce the commodity bundle $\mathbf{x}$ of finished goods (the *output*). Such a production process is said to be *linear*.

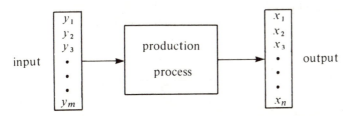

For example, suppose that the housewife of §1.33 is interested in producing x_1 kg of bread and x_2 kg of cake. Her output vector is therefore to be $\mathbf{x} = (x_1, x_2)^T$. In deciding how much of each type of ingredient will be required, she will need to consider her production process. If this is linear, her input vector $\mathbf{y} = (y_1, y_2, y_3, y_4)^T$ of required ingredients might, for example, be given by

$$\begin{pmatrix} y_1 \\ y_2 \\ y_3 \\ y_4 \end{pmatrix} = \begin{pmatrix} 1 & 1 \\ 0 & .2 \\ 0 & .3 \\ .1 & 0 \end{pmatrix} \begin{pmatrix} x_1 \\ x_2 \end{pmatrix}.$$

1.35 Price vectors

The co-ordinates of a *price vector*

$$\mathbf{p} = (p_1, p_2, \ldots, p_n)^T$$

list the prices at which the corresponding commodities can be bought or sold. Given the price vector $\mathbf{p}$, the *value* of a commodity bundle $\mathbf{x}$ is

$$\langle \mathbf{p}, \mathbf{x} \rangle = p_1 x_1 + p_2 x_2 + \ldots + p_n x_n.$$

This is the amount for which the commodity bundle may be bought or sold.
If $\mathbf{p}$ is a fixed price vector and c is a given quantity of money, the

commodity vectors **x** which lie on the hyperplane

$$\langle \mathbf{p}, \mathbf{x} \rangle = c$$

are those whose purchase costs precisely c.

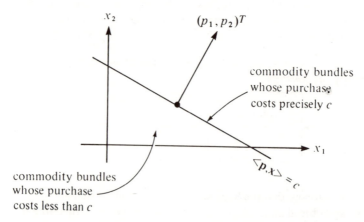

Alternatively, it may be that **x** is a fixed commodity bundle and that c is a given amount of money. In this case, the points **p** which lie on the hyperplane

$$\langle \mathbf{x}, \mathbf{p} \rangle = c$$

are the price vectors which ensure that the sale of **x** will realize an amount c.

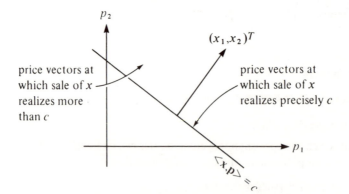

These two ways of looking at things are said to be *dual*. A buyer will be more interested in the former and a seller in the latter.

1.36 Linear programming

Suppose that, having acquired the commodity bundle **b** of ingredients, the housewife of §1.33 decides to sell the result **x** of her baking. If the relevant price vector is **p**, then the revenue she will acquire from the sale of **x** is

$$\langle \mathbf{p}, \mathbf{x} \rangle = \mathbf{p}^T \mathbf{x}.$$

Her problem is to choose **x** so as to *maximize* this revenue. But she cannot choose **x** freely. In particular, she cannot bake more than her stock **b** of ingredients allows. To bake **x**, she requires $\mathbf{y} = A\mathbf{x}$. Her choice of **x** is therefore restricted by the *constraint*

$$A\mathbf{x} \leqslant \mathbf{b}.$$

(Note that $\mathbf{c} \leqslant \mathbf{d}$ means that $c_1 \leqslant d_1, c_2 \leqslant d_2, c_3 \leqslant d_3$, etc.) Also, she cannot choose to bake negative quantities of bread or cakes. We therefore have the additional constraint

$$\mathbf{x} \geqslant \mathbf{0}.$$

The housewife's problem is therefore to find

$$\max \{\mathbf{p}^T \mathbf{x}\}$$

where **x** is subject to the *constraints*

$$A\mathbf{x} \leqslant \mathbf{b}$$
$$\mathbf{x} \geqslant \mathbf{0}.$$

Such a problem is called a *linear programming* problem. Observe how the use of matrix notation allows us to state the problem neatly and concisely. Written out in full, the problem takes the form: find

$$\max \{p_1 x_1 + p_2 x_2 + \ldots + p_n x_n\}$$

where $x_1, x_2, x_3, \ldots, x_n$ are subject to the constraints

$$a_{11} x_1 + a_{12} x_2 + \ldots + a_{1n} x_n \leqslant b_1$$
$$a_{21} x_1 + a_{22} x_2 + \ldots + a_{2n} x_n \leqslant b_2$$
$$\vdots$$
$$a_{m1} x_1 + a_{m2} x_2 + \ldots + a_{mn} x_n \leqslant b_n$$

and $x_1 \geqslant 0, x_2 \geqslant 0, \ldots, x_n \geqslant 0$.

1.37   Dual problem

A large bakery concern cannot compete with housewives in respect of the quality of its products. Instead, it proposes to buy up our housewife's stock of ingredients. What price vector $\mathbf{q}$ should it offer the housewife?

If the housewife bakes and sells $\mathbf{x}$, she will receive $\mathbf{p}^T\mathbf{x}$. If instead she sells the ingredients $\mathbf{y} = A\mathbf{x}$ at $\mathbf{q}$ she will receive $\mathbf{q}^T\mathbf{y} = \mathbf{q}^TA\mathbf{x} = (A^T\mathbf{q})^T\mathbf{x}$ (see exercise 1.15 (5)). Selling the ingredients at $\mathbf{q}$ is therefore the same as selling the results of her baking at prices $A^T\mathbf{q}$. For these prices to be more attractive than the market prices, we require that

$$A^T\mathbf{q} \geqslant \mathbf{p}.$$

We also, of course, require that

$$\mathbf{q} \geqslant \mathbf{0}.$$

Assuming that the bakery concern wishes to acquire the housewife's stock $\mathbf{b}$ at minimum cost, it therefore has to find

$$\min \{\mathbf{q}^T\mathbf{b}\}$$

where $\mathbf{q}$ is subject to the *constraints*

$$A^T\mathbf{q} \geqslant \mathbf{p}$$

$$\mathbf{q} \geqslant \mathbf{0}.$$

This linear programming problem is said to be *dual* to that of §1.36. It is more or less obvious that the minimum cost at which the bakery concern can buy up the housewife's ingredients is equal to the maximum revenue she can acquire by baking the ingredients and selling the results. This fact is the important *duality theorem* of linear programming. The prices $\mathbf{q}$ which solve the dual problem are called the *shadow prices* of the ingredients. These are the prices at which it is sensible to value the stock $\mathbf{b}$ given that an amount $\mathbf{x} = A\mathbf{y}$ of finished goods can be obtained costlessly from an amount $\mathbf{y}$ of stock and sold at prices $\mathbf{p}$.

1.38   Game theory

An $m \times n$ matrix A can be used as the *payoff matrix* in a *zero-sum game*. We interpret the m rows of A as possible strategies for the first player and the n columns of A as possible strategies for the second player. The choice of a row and column by the two players determines an entry of the matrix. The first player then *wins* a payoff equal to this entry and the second player *loses* an equal amount.

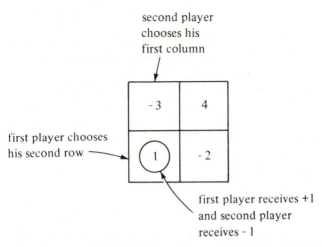

second player
chooses his
first column

first player chooses
his second row

first player receives +1
and second player
receives - 1

The game described above is called zero-sum because the sum of the payoffs to the two players is always zero. For such games we are interested in what happens when the players simultaneously act in such a way that each player's action is a 'best reply' to the action of the other. Neither player will then have cause to regret his choice of action.

It turns out that the players should consider using 'mixed strategies' – i.e. instead of simply choosing a strategy, they should assign probabilities to each of their strategies and leave chance to make the final decision. This has the advantage that the opponent cannot then possibly predict what the final choice will be.

We use an $m \times 1$ vector $\mathbf{p} = (p_1, p_2, \ldots, p_m)^T$ to represent a mixed strategy for the first player. The kth co-ordinate p_k represents the probability with which the kth row is to be chosen. Similarly, an $n \times 1$ vector $\mathbf{q} = (q_1, q_2, \ldots, q_n)^T$ represents a mixed strategy for the second player. If the two players independently choose the mixed strategies $\mathbf{p}$ and $\mathbf{q}$, then the expected gain to the first player (and hence the expected loss to the second player) is

$$\mathbf{p}^T A \mathbf{q}.$$

We illustrate this with the 2×2 payoff matrix considered above.

The expected gain to the first player is calculated by multiplying each payoff by the probability with which it occurs and then adding the results. The expected gain to the first player is therefore

$$(-3)p_1q_1 + 4p_1q_2 + 1p_2q_1 + (-2)p_2q_2$$

$$= (p_1, p_2) \begin{pmatrix} -3 & 4 \\ 1 & -2 \end{pmatrix} \begin{pmatrix} q_1 \\ q_2 \end{pmatrix} = \mathbf{p}^T A \mathbf{q}.$$

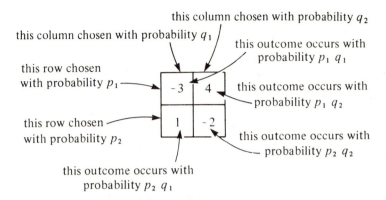

this column chosen with probability q_2

this column chosen with probability q_1

this outcome occurs with probability $p_1 q_1$

this row chosen with probability p_1

this outcome occurs with probability $p_1 q_2$

this row chosen with probability p_2

this outcome occurs with probability $p_2 q_2$

this outcome occurs with probability $p_2 q_1$

We assume that the first player chooses **p** in an attempt to maximize this quantity and the second player chooses **q** in an attempt to minimize it.

If each of the mixed strategies $\tilde{\mathbf{p}}$ and $\tilde{\mathbf{q}}$ is a best reply to the other, then

$$\tilde{\mathbf{p}}^T A \tilde{\mathbf{q}} \geqslant \mathbf{p}^T A \tilde{\mathbf{q}} \quad \text{(for all } \mathbf{p})$$

and

$$\tilde{\mathbf{p}}^T A \tilde{\mathbf{q}} \leqslant \tilde{\mathbf{p}}^T A \mathbf{q} \quad \text{(for all } \mathbf{q}).$$

A pair of mixed strategies $\tilde{\mathbf{p}}$ and $\tilde{\mathbf{q}}$ with these properties is called a *Nash equilibrium* of the game. Von Neumann proved that every matrix A has such a Nash equilibrium and that it is sensible to regard this as the *solution* of the game.

In the case of the 2 × 2 matrix given above, the first player should choose the mixed strategy $(.3, .7)^T$ and the second player should choose the mixed strategy $(.6, .4)^T$. It is instructive to check that each of these is a best reply to the other. The resulting expected gain to the first player is -0.2 — i.e. this is a good game for the second player!

2

Differentiation

This chapter introduces the idea of a partial derivative and explains the role that partial derivatives play in finding the rate of change of a function of several variables in a given direction. The chapter is vital for nearly everything which follows in this book. Some discussion of relevant material from the one variable case appears in §2.3. It may be that some of the notation in this revision section is unfamiliar, but those who find §2.3 a struggle to understand are probably reading the wrong book and would do better to turn to a more elementary text.

2.1 Real-valued functions

A *function* $f: \mathcal{X} \to \mathcal{Y}$ is a rule which assigns a unique y in the set $\mathcal{Y}$ to each x in the set $\mathcal{X}$. We denote the object in $\mathcal{Y}$ assigned to x by

$$y = f(x).$$

We denote the set of all real numbers by $\mathbb{R}$ and the set of all n-dimensional vectors by $\mathbb{R}^n$. In this chapter we shall consider functions

$$f: \mathbb{R}^n \to \mathbb{R}$$

– i.e. functions which assign to each $\mathbf{x} = (x_1, x_2, \ldots, x_n)^T$ a unique real number $f(\mathbf{x}) = f(x_1, x_2, \ldots, x_n)$.

Most of our examples will be for the cases $n = 1$ or $n = 2$. In these cases, the function $f: \mathbb{R}^n \to \mathbb{R}$ can be illustrated by drawing a *graph*.

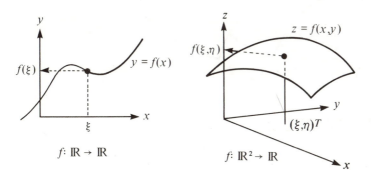

$f: \mathbb{R} \to \mathbb{R}$ $f: \mathbb{R}^2 \to \mathbb{R}$

In the case of a function $f : \mathbb{R}^2 \to \mathbb{R}$, the graph is a *surface* in three-dimensional space. This is not always easy to sketch. We therefore sometimes choose to think of the surface as a landscape. It is then natural to represent this landscape by drawing a *contour map*.

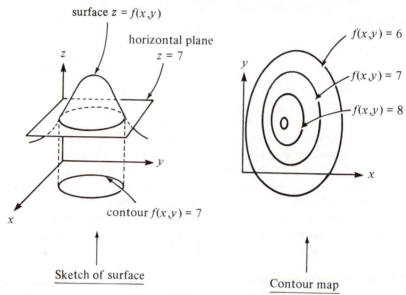

Sketch of surface   Contour map

One cannot, of course, draw a picture of the graph of a function $f : \mathbb{R}^3 \to \mathbb{R}$ but it is sometimes helpful to sketch a 'contour map' of such a function in three-dimensional space.

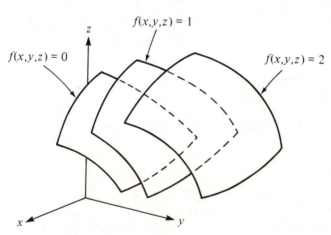

Contour map of $f \colon \mathbb{R}^3 \to \mathbb{R}$

2.2 Linear and affine functions

If M is an $m \times n$ matrix, the function $L: \mathbb{R}^n \to \mathbb{R}^m$ defined by

$$L(\mathbf{x}) = M\mathbf{x}$$

is said to be *linear*. If $\mathbf{c}$ is an $m \times 1$ vector, the function $A: \mathbb{R}^n \to \mathbb{R}^m$ defined by

$$A(\mathbf{x}) = M\mathbf{x} + \mathbf{c}$$

is said to be *affine*.

The graph of an affine function $A: \mathbb{R} \to \mathbb{R}$ is a *non-vertical* straight line. The graph of an affine function $A: \mathbb{R}^2 \to \mathbb{R}$ is a *non-vertical* plane.

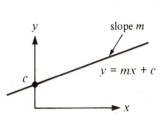

Affine function $A: \mathbb{R} \to \mathbb{R}$ defined by $y = A(x) = mx + c$

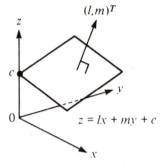

Affine function $A: \mathbb{R}^2 \to \mathbb{R}$ defined by

$$z = A(x, y) = (l, m)\begin{pmatrix} x \\ y \end{pmatrix} + c$$

i.e. $z = lx + my + c$

Note that a linear function is an affine function whose graph passes through the origin.

2.3 Derivatives[□]

In this section we shall revise the idea of the derivative of a function $f: \mathbb{R} \to \mathbb{R}$. The formula

$$\frac{dy}{dx} = \lim_{\delta x \to 0} \frac{\delta y}{\delta x}$$

will be familiar. This is a very useful piece of notation but it does have some drawbacks. The most notable of these is the fact that x is used *ambiguously* both as the variable with respect to which we are to differentiate *and* the point at which the derivative is to be evaluated.

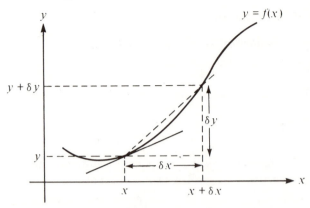

Where this ambiguity matters, we shall therefore prefer to use the more precise notation

$$f'(\xi)$$

for the *derivative* of the function f evaluated at the point ξ.

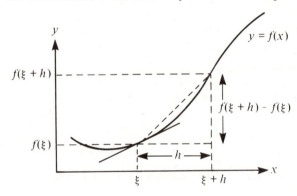

We have that

$$f'(\xi) = \lim_{h \to 0} \left\{ \frac{f(\xi + h) - f(\xi)}{h} \right\}.$$

We shall also use the notation $Df(x)$ for the derivative of f at the point x. Thus

$$Df(x) = \frac{d}{dx} \{f(x)\} = f'(x).$$

Example 2.4. □
If $y = x^2$, then

$$\frac{dy}{dx} = 2x.$$

Proof.

Put $f(x) = x^2$. Then

$$f'(\xi) = \lim_{h \to 0} \frac{f(\xi+h) - f(\xi)}{h} = \lim_{h \to 0} \frac{(\xi+h)^2 - \xi^2}{h}$$

$$= \lim_{h \to 0} \frac{\xi^2 + 2\xi h + h^2 - \xi^2}{h} = \lim_{h \to 0} (2\xi + h) = 2\xi.$$

By the same method, but using more complicated calculations, one can obtain the following results. It is essential that these be known by heart.

(i)	$\dfrac{d}{dx}(x^\alpha) = \alpha x^{\alpha-1}$
(ii)	$\dfrac{d}{dx}(e^x) = e^x$
(iii)	$\dfrac{d}{dx}(\log x) = \dfrac{1}{x}$
(iv)	$\dfrac{d}{dx}(\cos x) = -\sin x \;\; ; \;\; \dfrac{d}{dx}(\sin x) = \cos x.$
(v)	$\dfrac{d}{dx}(\tan x) = \sec^2 x = \dfrac{1}{\cos^2 x}$

Note that $\log x = \ln x$ will always mean the *natural* logarithm of x — i.e. $y = \log x$ if and only if $x = e^y$. Aside from these formula, it is also essential to know the following rules for manipulating derivatives:

(I)	$\dfrac{d}{dx}(Ay + Bz) = A\dfrac{dy}{dx} + B\dfrac{dz}{dx}$	(where A and B are constant).
(II)	$\dfrac{d}{dx}(yz) = z\dfrac{dy}{dx} + y\dfrac{dz}{dx}$	
(III)	$\dfrac{d}{dx}\left(\dfrac{y}{z}\right) = \dfrac{z\dfrac{dy}{dx} - y\dfrac{dz}{dx}}{z^2}$	
(IV)	$\dfrac{dz}{dx} = \dfrac{dz}{dy}\dfrac{dy}{dx}$	
(V)	$\dfrac{dx}{dy} = \left(\dfrac{dy}{dx}\right)^{-1}$	

Rules iv and v require some supplementary discussion. The discussion of v we postpone until a later chapter.

Rule iv is the rule for differentiating a 'function of a function'. It is also called the 'chain rule'. If $z = f(y)$ and $y = g(x)$, rule iv asserts that

$$\frac{d}{dx} f(g(x)) = f'(y)g'(x) = f'(g(x))g'(x).$$

Example 2.5$^\square$

$$\frac{d}{dx} \{\cos(x^2 + 2x + 1)\} = -(2x + 2)\sin(x^2 + 2x + 1).$$

We put $z = \cos y$ and $y = x^2 + 2x + 1$. Then

$$\frac{dz}{dy} = -\sin y; \quad \frac{dy}{dx} = 2x + 2.$$

Hence

$$\frac{dz}{dx} = \frac{dz}{dy}\frac{dy}{dx} = (-\sin y)(2x + 2)$$

$$= -(2x + 2)\sin(x^2 + 2x + 1).$$

Geometrically, $f'(\xi)$ is the *slope* (or *gradient*) of the tangent line to the graph $y = f(x)$ at the point $(\xi, f(\xi))^T$.

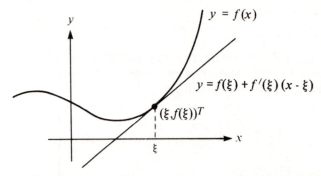

Since the tangent line has slope $f'(\xi)$ and passes through the point $(\xi, f(\xi))^T$, it therefore has equation

$$y = f(\xi) + f'(\xi)(x - \xi).$$

Finally, note that as x increases by h units from ξ to $\xi + h$, y increases by $f(\xi + h) - f(\xi)$ units. On average, y therefore increases by

$$\frac{f(\xi + h) - f(\xi)}{h}$$

units for each unit by which x increases. We say that

$$\frac{f(\xi + h) - f(\xi)}{h}$$

is the average *rate of increase* of y with respect to x as x increases from ξ to $\xi + h$.

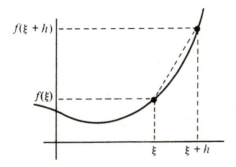

If we allow h to approach 0, we obtain that $f'(\xi)$ is the *instantaneous* rate of increase of y with respect to x at the point ξ. Economists call $f'(\xi)$ the *marginal* rate of increase.

Example 2.6.

A miller sells flour at a price which varies. The money in the miller's possession ($\$y$) and the amount of flour sold (x kg) are related by the formula $y = 1 + x^2$. At what price is the miller selling flour when he has sold 10 kg?

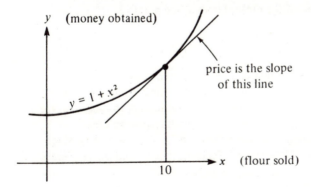

The price of flour is the amount of money received by the miller per unit of flour sold – i.e. it is the marginal rate of increase of y with respect to x. The price of flour when $x = 10$ kg is therefore $f'(10)$. But $f'(x) = 2x$ and hence the required price is

$20/kg.

Exercises 2.7

1.$^{\pi}$ Compute $f'(1)$ from first principles when

$$f(x) = x^3 + 2x + 3.$$

2.$^{\pi}$ Differentiate the following expressions:

(i) $x^4 - 3x^2 + 1$ (ii) $\sqrt{x}$ (iii) $\sqrt{(x^2 + 1)}$

(iv) $(x + 1)^{3/2}$ (v) $x \sin x$ (vi) $\dfrac{\tan x}{x}$.

3.$^{\pi}$ Obtain the following results where a and b are constants:

(i) $\dfrac{d}{dx} e^{(ax+b)} = a e^{ax+b}$

(ii) $\dfrac{d}{dx}(e^{-x}) = -e^{-x}$

(iii) $\dfrac{d}{dx}\{\log(ax + b)\} = \dfrac{a}{ax + b}$

(iv) $\dfrac{d}{dx}\{\cos(ax + b)\} = -a\sin(ax + b)$.

4.$^{\pi}$ Find the equations of the tangent lines at the point $(1, 1)^T$ to the following curves:

(i) $y = x^2$ (ii) $y = x^3$ (iii) $y = \{(x - 1)^2 + 1\}^{1/2}$.

5.$^{\pi}$ Money obtained ($\$y$) is related to amount sold (x kg) by $y = \{(x - 1)^2 + 1\}^{1/2}$. What is the price per kg when the amount sold is $x = 2$?

6.$^{\pi}$ Speed is the rate of increase of distance with respect to time. Acceleration is the rate of increase of speed with respect to time.

A car moves in a straight line. After t seconds it has reached a distance $x = t^3$ metres from its starting point. Find the speed v and the acceleration a at time $t = 3$ secs.

2.8   Partial derivatives

The partial derivatives of a function $f: \mathbb{R}^2 \to \mathbb{R}$ are defined by

$$\frac{\partial f}{\partial x} = \lim_{h \to 0} \left\{ \frac{f(x+h,y) - f(x,y)}{h} \right\}$$

$$\frac{\partial f}{\partial y} = \lim_{k \to 0} \left\{ \frac{f(x,y+k) - f(x,y)}{k} \right\}.$$

The computation of partial derivatives is easy. Simply differentiate with respect to one of the variables treating the other variable as a constant.

Examples 2.9

(i) If $z = x^2 y + y^3 x$, then

$$\frac{\partial z}{\partial x} = 2xy + y^3$$

$$\frac{\partial z}{\partial y} = x^2 + 3y^2 x.$$

(ii) If $z = \sin(xy + y^2)$, then

$$\frac{\partial z}{\partial x} = y \cos(xy + y^2)$$

$$\frac{\partial z}{\partial y} = (x + 2y) \cos(xy + y^2).$$

2.10   Notation

The first thing to observe is that the symbols

$$\partial, d, \delta$$

are all different and are used with different meanings. It is therefore important not to confuse these symbols.

An alternative notation for $\partial f/\partial x$ is f_x. This has the advantage that it allows one to distinguish between the variable with respect to which differentiation takes place and the point at which the partial derivative is evaluated. Thus

$$f_x(a, b)$$

means 'the partial derivative with respect to x evaluated at the point (a, b)'.

When a number of variables are in play, it is not always obvious which are to be held constant when differentiating partially. But it is of the *greatest* importance to be clear on this point. Where necessary, use the notation

$$\left(\frac{\partial f}{\partial x}\right)_y$$

to mean 'the partial derivative with respect to x keeping y constant'.

Example 2.11.
Consider the equations

$$\left.\begin{array}{l} u = x + y \\ v = x - y. \end{array}\right\}$$

From the first equation

$$\frac{\partial u}{\partial x} = 1.$$

If the equations are solved for x and y in terms of u and v, we obtain that

$$\left.\begin{array}{l} x = \frac{1}{2}(u + v) \\ y = \frac{1}{2}(u - v) \end{array}\right\}$$

and therefore that

$$\frac{\partial x}{\partial u} = \frac{1}{2}.$$

Note that

$$\frac{\partial u}{\partial x} \neq \left(\frac{\partial x}{\partial u}\right)^{-1}$$

This is not in the least surprising since *different* variables were held constant during the two differentiations. What has been shown is that

$$\left(\frac{\partial u}{\partial x}\right)_y = 1; \quad \left(\frac{\partial x}{\partial u}\right)_v = \frac{1}{2}.$$

It is instructive to calculate

$$\left(\frac{\partial u}{\partial x}\right)_v .$$

Eliminate y from the first set of equations. We obtain that

$$u = x + (x - v) = 2x - v$$

and therefore

$$\left(\frac{\partial u}{\partial x}\right)_v = 2$$

as one would expect.

2.12   Geometric interpretation

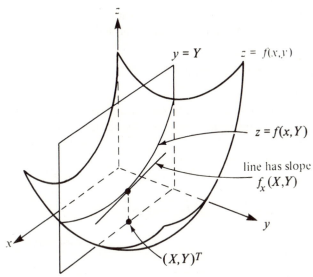

The picture shows the surface $z = f(x, y)$ sliced by the plane $y = Y$. On the plane $y = Y$, one can see the graph of the function $z = f(x, Y)$. The partial derivative $f_x(X, Y)$ is simply the slope of this graph at the point $x = X$.

The diagrams below illustrate both partial derivatives simultaneously:

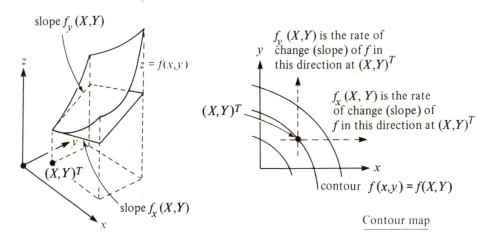

$f_y(X,Y)$ is the rate of change (slope) of f in this direction at $(X,Y)^T$

$f_x(X, Y)$ is the rate of change (slope) of f in this direction at $(X,Y)^T$

contour $f(x,y) = f(X,Y)$

Contour map

2.13   Tangent plane

If $Z = f(X, Y)$, then $(X, Y, Z)^T$ is a point on the surface $z = f(x, y)$. If the surface admits a non-vertical tangent plane at $(X, Y, Z)^T$, then we say that f is *differentiable* at $\boldsymbol{\xi} = (X, Y)^T$.

The general equation of a plane through the point $(X, Y, Z)^T$ is

$$u(x - X) + v(y - Y) + w(z - Z) = 0,$$

where $(u, v, w)^T$ is normal to the plane. We assume that this plane is non-vertical which means that $w \neq 0$ and hence we can rewrite the equation in the form

$$z = A(x, y) = l(x - X) + m(y - Y) + Z,$$

where $l = -u/w$ and $m = -v/w$.

For the plane to be tangent to the surface, our original function f and the affine function A must have the *same* partial derivatives at the point $(X, Y)^T$.

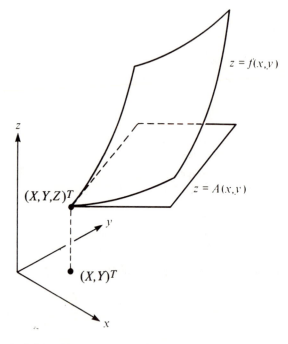

It follows that

$$\left.\begin{aligned} f_x(X, Y) &= A_x(X, Y) = l \\ f_y(X, Y) &= A_y(X, Y) = m. \end{aligned}\right\}$$

Hence, if f is differentiable at $(X, Y)^T$ its tangent plane must have equation

$$z - Z = f_x(X, Y)(x - X) + f_y(X, Y)(y - Y).$$

We usually write this in the less precise form

$$z - Z = \frac{\partial f}{\partial x}(x - X) + \frac{\partial f}{\partial y}(y - Y)$$

(keeping in mind that the partial derivatives are to be evaluated at the point $(X, Y)^T$).

Example 2.14.

Find the equation of the tangent plane to the surface $z = x^2 y + y^3 x$ at the point where $x = 1$ and $y = 2$.

At this point $z = 2 + 8 = 10$ and

$$\left. \begin{aligned} \frac{\partial z}{\partial x} &= 2xy + y^3 = 2 \times 2 + 8 = 12 \\[2ex] \frac{\partial z}{\partial y} &= x^2 + 3y^2 x = 1 \times 1 + 3 \times 4 \times 1 = 13. \end{aligned} \right\}$$

The tangent plane therefore has equation

$$z - 10 = 12(x - 1) + 13(y - 2).$$

A *normal* to the surface $z = x^2 y + y^3 x$ at $(1, 2, 10)^T$ is therefore $(12, 13, -1)^T$.

2.15   Gradient

The tangent plane to the surface $z = f(x, y)$ at the point $(X, Y, Z)^T$ has equation

$$z - Z = \frac{\partial f}{\partial x}(x - X) + \frac{\partial f}{\partial y}(y - Y).$$

If we slice the surface and its tangent plane by the horizontal plane $z = Z$, we find that the line

$$0 = \frac{\partial f}{\partial x}(x - X) + \frac{\partial f}{\partial y}(y - Y)$$

is tangent to the contour

$$Z = f(x, y)$$

at the point $(X, Y)^T$.

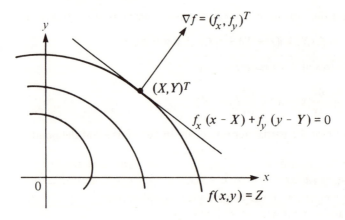

Contour map

It follows immediately that the vector

$$\nabla f = \left(\frac{\partial f}{\partial x}, \frac{\partial f}{\partial y}\right)^T$$

is normal to the contour $f(x,y) = Z$ at the point $(X, Y)^T$.
We call the vector ∇f the *gradient* of the function f.

Example 2.16.
Let $f(x,y) = x^2 y + y^3 x$. Then

$$\nabla f = \left(\frac{\partial f}{\partial x}, \frac{\partial f}{\partial y}\right)^T = (2xy + y^3,\ x^3 + 3y^2 x)^T.$$

It follows that the vector $(12, 13)^T$ is normal to the contour $x^2 y + y^3 x = 10$
at the point $(1, 2)^T$.

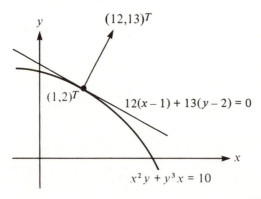

2.17   Chain rule

In §2.3 we considered the chain rule for functions $f: \mathbb{R} \to \mathbb{R}$ and $g: \mathbb{R} \to \mathbb{R}$.
If we write $z = f(x)$ and $x = g(t)$, the rule takes the form

$$\frac{dz}{dt} = \frac{dz}{dx}\frac{dx}{dt}.$$

We shall consider here a generalization to the case when $f: \mathbb{R}^2 \to \mathbb{R}$ and
$g: \mathbb{R} \to \mathbb{R}^2$.

We write $z = f(x, y)$ and $(x, y)^T = g(t)$. The latter formula means that
both x and y are functions of t – i.e. $x = g_1(t)$ and $y = g_2(t)$. We then have
that

$$z = f(g_1(t), g_2(t)).$$

The chain rule then asserts that

$$\frac{dz}{dt} = \frac{\partial z}{\partial x}\frac{dx}{dt} + \frac{\partial z}{\partial y}\frac{dy}{dt}.$$

Here $\partial z/\partial x$ means $(\partial z/\partial x)_y$ and $\partial z/\partial y$ means $(\partial z/\partial y)_x$.

An explanation of why the chain rule takes this form in this case will be
found in chapter 4. At this stage we shall simply confirm that the formula
gives the correct answer for the example which follows.

Example 2.18.

Let $z = x^2 y + y^3 x$, where $x = t^2$ and $y = t^3$. If we substitute for
x and y, we obtain that

$$z = t^4 t^3 + t^9 t^2 = t^7 + t^{11}.$$

Hence

$$\frac{dz}{dt} = 7t^6 + 11t^{10}.$$

We shall now confirm that the chain rule in the form introduced above gives
the same result. We have that

$$\frac{dz}{dt} = \frac{\partial z}{\partial x}\frac{dx}{dt} + \frac{\partial z}{\partial y}\frac{dy}{dt}$$

$$= (2xy + y^3)2t + (x^2 + 3y^2 x)3t^2$$

$$= (2t^5 + t^9)2t + (t^4 + 3t^8)3t^2$$

$$= 7t^6 + 11t^{10}.$$

2.19   Directional derivatives

Let $\mathbf{u} = (u_1, u_2)^T$ be a *unit* vector. We know from §1.23, that $\mathbf{x} = \boldsymbol{\xi} + t\mathbf{u}$ is a point on the straight line through $\boldsymbol{\xi}$ in the direction $\mathbf{u}$ whose distance from $\boldsymbol{\xi}$ is t.

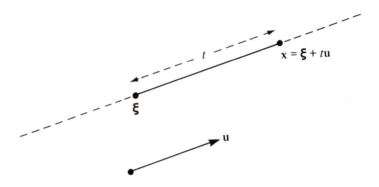

Suppose that $f: \mathbb{R}^2 \to \mathbb{R}$ and that $\boldsymbol{\xi}$ and $\mathbf{u}$ are given. Then $f(\boldsymbol{\xi} + t\mathbf{u})$ is a function of t and

$$\frac{d}{dt} \{ f(\boldsymbol{\xi} + t\mathbf{u}) \}$$

is the rate of increase of f with respect to t. If we evaluate this quantity at $t = 0$, we obtain the rate of increase of f in the direction $\mathbf{u}$ at the point $\boldsymbol{\xi}$. The chain rule is helpful here. Put $\boldsymbol{\xi} = (X, Y)^T$ and write $x = X + tu_1$, $y = Y + tu_2$. Then

$$\frac{d}{dt} \{ f(\boldsymbol{\xi} + t\mathbf{u}) \} = \frac{d}{dt} f(x, y) = \frac{\partial f}{\partial x} \frac{dx}{dt} + \frac{\partial f}{\partial y} \frac{dy}{dt}$$

$$= \frac{\partial f}{\partial x} u_1 + \frac{\partial f}{\partial y} u_2 = \langle \nabla f, \mathbf{u} \rangle.$$

We are interested in this quantity when $t = 0$ – i.e. when ∇f is evaluated at the point $(X, Y)^T$. We obtain that, if $\mathbf{u}$ is a *unit* vector and ∇f is evaluated at $\boldsymbol{\xi} = (X, Y)^T$, then

$$\langle \nabla f, \mathbf{u} \rangle$$

is the rate of increase of f in the direction $\mathbf{u}$ at the point $\boldsymbol{\xi}$.

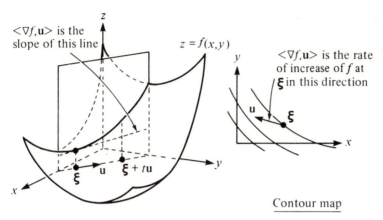

$\langle \nabla f, \mathbf{u} \rangle$ is the slope of this line

$z = f(x,y)$

$\langle \nabla f, \mathbf{u} \rangle$ is the rate of increase of f at $\boldsymbol{\xi}$ in this direction

Contour map

We call $\langle \nabla f, \mathbf{u} \rangle$ the *directional derivative* of f in the direction $\mathbf{u}$. If θ is the angle between the vectors ∇f and $\mathbf{u}$, then

$$\langle \nabla f, \mathbf{u} \rangle \ = \ \|\nabla f\| \cos \theta$$

(because $\|\mathbf{u}\| = 1$). Hence

$$|\langle \nabla f, \mathbf{u} \rangle| \ \leqslant \ \|\nabla f\|.$$

But

$$\left\langle \nabla f, \frac{\nabla f}{\|\nabla f\|} \right\rangle \ = \ \frac{\|\nabla f\|^2}{\|\nabla f\|} \ = \ \|\nabla f\|.$$

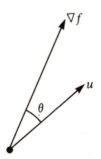

It follows that the direction in which ∇f points is the direction of maximum rate of increase of f at $\boldsymbol{\xi} = (X, Y)^T$ and that this maximum rate of increase is equal to $\|\nabla f\|$.

Example 2.20.
Let $f(x, y) = x^2 y + y^3 x$. We know from example 2.16 that

$$\nabla f \ = \ (12, 13)^T$$

at the point $(1, 2)^T$. What is the rate of increase of f at $(1, 2)^T$ in the direction $(3, 4)^T$?

Note that $(3, 4)^T$ is *not* a unit vector because $3^2 + 4^2 = 25$. But

$$(\tfrac{3}{5}, \tfrac{4}{5})^T$$

is a unit vector which points in the same direction. The rate of increase of f in this direction is therefore

$$\langle \nabla f, \mathbf{u} \rangle = 12 \cdot \tfrac{3}{5} + 13 \cdot \tfrac{4}{5} = \tfrac{88}{5} = 17.6.$$

The direction of maximum increase of f at $(1, 2)^T$ is simply $\nabla f = (12, 13)^T$ and the rate of increase in this direction is

$$\| \nabla f \| = \{12^2 + 13^2\}^{1/2} = \sqrt{(313)}.$$

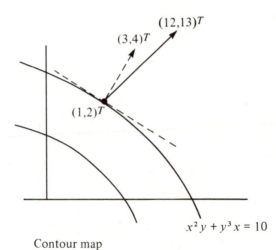

Contour map

2.21   More than two variables

All the above work applies equally well in the case of functions $f : \mathbb{R}^n \to \mathbb{R}$. We cannot draw pictures when $n > 2$, but we continue to use geometrical language. Thus the equation

$$z - Z = \frac{\partial f}{\partial x_1}(x_1 - X_1) + \frac{\partial f}{\partial x_2}(x_2 - X_2) + \ldots + \frac{\partial f}{\partial x_n}(x_n - X_n)$$

is the tangent hyperplane to the 'hypersurface' $z = f(x_1, x_2, \ldots, x_n)$ at the point $(X_1, X_2, \ldots, X_n, Z)^T$ provided that $Z = f(X_1, X_2, \ldots, X_n)$. Similarly,

$$\nabla f = \left(\frac{\partial f}{\partial x_1}, \frac{\partial f}{\partial x_2}, \ldots, \frac{\partial f}{\partial x_n}\right)^T$$

evaluated at $(X_1, X_2, \ldots, X_n)^T$ is the normal to the contour $Z = f(x_1, x_2, \ldots, x_n)$ at $(X_1, X_2, \ldots, X_n)^T$. The diagram illustrates the case $n = 3$.

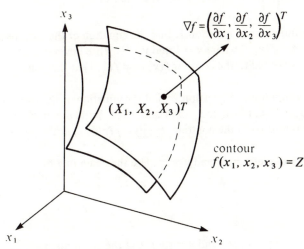

Finally, note that, if $\mathbf{u}$ is a unit vector, then

$$\langle \nabla f, \mathbf{u} \rangle$$

remains the rate of increase of f in the direction $\mathbf{u}$. In particular, the maximum rate of increase is $\|\nabla f\|$ in the direction of ∇f.

Exercises 2.22

1. Compute

$$\frac{\partial f}{\partial x} \quad \text{and} \quad \frac{\partial f}{\partial y}$$

in the following cases:

(i) $f(x, y) = x^2 + 3y^2$ (ii) $f(x, y) = \dfrac{x - y}{x + y}$

(iii) $f(x, y) = e^{xy^2}$ (iv) $f(x, y) = \cos(x^2 y)$.

2.* Compute

$$\frac{\partial f}{\partial x} \quad \text{and} \quad \frac{\partial f}{\partial y}$$

in the following cases:

(i) $f(x,y) = x + 2y$ (ii) $f(x,y) = x^2 y^5$

(iii) $f(x,y) = \sin\left(\dfrac{x}{y}\right)$ (iv) $f(x,y) = \log(y \tan x)$.

3. For each of the functions of question 1, find the equation of the tangent plane to the surface $z = f(x,y)$ where $x = 1$ and $y = 1$. Write down in each case a vector which is normal to the contour $f(x,y) = f(1, 1)$ at the point $(1, 1)^T$.

4.* For each of the functions of question 2, find the equation of the tangent plane to the surface $z = f(x,y)$ where $x = \pi/4$ and $y = 1$. Write down in each case a vector which is normal to the contour $f(x,y) = f(\pi/4, 1)$ at the point $(\pi/4, 1)^T$.

5. Let $z = \cos(x^2 y)$ and let $x = t^3$ and $y = t^2$. Use the chain rule to compute

$$\frac{dz}{dt}.$$

6.* Let $z = x^2 y^5$ and let $x = \cos t$ and $y = \sin t$. Use the chain rule to compute

$$\frac{dz}{dt}.$$

7. For each of the functions of question 1, find
   (a) the rate of increase at $(1, 1)^T$ in the direction $(-3, 4)^T$,
   (b) the direction of maximum rate of increase at $(1, 1)^T$. What is the rate of increase in this direction?

8.* For each of the functions of question 2, find
   (a) the rate of increase at $(\pi/4, 1)^T$ in the direction $(-1, 3)^T$,
   (b) the direction of maximum rate of increase at $(\pi/4, 1)^T$. What is the rate of increase in this direction?

9. Let $\mathbf{i} = (1, 0)^T$ and $\mathbf{j} = (0, 1)^T$. If $f: \mathbb{R}^2 \to \mathbb{R}$, prove that

$$\frac{\partial f}{\partial x} = \langle \nabla f, \mathbf{i} \rangle; \quad \frac{\partial f}{\partial y} = \langle \nabla f, \mathbf{j} \rangle.$$

Interpret these results.

10.* Let

$$\left. \begin{aligned} u &= x^2 + y^2 \\ v &= x^2 - y^2. \end{aligned} \right\}$$

Compute each of the following:

(i) $\left(\dfrac{\partial u}{\partial x}\right)_y$ (ii) $\left(\dfrac{\partial u}{\partial x}\right)_v$ (iii) $\left(\dfrac{\partial v}{\partial x}\right)_y$ (iv) $\left(\dfrac{\partial v}{\partial x}\right)_u$.

11. Let $f(x_1,x_2,x_3) = x_1^2 + x_2 x_3$. Find

$$\frac{\partial f}{\partial x_1}, \frac{\partial f}{\partial x_2} \quad \text{and} \quad \frac{\partial f}{\partial x_3}.$$

What is the tangent hyperplane to the hypersurface $z = x_1^2 + x_2 x_3$ where $x_1 = 1, x_2 = 0$ and $x_3 = 2$. Find the normal and the tangent plane to the contour $1 = x_1^2 + x_2 x_3$ at $(1, 0, 2)^T$. What is the rate of increase of f in the direction $(2, -1, 2)^T$ at the point $(1, 0, 2)^T$?

12.* Let $f(x_1,x_2,x_3) = x_1 x_2^2 x_3^3$. Find

$$\frac{\partial f}{\partial x_1}, \frac{\partial f}{\partial x_2} \quad \text{and} \quad \frac{\partial f}{\partial x_3}.$$

What is the tangent hyperplane to the hypersurface $z = x_1 x_2^2 x_3^3$ where $x_1 = 1, x_2 = 1$ and $x_3 = 1$. Find the normal and the tangent plane to the contour $1 = x_1 x_2^2 x_3^3$ at $(1, 1, 1)^T$. What is the rate of increase of f in the direction $(1, 2, 3)^T$ at the point $(1, 1, 1)^T$?

13. Let $f: \mathbb{R}^2 \to \mathbb{R}$ be homogeneous of degree α. This means that, for all x, y and λ,

$$f(\lambda x, \lambda y) = \lambda^\alpha f(x, y).$$

Put $z = f(x, y)$ and $x = tX, y = tY$. Use the chain rule to prove that

$$\alpha t^{\alpha-1} f(X, Y) = X f_x(tX, tY) + Y f_y(tX, tY).$$

Deduce that

$$x\frac{\partial f}{\partial x} + y\frac{\partial f}{\partial y} = \alpha f \qquad \text{(Euler's theorem)}.$$

SOME APPLICATIONS (OPTIONAL)

2.23   Indifference curves

It is often convenient to describe the preferences of a consumer over a set of commodity bundles (§1.33) with a real-valued utility function u. We then interpret

$$u(\mathbf{x}) < u(\mathbf{y})$$

as meaning that commodity bundle **y** is preferred to commodity bundle **x**.

The contours of such a utility function are called *indifference curves* because the consumer is indifferent between any two bundles on the same contour.

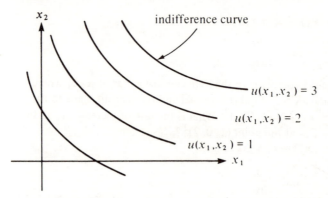

If the utility function is differentiable, its gradient indicates the direction in which utility increases fastest — i.e. the direction in which the consumer would most like to go. However, the consumer will be happy to move in any direction **v** for which $\langle \nabla u, \mathbf{v} \rangle \geq 0$. In our diagram, this means that points above an indifference curve are preferred to points on an indifference curve.

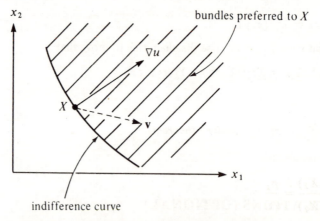

It is not unusual for a consumer to be constrained in his choice of a commodity bundle by the amount c of money in his possession. If the relevant price vector is **p**, he will only be able to purchase those bundles **x** for which

$$\langle \mathbf{p}, \mathbf{x} \rangle \leq c$$

(see §1.35). We call this inequality the consumer's *budget constraint* and the set of **x** which satisfies this constraint is the consumer's budget set. What **X** in the budget set maximizes the consumer's utility?

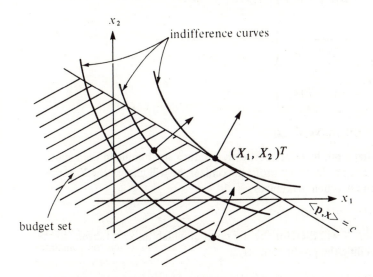

The only point in the budget set where the consumer cannot improve his position without going outside the budget set is $(X_1, X_2)^T$. Observe that, at this point, the gradient ∇u points in the same direction as the normal **p** to the hyperplane $\langle \mathbf{p}, \mathbf{x} \rangle = c$. This means that, for some scalar α,

$$\nabla \mathbf{u} = \alpha \mathbf{p}.$$

In the two-dimensional case, this can be written as

$$\left. \begin{aligned} u_{x_1}(X_1, X_2) &= \alpha p_1 \\ u_{x_2}(X_1, X_2) &= \alpha p_2. \end{aligned} \right\}$$

Eliminating α, we obtain that

$$\frac{u_{x_1}(X_1, X_2)}{u_{x_2}(X_1, X_2)} = \frac{p_1}{p_2}$$

In the special case when $u(x_1, x_2) = x_1^a x_2^b$, this equation reduces to

$$\frac{a X_1^{a-1} X_2^b}{b X_1^a X_2^{b-1}} = \frac{p_1}{p_2}$$

i.e.

$$p_2 X_2 = \frac{b}{a} p_1 X_1.$$

We also have that $(X_1, X_2)^T$ lies on $\langle \mathbf{p}, \mathbf{x} \rangle = c$ – i.e. $p_1 x_1 + p_2 x_2 = c$. Thus

$$c = p_1 X_1 + p_2 X_2 = p_1 X_1 + \frac{b}{a} p_1 X_1$$

and so

$$\left.\begin{array}{l} X_1 = \dfrac{ac}{(a+b)p_1}. \\[1em] X_2 = \dfrac{bc}{(a+b)p_2}. \end{array}\right\}$$

Therefore

2.24   Profit maximization

Suppose that a producer can costlessly produce any bundle in the region shaded. If the producer is a profit maximizer, he will choose the $\mathbf{X}$ in the production set which maximizes

$$u(\mathbf{x}) = \langle \mathbf{p}, \mathbf{x} \rangle = \mathbf{c}.$$

The point $\mathbf{X}$ will therefore be the point at which $\langle \mathbf{p}, \mathbf{x} \rangle$ is tangent to the curve bounding the production set.

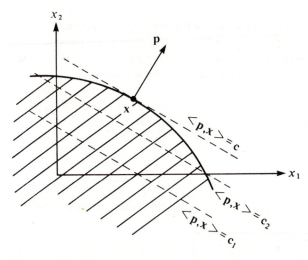

2.25   Contract curve

Robinson Crusoe and Man Friday wish to trade in wheat and fish. Robinson Crusoe begins with a commodity bundle $(\mathcal{W}, 0)$ representing a quantity $\mathcal{W}$ of wheat and no fish. Man Friday begins with a commodity bundle $(0, \mathcal{F})$ representing a quantity $\mathcal{F}$ of fish and no wheat. We call these bundles their respective endowments.

We denote Robinson's utility function by u and Friday's by v. If the result of the trading is that Robinson receives commodity bundle (w, f), then Friday will receive $(\mathscr{W} - w, \ \mathscr{F} - f)$. From the trade in which Robinson receives (w, f), it follows that Robinson will derive utility $u(w, f)$ and Friday will derive utility $V(w, f) = v(\mathscr{W} - w, \ \mathscr{F} - f)$.

The diagram drawn below is called the *Edgeworth box*. A point (w, f) in this box represents a trade in which Robinson receives the bundle (w, f) and Friday receives $(\mathscr{W} - w, \ \mathscr{F} - f)$. The firm curves represent Robinson's indifference curves (i.e. $u(w, f) = $ constant) and the broken curves represent Friday's indifference curves (i.e. $V(w, f) = $ constant)

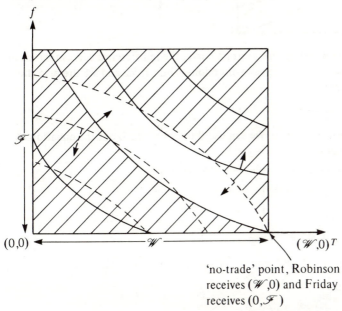

'no-trade' point, Robinson receives $(\mathscr{W}, 0)$ and Friday receives $(0, \mathscr{F})$

The points in the shaded region represent trades on which it is impossible that they will agree because one or the other receives less than the utility he enjoys at the 'no-trade' point. The traders will also not agree on the trades P or P' in the accompanying diagram because *both* traders prefer the trade Q.

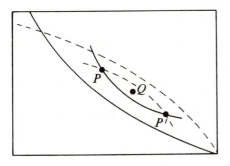

The only trades not ruled out by these considerations are those in the diagram below. The curve on which these lie is called the *contract curve*.

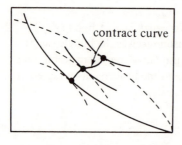
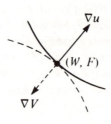

At points (W, F) on the contract curve, the indifference curves touch and their normals point in opposite directions. Thus, if ∇u and ∇V are evaluated at (W, F), then

$$\nabla V = -\alpha \nabla u$$

for some positive scalar α. We can use this fact to work out the equation of the contract curve.

Consider, for example, the case when

and
$$u(w, f) = wf^2$$
$$v(w, f) = w^2 f.$$

Let $\mathscr{W} = 2$ and $\mathscr{F} = 1$. Then

$$V(w, f) = v(2 - w, 1 - f) = (2 - w)^2 (1 - f).$$

We have that

$$\nabla u = (f^2, 2wf)$$
$$\nabla V = (-2(2 - w)(1 - f), -(2 - w)^2).$$

Since $\nabla V = -\alpha \nabla u$ at the point (W, F), we obtain that

$$-2(2 - W)(1 - F) = -\alpha F^2$$
$$-(2 - W)^2 = -\alpha 2WF.$$

Hence
$$\frac{2(1 - F)}{(2 - W)} = \frac{F}{2W}$$

$$4W - 4WF = 2F - FW$$

and so the contract curve is a segment of the curve with equation $3wf - 4w + 2f = 0$.

3

Stationary points

This chapter is about the use of first order partial derivatives in optimization. The one variable case (considered in §3.1, 3.2 and §3.13) is worth some careful attention since the discussion makes it clear that there is more to optimization than simply setting derivatives equal to zero. Example 3.14 is particularly instructive. The discussion of constained stationary points given in §3.15 and §3.17 might perhaps be omitted at a first reading. (Such sections are marked with the symbol †.) However, Lagrange's method, described in §3.18, is too widely used to be left out and it is important to be able to solve straightforward problems with this technique (even if the explanation of why the method works may not seem crystal clear). Note, incidentally, that example 3.21 does not qualify as a straightforward problem in this sense.

3.1 The one variable case[¤]

Let $f: \mathbb{R} \to \mathbb{R}$. A point ξ for which $f'(\xi) = 0$ is called a *stationary point* of the function. Recall from §2.3 that the tangent line to $y = f(x)$ at the point where $x = \xi$ has equation

$$y = f(\xi) + f'(\xi)(x - \xi).$$

It follows that ξ is a stationary point for f if and only if f has a *horizontal* tangent line where $x = \xi$.

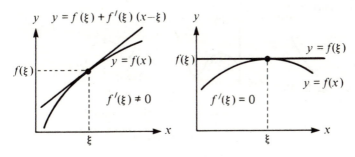

The stationary points of 'well-behaved' functions fall into three classes:
  (i) *local maxima*
  (ii) *local minima*
  (iii) *points of inflexion.*

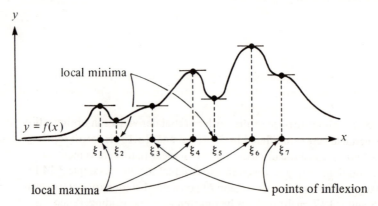

The word 'stationary' arises from the fact that, if a tiny ball bearing were to be balanced on the graph of $y = f(x)$ at any point where the tangent line is horizontal, then the ball bearing would not move. At any other point on the graph it would immediately begin to roll up or down.

3.2   Optimization

Suppose that $f: \mathbb{R} \to \mathbb{R}$. A common problem is to find a point ξ at which $f(x)$ achieves its maximum value – i.e.

$$f(\xi) = \max_{x} f(x).$$

Such a point ξ is called a *global maximum* for f. A function need *not* have a global maximum. However, if f does have a global maximum and f is differentiable, then the *global* maximum must be one of the *local* maxima.

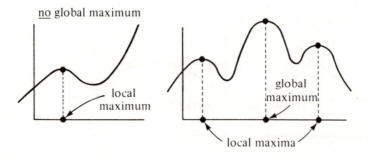

A helpful way to start on finding a global maximum of a function is therefore to find the stationary points of the function. Amongst these stationary points are the local maxima and hence the global maximum (if this exists).

Assuming that a global maximum exists, how does one decide which of the stationary points it is? A simple-minded (but not to be despised) method, is to evaluate $f(x)$ at each stationary point and to see at which of these points $f(x)$ is largest. One can often avoid some or all of these calculations by sketching a graph of $y = f(x)$. It is, in any case, *always* a good idea to sketch a graph of $y = f(x)$ whenever this is practicable. Amongst other things, it helps avoid the error of identifying a global maximum when none exists. Finally, the value of second derivative at the stationary points is often useful. This point is discussed in chapter 5.

Everything which has been said above about global maxima applies equally well, of course, to *global minima*.

Example 3.3.
Let $f: \mathbb{R} \to \mathbb{R}$ be defined by

$$f(x) = 3 + 7x - 5x^2 .$$

Then

$$f'(x) = 7 - 10x .$$

The stationary points are therefore found by solving the equation

$$7 - 10x = 0$$

i.e.

$$x = \tfrac{7}{10} .$$

It is not hard to sketch the graph of $y = f(x)$ in this case. It is apparent from the graph that we have located a global maximum.

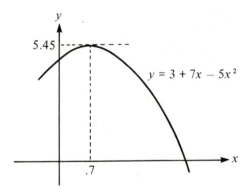

Observe that

$$\max_{x} f(x) = f(\tfrac{7}{10}) = 3 + 7(\tfrac{7}{10}) - 5(\tfrac{7}{10})^2$$
$$= 3 + \tfrac{49}{100}(10 - 5) = 3 + \tfrac{245}{100} = 5.45.$$

The maximum value of $f(x)$ is therefore 5.45.

Example 3.4.
Let $f: \mathbb{R} \to \mathbb{R}$ be defined by

$$f(x) = 2 + (x - 2)^3.$$

Then

$$f'(x) = 3(x - 2)^2.$$

The stationary points are therefore found by solving the equation

$$3(x - 2)^2 = 0$$

i.e.

$$x = 2.$$

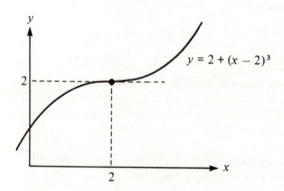

This stationary point is a *point of inflexion*. Observe that, when $x > 2$, $2 + (x - 2)^3 > 2$ and, when $x < 2$, $2 + (x - 2)^3 < 2$.

Note also that this function has *no* global maximum and *no* global minimum.

Example 3.5.
Let $f: \mathbb{R} \to \mathbb{R}$ be defined by

$$f(x) = x^3 - 3x^2 + 2x.$$

Then

$$f'(x) = 3x^2 - 6x + 2.$$

The stationary points are therefore found by solving the equation

$$3x^2 - 6x + 2 = 0$$

i.e.

$$x = \frac{6 \pm \sqrt{(36 - 24)}}{6} = 1 \pm \frac{1}{6}\sqrt{(12)}$$

$$x = 1 \pm \frac{1}{\sqrt{3}}.$$

Because $x^3 - 3x^2 + 2x = x(x^2 - 3x + 2) = x(x - 1)(x - 2)$, it is fairly easy to sketch the graph of $y = f(x)$. From this graph it is apparent that $\{1 - (1/\sqrt{3})\}$ is a *local maximum* and that $\{1 + (1/\sqrt{3})\}$ is a *local minimum*.

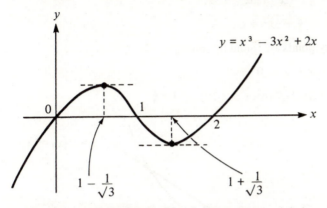

It is important to note that this function has *no* global maximum. There is *no* largest value of $f(x)$. The fact that a function has just one *local* maximum therefore does *not* guarantee that this local maximum is a global maximum. Precisely the same remarks apply in the case of minima.

Example 3.6.

A producer is assumed to seek to maximize profit. If a quantity x of his product is sold at price p (per unit quantity) his *profit* is given by

$$\pi(x) = R(x) - C(x),$$

where $R(x) = px$ is the *revenue* obtained from selling x and $C(x)$ is the *cost* of producing x.

Stationary points are obtained by solving the equation

$$\pi'(x) = R'(x) - C'(x) = 0$$

i.e.

$$R'(x) = C'(x).$$

Recalling that economists use the word 'marginal', to describe the derivative of a quantity, we obtain the 'economic law' that profit is maximized when

marginal revenue = marginal cost.

Note, however, that a capitalist who blindly sets marginal revenue equal to marginal cost could equally well be *minimizing* profit. The 'law' must therefore be applied with some caution.

Consider the case of 'perfect competition'. Here the producer is assumed to have such a small share of the market that variations in his output have a negligible effect on the price p of his product – i.e. p may be regarded as constant.

As an example, we take $p = \frac{4}{3}$ and $C(x) = x^3 - 3x^2 + 3x$. Thus

$$\pi(x) = \tfrac{4}{3}x - x^3 + 3x^2 - 3x.$$

Then

$$\pi'(x) = -\tfrac{5}{3} - 3x^2 + 6x.$$

The stationary points are found by solving

$$9x^2 - 18x + 5 = 0$$

i.e.

$$x = 1 \pm 18^{-1}\sqrt{(18^2 - 9 \times 20)} = 1 \pm \tfrac{2}{3}.$$

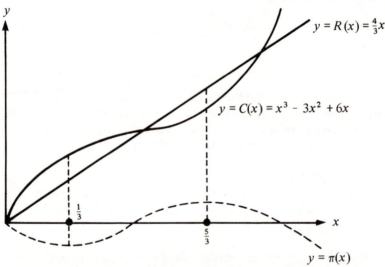

An examination of the diagram shows that it is the larger of these stationary points (i.e. $x = \frac{5}{3}$) which yields the required maximum – i.e. profit is maximized at $x = \frac{5}{3}$. However, marginal revenue = marginal cost also at the point $x = \frac{1}{3}$ where profit has a local minimum.

3.7   The two variable case

Let $f: \mathbb{R}^2 \to \mathbb{R}$. For such functions, $\boldsymbol{\xi}$ is a stationary point if and only if $z = f(x, y)$ has a *horizontal* tangent plane where $(x, y)^T = \boldsymbol{\xi}$.

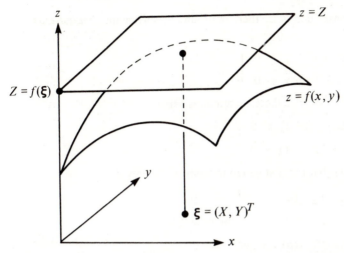

If $\boldsymbol{\xi} = (X, Y)^T$ and $Z = f(X, Y)$, then we know from §2.13 that a differentiable function has tangent plane

$$z - Z = f_x(X, Y)(x - X) + f_y(X, Y)(y - Y)$$

at the point in question. For this to be horizontal, it must take the form $z = Z$. Thus the condition for $\boldsymbol{\xi} = (X, Y)^T$ to be a stationary point is that

$$\left.\begin{aligned} f_x(X, Y) &= 0 \\ f_y(X, Y) &= 0. \end{aligned}\right\}$$

and

To find the stationary points of a function $f: \mathbb{R}^2 \to \mathbb{R}$, one must therefore find *all* solutions of the simultaneous equations

$$\left.\begin{aligned} \frac{\partial f}{\partial x} &= 0 \\ \frac{\partial f}{\partial y} &= 0. \end{aligned}\right\}$$

Example 3.8.
Let $f: \mathbb{R}^2 \to \mathbb{R}$ be defined by

$$f(x, y) = x^2 y + y^3 x - xy.$$

Then

$$\left.\begin{aligned} \frac{\partial f}{\partial x} &= 2xy + y^3 - y \\ \frac{\partial f}{\partial y} &= x^2 + 3y^2 x - x. \end{aligned}\right\}$$

The stationary points are therefore found by solving the *simultaneous* equations

$$2xy + y^3 - y = 0 \atop x^2 + 3y^2x - x = 0. \Big\}$$

It is helpful to factorize these equations as below:

$$y(2x + y^2 - 1) = 0 \atop x(x + 3y^2 - 1) = 0. \Big\}$$

It is then clear that the first equation holds if and only if

(a) $y = 0$ OR   (b) $2x + y^2 - 1 = 0$

and the second equation holds if and only if

(c) $x = 0$ OR   (d) $x + 3y^2 - 1 = 0$.

All solutions to the simultaneous equations are found by taking these latter equations in pairs — i.e.

(i) a and c
(ii) a and d
(iii) b and c
(iv) b and d.

Notice how easy it would be to lose one or more of the solutions by being careless at this stage. A systematic approach is therefore essential.

(i) *a and c*

$$y = 0 \atop x = 0. \Big\}$$

These equations have the unique solution $(0, 0)^T$.

(ii) *a and d*

$$y = 0 \atop x + 3y^2 - 1 = 0. \Big\}$$

Substituting $y = 0$ in $x + 3y^2 - 1 = 0$ yields $x = 1$. The equations therefore have the unique solution $(1, 0)^T$.

(iii) *b and c*

$$2x + y^2 - 1 = 0 \atop x = 0.. \Big\}$$

substituting $x = 0$ in $2x + y^2 - 1 = 0$ yields $y^2 = 1 - $ i.e. $y = \pm 1$. The equations therefore have two solutions, namely, $(0, 1)^T$ and $(0, -1)^T$.

(iv) $\underline{c}$ and $\underline{d}$

$$2x + y^2 - 1 = 0$$

$$x + 3y^2 - 1 = 0.$$

Multiply the first equation by 3 and then subtract the second equation. This yields $5x - 2 = 0$ – i.e. $x = \frac{2}{5}$. Substitute this result in the first equation. This gives $y^2 = \frac{1}{5}$ – i.e. $y = \pm 1/\sqrt{5}$. The equations therefore have two solutions, namely, $(\frac{2}{5}, 1/\sqrt{5})^T$ and $(\frac{2}{5}, -1/\sqrt{5})^T$.

The full list of stationary points is

$$(0,0)^T, (1,0)^T, (0,1)^T, (0,-1)^T, \left(\frac{2}{5}, \frac{1}{\sqrt{5}}\right)^T, \left(\frac{2}{5}, \frac{-1}{\sqrt{5}}\right)^T.$$

The stationary points of 'well-behaved' functions $f: \mathbb{R}^2 \to \mathbb{R}$ fall into three classes

(i) *local maxima*

(ii) *local minima*

(iii) *saddle points.*

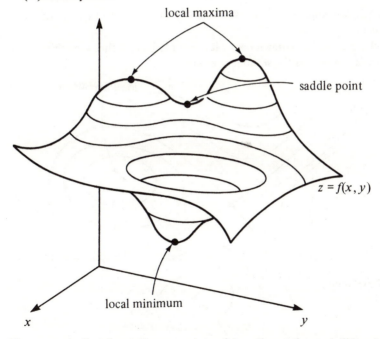

Observe again that the stationary points of f are those where a ball bearing can be balanced on the surface. Saddle points are so-called because the surface looks like a saddle at such a point. It is important to note that saddle points are a very common feature of a surface $z = f(x, y)$.

We shall discuss how one goes about classifying the stationary points of a function $f:\mathbb{R}^2 \to \mathbb{R}$ in chapter 5. For the moment, we shall only observe that, if one is seeking to find a global maximum or minimum of a function $f:\mathbb{R}^2 \to \mathbb{R}$, then it is a good idea to begin by finding the stationary points of f.

3.9 Gradient and stationary points

A stationary point of a function $f:\mathbb{R}^2 \to \mathbb{R}$ occurs when

$$\frac{\partial f}{\partial x} = 0$$

and

$$\frac{\partial f}{\partial y} = 0.$$

An alternative way of expressing this is by saying that, at a stationary point,

$$\nabla f = \left(\frac{\partial f}{\partial x}, \frac{\partial f}{\partial y}\right) = \mathbf{0}.$$

From §2.19 it follows that, at a stationary point, the rate of change of f in *all* directions is zero.

In the contour map drawn below it is not hard to see that something peculiar is going on at the stationary points.

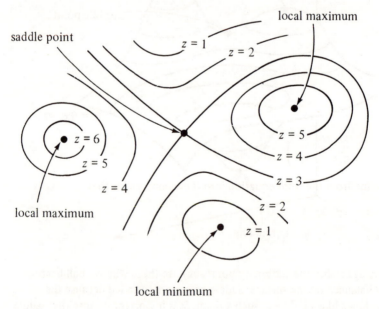

3.10 More than two variables

Given a function $f: \mathbb{R}^n \to \mathbb{R}$, one cannot draw pictures of tangent hyperplanes when $n > 2$. But one can still define a stationary point to be a point at which the rate of change of f in all directions is zero – i.e. a point at which

$$\nabla f = 0.$$

To find the stationary points of a function $f: \mathbb{R}^n \to \mathbb{R}$, one must therefore solve the *simultaneous* equations

$$\left. \begin{array}{l} \dfrac{\partial f}{\partial x_1} = 0 \\[2mm] \dfrac{\partial f}{\partial x_2} = 0 \\[1mm] \quad \vdots \\[1mm] \dfrac{\partial f}{\partial x_n} = 0. \end{array} \right\}$$

Example 3.11.
Let $f: \mathbb{R}^3 \to \mathbb{R}$ be defined by

$$f(x,y,z) = x^2 + yz + 3z^3 + y^4.$$

Then
$$\frac{\partial f}{\partial x} = 2x; \quad \frac{\partial f}{\partial y} = z + 4y^3; \quad \frac{\partial f}{\partial z} = y + 9z^2.$$

The stationary points are therefore found by solving the simultaneous equations

$$\left. \begin{array}{r} 2x = 0 \\ z + 4y^3 = 0 \\ y + 9z^2 = 0. \end{array} \right\}$$

Substituting from the third equation into the second yields that

$$z + 4(-9z^2)^3 = 0$$

i.e.
$$z = 0 \quad \underline{\text{OR}} \quad 1 - 4 \times 9^3 z^5 = 0$$

$$z^5 = \frac{1}{4 \times 9^3} = \frac{1}{12 \times 3^5}$$

$$z = \{\tfrac{1}{12}\}^{1/5} \tfrac{1}{3}.$$

We therefore obtain two stationary points, namely $(0, 0, 0)^T$ and

$$(0, -\{\tfrac{1}{12}\}^{2/5}, \tfrac{1}{3}\{\tfrac{1}{12}\}^{1/5})^T.$$

Exercises 3.12

1.$^\Pi$ Find the stationary points of the function $f: \mathbb{R} \to \mathbb{R}$ defined by

$$f(x) = e^{2x} + e^{-3x}.$$

Sketch the graph of $y = f(x)$ and hence classify the stationary points.

2.$^{*\Pi}$ Find the stationary points of the functions $f: \mathbb{R} \to \mathbb{R}$ defined by

$$\text{(i)} \quad f(x) = \frac{x^2 + x - 1}{x^2 + 1} \qquad \text{(ii)} \, f(x) = x^3 e^{-x^2}.$$

Sketch the graphs and hence classify the stationary points.

3.$^\Pi$ Show that the function $f: \mathbb{R} \to \mathbb{R}$ defined by

$$f(x) = x \sin x$$

has an infinite number of stationary points. Discuss the nature of the stationary point at $x = 0$. Does the function have a global minimum?

4.$^{*\Pi}$ Find the value of $x > 0$ which maximizes $\log x - x$.

5. By making the change of variable $X = x - 1$, $Y = y - 2$, show that the function $f: \mathbb{R}^2 \to \mathbb{R}$ defined by

$$f(x, y) = 2x + 4y - x^2 - y^2 - 3$$

achieves a global maximum of 2 at $(x, y)^T = (1, 2)^T$. Check that $(1, 2)^T$ is a stationary point of the function.

6.* Find all stationary points of the functions defined by

(i) $f(x, y) = x^2 y + y^3 x - xy^2$
(ii) $f(x, y) = e^{x+y}(x^2 - 2xy + 3y^2)$
(iii) $f(x, y) = y^3 + 3x^2 y - 3x^2 - 3y^2 + 2$
(iv) $f(x, y) = 8x^2 y - x^3 y - 5y^3 x$
(v) $f(x, y, z) = x^2 y + y^2 z + z^2 x.$

3.13   Constrained optimization

We have seen that a differentiable function $f: \mathbb{R} \to \mathbb{R}$ which has a global maximum achieves this maximum at a stationary point. In many problems, however, one is not particularly interested in finding the maximum of $f(x)$ for *all* values of x. Instead, one may have various *constraints* on the values which x can take. In this case the object will be to maximize $f(x)$ subject to these constraints.

Consider, for example, the problem of maximizing $f(x)$ subject to the constraint

$$a \leqslant x \leqslant b.$$

In this case, either the maximum will be achieved at a point ξ satisfying $a < \xi < b$ or else the maximum will be achieved at a or b.

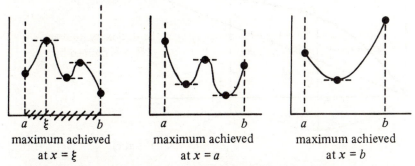

maximum achieved   maximum achieved   maximum achieved
at $x = \xi$ at $x = a$ at $x = b$

Observe that, *if* the maximum is achieved at $x = \xi$ where $a < \xi < b$, then ξ will be a stationary point of f. One way of proceeding is therefore to evaluate $f(\xi)$ for each stationary point ξ satisfying $a < \xi < b$ and *also* to evaluate $f(a)$ and $f(b)$. Then

$$\max_{a \leqslant x \leqslant b} f(x)$$

will be the largest of all these numbers.

Example 3.14.

Consider example 3.6 again but with $p = \frac{1}{3}$. This is actually a *constrained* optimization problem because it is impossible to produce a *negative* value of x. We are therefore really interested in

$$\max_{x \geqslant 0} \pi(x)$$

where $\pi(x) = R(x) - C(x) = \frac{1}{3}x - x^3 + 3x^2 - 3x$.

The stationary points are found by solving

$$\pi'(x) = \tfrac{1}{3} - 3x^2 + 6x - 3 = 0$$

i.e.

$$x^2 - 2x + (1 - \tfrac{1}{9}) = 0$$

i.e.

$$x = 1 \pm \sqrt{(1 - 8/9)} = 1 \pm \tfrac{1}{3}$$

The stationary points are therefore $x = \frac{2}{3}$ and $x = \frac{4}{3}$. Observe that $\pi(\frac{2}{3}) = -\frac{20}{27}$ and $\pi(\frac{4}{3}) = -\frac{16}{27}$. But one *cannot* deduce that the maximum is achieved at

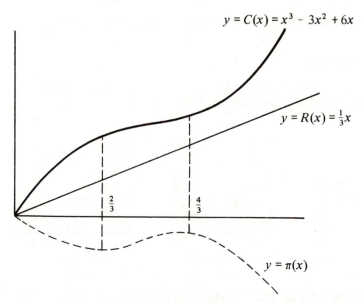

$$y = C(x) = x^3 - 3x^2 + 6x$$

$$y = R(x) = \tfrac{1}{3}x$$

$$\tfrac{2}{3}$$ $$\tfrac{4}{3}$$

$$y = \pi(x)$$

$x = \frac{4}{3}$. We also need to note the fact that $\pi(0) = 0$. It follows that the maximum is achieved at $x = 0$ – i.e. no production at all is the optimal course of action.

Similar considerations apply when seeking to maximize a function $f: \mathbb{R}^2 \to \mathbb{R}$ subject to the constraint that $(x, y)^T$ lie in some specified region S in $\mathbb{R}^2$. Assuming the maximum exists, it is *either* attained at an interior point of S *or* at a boundary point of S. If the maximum is attained at an interior point $\boldsymbol{\xi}$ then $\boldsymbol{\xi}$ is a stationary point of f but this need *not* be true if $\boldsymbol{\xi}$ is a boundary point of S.

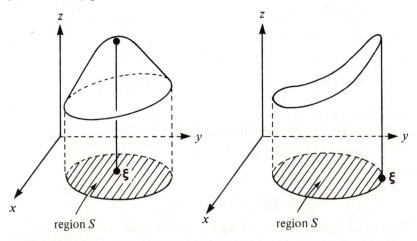

maximum attained at interior point $\boldsymbol{\xi}$ maximum attained at boundary point $\boldsymbol{\xi}$

3.15   Constrained stationary points†

Suppose that $f: \mathbb{R}^2 \to \mathbb{R}, g: \mathbb{R}^2 \to \mathbb{R}$ and we are seeking to evaluate

$$\max f(x,y)$$

subject to the constraint

$$g(x,y) = 0.$$

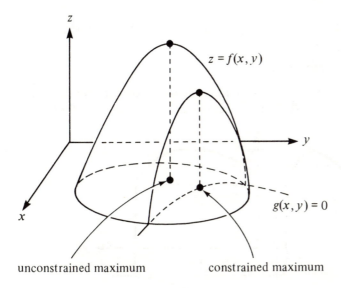

| unconstrained maximum | constrained maximum |

It may be, for example, that $g(x,y) = 0$ is the equation of the curve which bounds the region S in $\mathbb{R}^2$. The solution to our optimization problem will then provide the maximum value of $f(x,y)$ for $(x,y)^T$ on the boundary of S.

The most straightforward way of tackling this problem is to begin by solving the equation $g(x,y) = 0$ to obtain y in terms of x. Suppose that the formula which results from the solution of $g(x,y) = 0$ is

$$y = h(x).$$

Then we know that the points which lie on the curve $g(x,y) = 0$ are those of the form $(x, h(x))^T$.

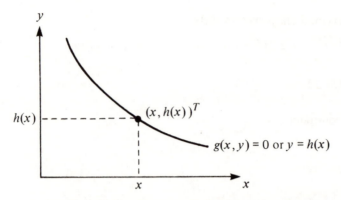

Our constrained optimization problem therefore reduces to finding the maximum of

$$\phi(x) = f(x, h(x))$$

for all x. But this is a one variable *unconstrained* problem. If ξ is a stationary point for ϕ, we shall call $(\xi, h(\xi))^T$ a constrained stationary point for f.

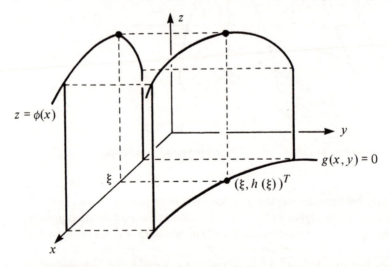

The same method can be used when $f : \mathbb{R}^3 \to \mathbb{R}, g : \mathbb{R}^3 \to \mathbb{R}$ and we are seeking to find

$$\max f(x, y, z)$$

subject to the constraint

$$g(x, y, z) = 0.$$

One can solve $g(x, y, z) = 0$ to obtain $z = h(x, y)$ and then examine the stationary points of

$$\phi(x, y) = f(x, y, h(x, y)).$$

The method can also be applied when there are several constraints. Suppose, for example, that $f: \mathbb{R}^3 \to \mathbb{R}, g_1: \mathbb{R}^3 \to \mathbb{R}, g_2: \mathbb{R}^3 \to \mathbb{R}$ and we are seeking to find

$$\max f(x, y, z)$$

subject to the constraints

$$\left. \begin{array}{l} g_1(x, y, z) = 0 \\ g_2(x, y, z) = 0. \end{array} \right\}$$

One can then solve the equations $g_1(x, y, z) = 0$ and $g_2(x, y, z) = 0$ simultaneously to obtain y *and* z as functions of x. Suppose that $y = h_1(x)$ and $z = h_2(x)$. The next step is then to examine the stationary points of

$$\phi(x) = f(x, h_1(x), h_2(x)).$$

Example 3.16.
Find

$$\min (3x^2 - 2y^2)$$

subject to the constraints

$$\left. \begin{array}{l} x + y \leqslant 2 \\ x - y \leqslant 0. \end{array} \right\}$$

The region S in $\mathbb{R}^2$ defined by these constraints is illustrated below:

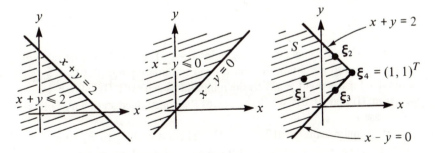

Assuming that the minimum exists, there are four different types of point in S at which the minimum could possibly be achieved.

(i) <u>Minimum achieved at a point like $\boldsymbol{\xi}_1$</u>. Then $\boldsymbol{\xi}_1$ will be an unconstrained stationary point of f. Such points are found by solving the simultaneous equations

$$\left.\begin{array}{l}\dfrac{\partial f}{\partial x} = 6x = 0 \\[3mm] \dfrac{\partial f}{\partial y} = -4y = 0.\end{array}\right\}$$

The only possibility is therefore $(0, 0)^T$. But this is *not* an interior point of S and so this case may be eliminated.

(ii) <u>Minimum achieved at a point like $\boldsymbol{\xi}_2$</u>. Then $\boldsymbol{\xi}_2$ will be a stationary point of f subject to the constraint $x + y = 2$. To find these stationary points, we can solve $x + y = 2$ for y in terms of x to obtain $y = 2 - x$. Then substitute the result in $f(x, y) = 3x^2 - 2y^2$. This yields

$$\phi(x) = 3x^2 - 2(2 - x)^2$$
$$= x^2 + 8x - 8.$$

But

$$\phi'(x) = 2x + 8$$

and hence ϕ has the single stationary point $x = -4$. The corresponding value of y is $y = 2 - (-4) = 6$ and so we obtain the constrained stationary point $(-4, 6)^T$ as a possibility for the point at which the minimum is achieved.

(iii) <u>Minimum achieved at a point like $\boldsymbol{\xi}_3$</u>. Then $\boldsymbol{\xi}_3$ will be a stationary point of f subject to the constraint $x - y = 0$. To find these stationary points, we can solve $x - y = 0$ for y in terms of x to obtain $y = x$. Then substitute the result in $f(x, y) = 3x^2 - 2y^2$. This yields

$$\phi(x) = 3x^2 - 2x^2 = x^2$$
$$\phi'(x) = 2x$$

and hence ϕ has the single stationary point $x = 0$. The corresponding value of y is $y = 0$, and so we obtain the constrained stationary point $(0, 0)^T$ as a possibility for the point at which the minimum is achieved. (It follows, in fact, from (i) that $(0, 0)^T$ would be one of the constrained stationary points in this case.)

(iv) <u>Minimum achieved at $\boldsymbol{\xi}_4 = (1, 1)^T$</u>. There is nothing to say here except that $(1, 1)^T$ is a possibility for the point at which the minimum is achieved.

We next observe that

$$f(-4, 6) = 48 - 72 = -24$$
$$f(0, 0) = 0$$
$$f(1, 1) = 1.$$

The minimum we are seeking is therefore -24 and this is achieved at the point $(-4, 6)^T$.

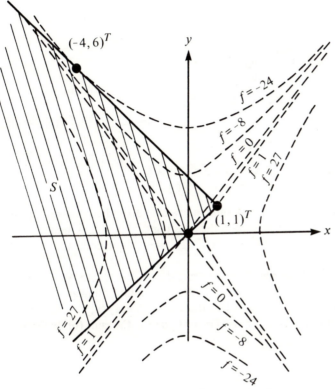

The contour map reveals a number of instructive facts:

(1) At the minimizing point $(-4, 6)^T$, the contour $3x^2 - 2y^2 = -24$ is tangent to the line $x + y = 2$.

(2) The point $(1, 1)^T$ is only a *local* maximum for the problem. The function has *no* maximum on the whole set S.

(3) The point $(0, 0)^T$ corresponds to a saddle point.

3.17   Constraints and gradients†

Suppose that $f: \mathbb{R}^2 \to \mathbb{R}$ and $g: \mathbb{R}^2 \to \mathbb{R}$ and that we are seeking to evaluate

$$\max f(x, y)$$

subject to the constraint

$$g(x,y) = 0.$$

The contour map below makes it clear that, if the maximum value of f subject to the constraint is M, then the contour $f(x,y) = M$ will be tangent to $g(x,y) = 0$ and the maximum will be achieved at the point ξ of tangency.

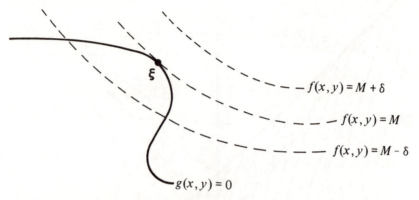

We encountered this phenomenon in example 3.16 when considering the minimum of $3x^2 - 2y^2$ subject to the constraint $x + y = 2$. A minimum of -24 is achieved at $(-4, 6)^T$ and $3x^2 - 2y^2 = -24$ is tangent to $x + y = 2$ at the point $(-4, 6)^T$.

One can exploit this phenomenon using the work of §2.15. We know that ∇f is a normal to the contour $f(x,y) = M$ at the point ξ. Similarly, ∇g is a normal to the contour $g(x,y) = 0$ at the point ξ. Since $f(x,y) = M$ touches $g(x,y) = 0$, it follows that ∇f and ∇g must point in the same (or opposite) directions.

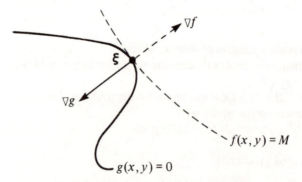

We conclude that ∇f must be a scalar multiple of ∇g at the point ξ and thus

$$\nabla f + \lambda \nabla g = 0$$

for some scalar λ. Thus

$$\left. \begin{array}{l} \dfrac{\partial f}{\partial x} + \lambda \dfrac{\partial g}{\partial x} = 0 \\[2mm] \dfrac{\partial f}{\partial y} + \lambda \dfrac{\partial g}{\partial y} = 0. \end{array} \right\}$$

If we apply this result in the case when $f(x,y) = 3x^2 - 2y^2$ and $g(x,y) = x + y - 2$, we obtain that

$$6x + \lambda = 0$$
$$-4y + \lambda = 0.$$

Subtracting these equations yields that

$$6x + 4y = 0.$$

In addition, we have the equation

$$x + y = 2$$

since ξ must satisfy the constraint $g(x, y) = 0$. Solving the simultaneous equations

$$\left. \begin{array}{l} 6x + 4y = 0 \\ x + y = 2 \end{array} \right\}$$

yields the result $(x, y)^T = (-4, 6)^T$ as obtained in example 3.16.

Lagrange's method, which we describe in the next section, is a simple generalization of this technique.

3.18   Lagrange's method

Suppose that $f : \mathbb{R}^n \to \mathbb{R}$ and that $g_1 : \mathbb{R}^n \to \mathbb{R}, g_2 : \mathbb{R}^n \to \mathbb{R}, \ldots, g_m : \mathbb{R}^n \to \mathbb{R}$. We have to consider the problem of evaluating the maximum or minimum of

$$f(x_1, x_2, \ldots, x_n)$$

subject to the constraints

$$\left. \begin{array}{l} g_1(x_1, x_2, \ldots, x_n) = 0 \\ g_2(x_1, x_2, \ldots, x_n) = 0 \\ \quad \vdots \\ g_m(x_1, x_2, \ldots, x_n) = 0. \end{array} \right\}$$

In §3.15, we discussed how this problem can be approached by solving the constraint equations for $x_{n-m+1}, x_{n-m+2}, \ldots, x_n$ in terms of $x_1, x_2, \ldots, x_{n-m}$. One may then substitute for $x_{n-m+1}, x_{n-m+2}, \ldots, x_n$ in $f(x_1, x_2, \ldots, x_n)$ and hence obtain an *unconstrained* problem in the $n-m$ variables $x_1, x_2, \ldots, x_{n-m}$.

This method is not to be despised but it needs little imagination to see that practical difficulties are liable to intrude.

In §3.17, we examined an alternative approach. Lagrange's method is the general version of this. It may seem somewhat perverse, but we begin by introducing m *new* variables

$$\lambda_1, \lambda_2, \ldots, \lambda_m$$

(one for each constraint equation). We then form the *Lagrangian* $L: \mathbb{R}^{n+m} \to \mathbb{R}$ defined by

$$L(x_1, x_2, \ldots, x_n, \lambda_1, \ldots, \lambda_m) = f + \lambda_1 g_1 + \lambda_2 g_2 + \ldots + \lambda_m g_m$$

and compute the *stationary points* of L. If

$$(\tilde{x}_1, \tilde{x}_2, \ldots, \tilde{x}_n)^T$$

is a constrained stationary point for f, then

$$(\tilde{x}_1, \tilde{x}_2, \ldots, \tilde{x}_n, \tilde{\lambda}_1, \tilde{\lambda}_2, \ldots, \tilde{\lambda}_m)^T$$

is an unconstrained stationary point for L (for some $\tilde{\lambda}_1, \tilde{\lambda}_2, \ldots, \tilde{\lambda}_m$).

Example 3.19.

Find the minimum distance from the point $(0, 0, 0)^T$ to the plane $lx + my + nz = p$. We shall assume that $l^2 + m^2 + n^2 = 1$ – i.e. that $(l, m, n)^T$ is a unit vector. We then know from §1.27 that the answer should be p.

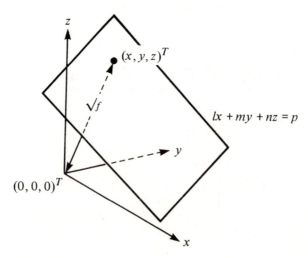

In analytical terms, the problem is to minimize

$$f(x, y, z) = x^2 + y^2 + z^2$$

subject to the constraint

$$lx + my + nz = p.$$

Method 1. We use the constraint to eliminate z. Then

$$\phi(x, y) = f\left(x, y, \frac{p - lx - my}{n}\right) = x^2 + y^2 + \frac{1}{n^2}(p - lx - my)^2.$$

We then seek the unconstrained stationary points of this expression. Since

$$\left.\begin{aligned}
\frac{\partial \phi}{\partial x} &= 2x + \frac{2}{n^2}(p - lx - my)(-l) \\
\frac{\partial \phi}{\partial y} &= 2y + \frac{2}{n^2}(p - lx - my)(-m),
\end{aligned}\right\}$$

the stationary points are found by solving

$$\left.\begin{aligned}
2x - \frac{2l}{n^2}(p - lx - my) &= 0 \\
2y - \frac{2m}{n^2}(p - lx - my) &= 0.
\end{aligned}\right\}$$

These simplify to

$$x(n^2 + l^2) + lmy = lp$$

$$lmx + y(n^2 + m^2) = mp$$

$$\begin{pmatrix} n^2 + l^2 & lm \\ lm & n^2 + m^2 \end{pmatrix}\begin{pmatrix} x \\ y \end{pmatrix} = \begin{pmatrix} lp \\ mp \end{pmatrix}$$

$$\begin{pmatrix} x \\ y \end{pmatrix} = \begin{pmatrix} n^2 + l^2 & lm \\ lm & n^2 + m^2 \end{pmatrix}^{-1}\begin{pmatrix} lp \\ mp \end{pmatrix} =$$

$$= \frac{p}{n^2(l^2 + m^2 + n^2)}\begin{pmatrix} n^2 + m^2 & -lm \\ -lm & n^2 + l^2 \end{pmatrix}\begin{pmatrix} l \\ m \end{pmatrix}.$$

Here we have used Cramer's rule (§1.13). Thus, recalling that $l^2 + m^2 + n^2 = 1$, we obtain that

$$x = lp \quad \text{and} \quad y = mp.$$

Substituting these values in $z = (p - lx - my)/n$ we obtain that the corresponding value of z is np. Thus there is a single constrained stationary point, namely $(lp, mp, np)^T$. Since the geometry guarantees the existence of a minimum, this must therefore be attained at $(lp, mp, np)^T$ and the required minimum value is

$$\sqrt{f} = \{(lp)^2 + (mp)^2 + (np)^2\}^{1/2} = p.$$

Method 2. We use Lagrange's method. The Lagrangian is

$$L = x^2 + y^2 + z^2 + \lambda(lx + my + nz - p).$$

To find the stationary points, we have to solve

(1) $\quad \dfrac{\partial L}{\partial x} = 2x + \lambda l = 0 \quad \text{i.e.}\ x = -\tfrac{1}{2}\lambda l$

(2) $\quad \dfrac{\partial L}{\partial y} = 2y + \lambda m = 0 \quad \text{i.e.}\ y = -\tfrac{1}{2}\lambda m$

(3) $\quad \dfrac{\partial L}{\partial z} = 2z + \lambda n = 0 \quad \text{i.e.}\ z = -\tfrac{1}{2}\lambda n$

(4) $\quad \dfrac{\partial L}{\partial \lambda} = lx + my + nz - p = 0.$

Note that the final equation is just the constraint equation again.

Substitute $x = -\tfrac{1}{2}\lambda l, y = -\tfrac{1}{2}\lambda m$ and $z = -\tfrac{1}{2}\lambda n$ in equation (4). We obtain that

$$-\tfrac{1}{2}\lambda(l^2 + m^2 + n^2) = p$$

i.e.

$$\lambda = -2p.$$

It follows that

$$x = lp, \quad y = mp, \quad z = np$$

and hence the result.

Example 3.20.
Evaluate

$$\max\{xy\}$$

subject to the constraint

$$x^2 + 4y^2 = 1.$$

The Lagrangian is

$$L = xy + \lambda(x^2 + 4y^2 - 1)$$

and

(1) $\quad \dfrac{\partial L}{\partial x} = y + 2\lambda x = 0 \left.\vphantom{\dfrac{\partial L}{\partial x}}\right\} \quad -2\lambda = \dfrac{y}{x}$

(2) $\quad \dfrac{\partial L}{\partial y} = x + 8\lambda y = 0 \left.\vphantom{\dfrac{\partial L}{\partial y}}\right\} \quad -\dfrac{1}{8\lambda} = \dfrac{y}{x}$

(3) $\quad \dfrac{\partial L}{\partial \lambda} = x^2 + 4y^2 - 1 = 0.$

From equations (1) and (2) we obtain that

$$\lambda^2 = \tfrac{1}{16}$$
$$\lambda = \pm \tfrac{1}{4}.$$

It follows that

$$y = \pm \tfrac{1}{2}x.$$

We substitute this result in equation (3) (the constraint equation) which has not yet been used. Then

$$x^2 + 4(\tfrac{1}{2}x)^2 = 1$$
$$x = \pm \frac{1}{\sqrt{2}}.$$

We therefore obtain four constrained stationary points, namely $(1/\sqrt{2}, 1/2\sqrt{2})^T, (-1/\sqrt{2}, 1/2\sqrt{2})^T, (1/\sqrt{2}, -1/2\sqrt{2})^T, (-1/\sqrt{2}, -1/2\sqrt{2})^T$. The value of xy at the first two of these points is $\tfrac{1}{4}$ and the value of xy at the second two is $-\tfrac{1}{4}$. The required maximum is therefore $\tfrac{1}{4}$.

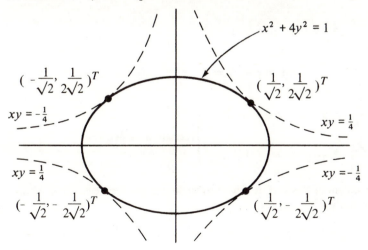

Example 3.21.
This is a considerably harder problem. Find

max {*xyz*}

subject to the constraints

$$x + y + z = a$$
$$xy + yz + zx = b.$$

One may regard this as the problem of finding the maximum volume of a brick when the sum of the lengths of the edges and the sum of the areas of the sides are given.

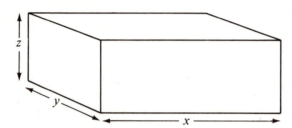

The Lagrangian is

$$L = xyz + \lambda(x + y + z - a) + \mu(xy + yz + zx - b).$$

From this we obtain the simultaneous equations

(1) $\dfrac{\partial L}{\partial x} = yz + \lambda + \mu(y + z) = 0$

(2) $\dfrac{\partial L}{\partial y} = zx + \lambda + \mu(z + x) = 0$

(3) $\dfrac{\partial L}{\partial z} = xy + \lambda + \mu(x + y) = 0$

(4) $\dfrac{\partial L}{\partial \lambda} = x + y + z - a = 0$

(5) $\dfrac{\partial L}{\partial \mu} = xy + yz + zx - b = 0.$

(constraint equations)

In a problem of this sort, it is wise to try and exploit the *symmetry* of the situation. Adding (1), (2) and (3) gives

$$(yz + zx + xy) + 3\lambda + 2\mu(x + y + z) = 0$$

i.e.

(6) $b + 3\lambda + 2a\mu = 0.$

Similarly, taking $x(1) + y(2) + z(3)$ yields that,

(7) $3f + a\lambda + 2b\mu = 0.$

Next consider (1)–(2) – i.e.

$$z(y - x) + \mu(y - x) = 0.$$

This implies that $x = y$ OR $z = -\mu$. Symmetric considerations lead to the equations

(8) $x = y$ OR $z = -\mu$

(9) $y = z$ OR $x = -\mu$

(10) $z = x$ OR $y = -\mu.$

From (8), (9) and (10), we obtain that

EITHER	(i) $x = y = z$
OR	(ii) $x = y = -\mu$
OR	(iii) $y = z = -\mu$
OR	(iv) $z = x = -\mu.$

Case (i). If $x = y = z$, then equations (4) and (5) show that $3x = a$ and $3x^2 = b$. Thus case (i) can hold only if $a^2 = 3b$ and then $x = a/3, y = a/3$ and $z = a/3$. Thus $f = (a/3)^3$.

Case (ii). If $x = y = -\mu$, then we may substitute in (3) and obtain

(11) $\mu^2 + \lambda - 2\mu^2 = 0$ – i.e. $\lambda = \mu^2$

and so, from (6) and (7),

(12) $b + 3\mu^2 + 2a\mu = 0$

(13) $3f + a\mu^2 + 2b\mu = 0.$

Thus,

$$ab + 3a\mu^2 + 2a^2\mu = 0$$

$$9f + 3a\mu^2 + 6b\mu = 0$$

(14) $(ab - 9f) + 2\mu(a^2 - 3b) = 0.$

Either $a^2 = 3b$, and so equation (14) asserts that $f = \frac{1}{9}ab = \frac{1}{27}a^3$ as in case (i), or else $a^2 \neq 3b$ and

$$\mu = \frac{9f - ab}{2(a^2 - 3b)}.$$

But, from (12),

$$\mu = \frac{-a \pm \sqrt{(a^2 - 3b)}}{3}$$

and thus

$$\frac{9f - ab}{2(a^2 - 3b)} = \frac{-a \pm \sqrt{(a^2 - 3b)}}{3}$$

$$f = \tfrac{1}{9}ab - \tfrac{2}{27}(a^2 - 3b)\{a \pm \sqrt{(a^2 - 3b)}\}.$$

By symmetry, cases (iii) and (iv) will, of course, lead to the same values for f. We may conclude that the problem only makes sense when $a^2 \geqslant 3b$ and in this case the maximum is

$$\tfrac{1}{9}ab - \tfrac{2}{27}(a^2 - 3b)\{a - \sqrt{(a^2 - 3b)}\}.$$

(One may ask about the minimum value of f. If we insist that $x \geqslant 0$, $y \geqslant 0, z \geqslant 0$, this will be achieved when $x = 0$. We then require that $y + z = a$ and $yz = b$. Thus $y = \tfrac{1}{2}\{a + \sqrt{(a^2 - 4b)}\}$ and $z = \tfrac{1}{2}\{a - \sqrt{(a^2 - 4b)}\}$. The constrained stationary point corresponding to $f = \tfrac{1}{9}ab - \tfrac{2}{27}(a^2 - 3b)\{a + \sqrt{(a^2 - 3b)}\}$ yields neither a maximum nor a minimum.)

Exercises 3.22

1. Find the stationary points of the function $f: \mathbb{R} \to \mathbb{R}$ defined by

$$f(x) = 6 - x - x^2.$$

Sketch the graph of the function and compute the following quantities:

(i) $\max\limits_{-3 \leqslant x \leqslant 2} f(x)$ (ii) $\min\limits_{-3 \leqslant x \leqslant 2} f(x)$

(iii) $\max\limits_{1 \leqslant x \leqslant 3} f(x)$ (iv) $\min\limits_{1 \leqslant x \leqslant 3} f(x)$.

2.* Let $f: \mathbb{R} \to \mathbb{R}$ be defined by

$$f(x) = x^3 - 3x^2 + 2x.$$

Compute the following quantities.

(i) $\max\limits_{0 \leqslant x \leqslant 1} f(x)$ (ii) $\min\limits_{0 \leqslant x \leqslant 1} f(x)$

(iii) $\max\limits_{0 \leqslant x \leqslant 2} f(x)$ (iv) $\min\limits_{0 \leqslant x \leqslant 2} f(x)$

(v) $\max_{0 \leqslant x \leqslant 3} f(x)$ (vi) $\min_{0 \leqslant x \leqslant 3} f(x)$.

[Example 3.5 will be helpful.]

3. A quantity x of a product realizes revenue $R(x) = px$ but costs $C(x) = x^3 - 3x^2 + 3x$ to produce (see example 3.6). The profit $\pi(x)$ is given by $\pi(x) = R(x) - C(x)$. Prove that

(i) If $0 \leqslant p < \frac{3}{4}$, then

$$\max_{x \geqslant 0} \pi(x) = \pi(0) = 0$$

(ii) If $p \geqslant \frac{3}{4}$, then

$$\max_{x \geqslant 0} \pi(x) = \pi(\xi)$$

where $\xi = 1 + \sqrt{(p/3)}$.

4.* A quantity x of a product realises revenue $R(x) = px$ but costs $C(x)$ to produce. The profit $\pi(x)$ is given by $\pi(x) = R(x) - C(x)$. For each $p \geqslant 0$, discuss the problem of finding the value of $x \geqslant 0$ which maximizes profit in the two cases

(i) $C(x) = x^2$ (ii) $C(x) = \sqrt{x}$.

5. Find all stationary points of the function $f : \mathbb{R}^2 \to \mathbb{R}$ defined by $f(x, y) = (x - 2)^2 + (y - 3)^2$.

(i) Determine

$$\min f(x, y)$$

subject to the constraint

$$2x + y \leqslant 2.$$

(ii) Determine

$$\min f(x, y)$$

subject to the constraint

$$3x + 2y \geqslant 6.$$

6.* Find all stationary points of the function $f : \mathbb{R}^2 \to \mathbb{R}$ defined by $f(x, y) = x^3 + y^3 - 3xy^2$.

Let S be the region in $\mathbb{R}^2$ consisting of the points inside and on the boundary of the triangle with vertices $(0, 0)^T$, $(0, 1)^T$ and $(1, 0)^T$. Find the maximum and minimum values of $f(x, y)$ subject to the constraint that $(x, y)^T$ lie in S.

7. Find the maximum and minimum values of $f = x^2 + y^2$ subject to the constraint

$$5x^2 + 6xy + 5y^2 = 8.$$

8.* Find the maximum and minimum values of $f = xy$ subject to the constraint

$$2x^2 + y^2 \leqslant 1.$$

9. Find the rectangle of largest area which has given perimeter $p > 0$.

10.* Find the maximum and minimum values of

$$f = lx + my + nz$$

subject to the constraint

$$\frac{x^2}{a^2} + \frac{y^2}{b^2} + \frac{z^2}{c^2} = 1.$$

11. Find the minimum distance between the curves $xy = 1$ and $x + 2y = 1$ — i.e. minimize

$$(x - X)^2 + (y - Y)^2$$

subject to the constraints

$$\left. \begin{array}{r} xy = 1 \\ X + 2Y = 1. \end{array} \right\}$$

12.* Find the minimum distance from the point $(0, 0, 0)^T$ to the line of intersection of the planes

$$\left. \begin{array}{r} x + 2y - 2z = 3 \\ 2x + y + 2z = 6. \end{array} \right\}$$

13. Find the constrained stationery points for xyz when the constraints are

$$\left. \begin{array}{r} x + 2y + 3z = 1 \\ 3x + 2y + z = 1. \end{array} \right\}$$

At which of these points is xyz largest? Is this a point at which xyz is maximised subject to the constraints?

14.* Show that the maximum and minimum values of $f = x^2 + y^2 + z^2$ subject to the constraints

$$\left. \begin{array}{r} a^2 x^2 + b^2 y^2 + c^2 z^2 = 1 \\ lx + my + nz = 0 \end{array} \right\}$$

satisfy the equation

$$\frac{l^2}{1 - a^2 f} + \frac{m^2}{1 - b^2 f} + \frac{n^2}{1 - c^2 f} = 0.$$

15. Suppose that an electricity generating company meets all demand made on it for the supply of electricity (i.e. no 'brown-outs') but is able to control demand by fixing the price. Suppose that demand x is related to price p by the formula $xp^2 = 1$. If the cost of meeting demand x is $C(x) = 2x^2$, what price maximizes profit $\pi(x) = px - C(x)$ and what will be the demand at this price?

16.* If the company need not meet all demand in the previous problem, we have to maximize $\pi(x)$ subject to $x \geqslant 0, p \geqslant 0, xp^2 \leqslant 1$. What is the result?

SOME APPLICATIONS (OPTIONAL)

3.23 The Nash bargaining problem

Suppose that S is a convex region in $\mathbb{R}^2$ and that s is a specified point in S. Two individuals negotiate over which point in S should be adopted. If they agree on $\mathbf{x} = (x_1, x_2)^T$, then the first individual receives a utility of x_1 units and the second a utility of x_2 units. On which point of S should they agree? If they *fail* to agree, it is to be understood that the result will be the 'status quo' point s.

The '*Nash bargaining solution*' for this problem is the point $\boldsymbol{\xi}$ at which

$$\max_{x} \; (x_1 - s_1)(x_2 - s_2)$$

is achieved subject to the constraints that $\mathbf{x}$ lie in the region S and $x_1 \geqslant s_1$, $x_2 \geqslant s_2$. Note that $\boldsymbol{\xi}$ is always a point on the boundary of S.

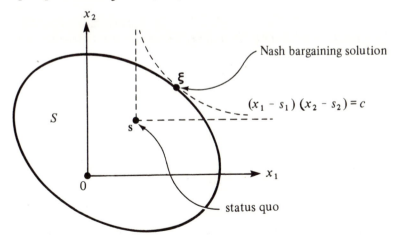

We shall not discuss the reasons which led Nash to propose $\boldsymbol{\xi}$ as the solution of the problem. Instead we shall consider some examples.

(i) Consider, to begin with, the case when S is the set of points inside or on the ellipse

$$x^2 + 4y^2 = 1$$

and $\mathbf{s} = (0, 0)^T$. Example 3.20 then applies and we obtain that the Nash bargaining solution is $\boldsymbol{\xi} = (1/\sqrt{2}, 1/2\sqrt{2})^T$. Thus the first individual receives $1/\sqrt{2}$ and the second $1/2\sqrt{2}$.

(ii) Suppose that S is the quadrilateral with vertices $(0, 0)^T$, $(0, 2)^T$, $(4, 1)^T$, $(5, 0)^T$ and $\mathbf{s} = (1, 0)^T$.

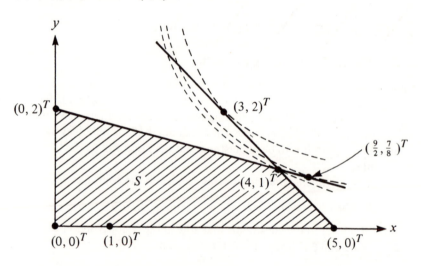

We begin by finding the point at which $(x - 1)y$ is maximized subject to the constraint $x + 4y = 8$. For this purpose, we evaluate the stationary points of

$$\phi(y) = (8 - 4y - 1)y = 7y - 4y^2$$

$$\phi'(y) = 7 - 8y$$

and hence a unique stationary point occurs when $y = \frac{7}{8}$. But the point $(\frac{9}{2}, \frac{7}{8})^T$ does not lie in the region S.

Next we find the point at which $(x - 1)y$ is maximized subject to $x + y = 5$.

$$\phi(y) = (5 - y - 1)y = (4 - y)y = 4y - y^2$$

$$\phi'(y) = 4 - 2y.$$

Hence a unique stationary point occurs when $y = 2$. But the point $(3, 2)^T$ does not lie in the region S.

We conclude that the Nash bargaining solution is $\boldsymbol{\xi} = (4, 1)^T$.

3.24   Inventory control

(i) At regular intervals, a firm orders a quantity x of a commodity which is placed in stock. This stock is depleted at a constant rate until none remains whereupon the firm immediately restocks with quantity x. The graph below shows how the amount in stock varies with time.

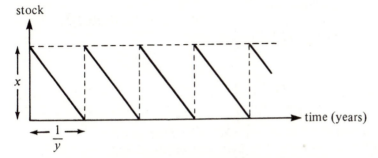

The firm requires X units of the commodity each year and, on average, the firm orders the commodity with a frequency of y times in each year. If the requirement is to be met, it is therefore necessary that

$$xy = X.$$

The cost of holding one unit of the commodity in stock for a year is d. Since the average amount of the commodity held in stock is $\frac{1}{2}x$, it follows that the yearly holding cost is $\frac{1}{2}xd$. The cost of reordering the commodity is e. The yearly reordering cost is therefore ey.

The problem that firm faces is to minimize cost

$$C(x, y) = \tfrac{1}{2}xd + ey$$

subject to the constraint

$$xy = X.$$

The Lagrangian is

$$L = \tfrac{1}{2}xd + ey + \lambda(xy - X)$$

(1) $\dfrac{\partial L}{\partial x} = \dfrac{1}{2}d + \lambda y = 0$

(2) $\dfrac{\partial L}{\partial y} = e + \lambda x = 0$

(3) $\dfrac{\partial L}{\partial \lambda} = xy - X = 0.$

From (1) and (2) we obtain that

$$\tfrac{1}{2}ed = \lambda^2 xy = \lambda^2 X.$$

Hence

$$\frac{1}{\lambda} = \pm \sqrt{\left(\frac{2X}{ed}\right)}$$

and the required solutions are

$$x = -\frac{e}{\lambda} = \sqrt{\left(\frac{2eX}{d}\right)} \\ y = -\frac{d}{2\lambda} = \sqrt{\left(\frac{dX}{2e}\right)}.$$

The firm should therefore order $\{2eXd^{-1}\}^{1/2}$ of the commodity $\{dX(2e)^{-1}\}^{1/2}$ times a year. The yearly cost will then be

$$\sqrt{(2deX)}.$$

(ii) Consider the previous problem but with the assumption that the firm does *not* immediately restock when its inventory is reduced to zero but waits until the demand for the commodity reaches a level of z before restocking with quantity $x + z$. Of this, z is processed immediately and the remaining x is depleted at a constant rate as before. Observe that $(x + z)y = X$.

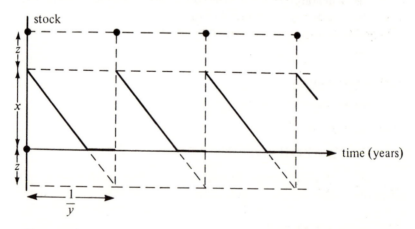

Since stock is only held for

$$\left(\frac{x}{x+z}\right)$$

of the year (under the new arrangements), the yearly holding cost becomes

$$\frac{1}{2}xd\left(\frac{x}{x+z}\right).$$

The yearly reordering cost remains ey but we must now also include a cost representing the penalty (e.g. overtime or customer dissatisfaction) resulting from the fact that a backlog is periodically built up which then has to be eliminated in a sudden flurry of activity. If it costs f to instantaneously clear one unit of backlog, then the yearly clearing cost will be

$$fyz.$$

The new problem is therefore to minimize

(1) $\quad C(x,y,z) = \frac{1}{2}xd\left(\frac{x}{x+z}\right) + ey + fyz$

subject to the constraint $(x+z)y = X$. Make the substitutions $x + z = Xy^{-1}$ and $z = Xy^{-1} - x$. We obtain

(2) $\quad D(x,y) = \frac{1}{2}dX^{-1}x^2y + ey + fX - fyx.$

The stationary points of this function are found by solving

(3) $\quad \dfrac{\partial D}{\partial x} = dX^{-1}xy - fy = 0$

(4) $\quad \dfrac{\partial D}{\partial y} = \dfrac{1}{2}dX^{-1}x^2 + e - fx = 0.$

From (3) we obtain that $y = 0$ or $x = fd^{-1}X$. The first possibility is incompatible with the constraint $(x + z)y = X$. Substituting the second possibility in (4), we obtain that

(5) $\quad \frac{1}{2}dX^{-1}(fd^{-1}X)^2 + e - f(fd^{-1}X) = 0$

i.e.

$$e = \frac{1}{2}f^2d^{-1}X.$$

If this equation happens not to hold, it then follows that (1) has *no* constrained stationary points. Note, however, that we have neglected to take account so far of the fact that $x \geqslant 0$, $y \geqslant 0$ and $z \geqslant 0$. It may therefore be that the minimum we are seeking occurs when $x = 0$ or $z = 0$ ($y = 0$ does not satisfy $(x + z)y = X$).

If $z = 0$, we are back with problem (i). If $x = 0$, we have to consider

$$C(0, y, Xy^{-1}) = ey + fX.$$

Since $y = 0$ is not admissible, this quantity has no minimum but we can make the quantity as close to fX as we choose by taking y sufficiently close to 0.

As a result of this analysis, we can distinguish two cases:

(a) $\sqrt{(2deX)} \leqslant fX$. In this case a minimum is obtained by taking $z = 0$ and choosing x and y as in problem (i).

(b) $\sqrt{(2deX)} > fX$. In this case *no* minimum exists. However, the firm can make its cost as close to fX as it chooses by taking $x = 0$ and making y very small (but positive). The corresponding value of z (i.e. Xy^{-1}) will then be very large. In this case, the penalty costs are so small compared with the other costs that the firm should hold nothing in stock and delay production for as long as possible.

Note incidentally that, when equation (5) holds, case (a) applies. Substituting $e = \frac{1}{2}f^2d^{-1}X$ and $x = fd^{-1}X$ in (2), we obtain that

$$D(x,y) = \tfrac{1}{2}dX^{-1}(fd^{-1}X)^2y + \tfrac{1}{2}f^2d^{-1}Xy + fX - fy(fd^{-1}X)$$

$$= fX.$$

4

Vector functions

We have seen in earlier chapters that partial derivatives are very useful objects. However, they do not provide, in themselves, an adequate generalization of the idea of the derivative of a real-valued function of one real variable. In particular, the rules for manipulating real derivatives take a rather complicated form when generalized to the vector case if expressed in terms of partial derivatives. In this chapter we therefore introduce the idea of the 'total derivative' of a vector function. This object is defined to be a matrix and an understanding of the material covered in this chapter requires a good knowledge of the matrix algebra surveyed in chapter 1. (Note especially § 1.31.) Although it is quite demanding at first to think systematically in terms of matrices and vectors, the payoff is quite considerable. The theory is elegant, the formulae are easily remembered and the actual amount that has to be written down is very much less than would otherwise be necessary. Some effort with this chapter will therefore be amply repaid.

4.1 Vector-valued functions

In this chapter we shall study functions

$$f : \mathbb{R}^n \to \mathbb{R}^m.$$

Such a function assigns to each $\mathbf{x} \in \mathbb{R}^n$ a unique vector $\mathbf{y} \in \mathbb{R}^m$. We write

$$\mathbf{y} = f(\mathbf{x}).$$

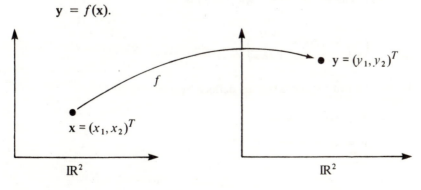

The diagram illustrates the case $f: \mathbb{R}^2 \to \mathbb{R}^2$. Since space is three-dimensional, there is no point of course in trying to draw the graph of such a function.

Suppose that

$$\mathbf{x} = \begin{pmatrix} x_1 \\ x_2 \\ \vdots \\ x_n \end{pmatrix} \quad \text{and} \quad \mathbf{y} = \begin{pmatrix} y_1 \\ y_2 \\ \vdots \\ y_m \end{pmatrix}.$$

Then the vector equation $\mathbf{y} = f(\mathbf{x})$ can be written out as a list of m *real* equations as below.

$$\left. \begin{aligned} y_1 &= f_1(x_1, x_2, \ldots, x_n) \\ y_2 &= f_2(x_1, x_2, \ldots, x_n) \\ &\vdots \\ y_m &= f_m(x_1, x_2, \ldots, x_n). \end{aligned} \right\}$$

A vector-valued function can therefore be studied by looking at the m *real*-valued functions $f_1, f_2, \ldots, f_m$. We shall say that $f_1, f_2, \ldots, f_m$ are the *component* functions of f.

Examples 4.2
(i) The equations

$$\left. \begin{aligned} x &= \cos t \\ y &= \sin t \end{aligned} \right\}$$

define a function $f: \mathbb{R}^1 \to \mathbb{R}^2$. We may write

$$\begin{pmatrix} x \\ y \end{pmatrix} = f(t) = \begin{pmatrix} \cos t \\ \sin t \end{pmatrix}.$$

The component functions of f are defined by

$$\left. \begin{aligned} x &= f_1(t) = \cos t \\ y &= f_2(t) = \sin t. \end{aligned} \right\}$$

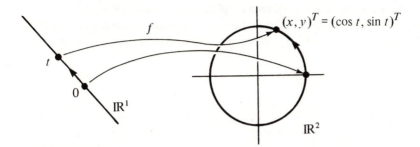

As t varies over the set of real numbers, the point $(x, y)^T = (\cos t, \sin t)^T$ describes a circle in $\mathbb{R}^2$. In a similar way, any function $g : \mathbb{R}^1 \to \mathbb{R}^m$ defines a curve in $\mathbb{R}^m$.

(ii) The equations

$$\left.\begin{aligned} u &= x^2 + y^2 \\ v &= x^2 - y^2 \end{aligned}\right\}$$

define a function $f : \mathbb{R}^2 \to \mathbb{R}^2$. We may write

$$\begin{pmatrix} u \\ v \end{pmatrix} = f(x, y) = \begin{pmatrix} x^2 + y^2 \\ x^2 - y^2 \end{pmatrix}.$$

The component functions of f are defined by

$$\left.\begin{aligned} u &= f_1(x, y) = x^2 + y^2 \\ v &= f_2(x, y) = x^2 - y^2. \end{aligned}\right\}$$

To illustrate a function $f : \mathbb{R}^2 \to \mathbb{R}^2$ one can draw contour maps of the real-valued component functions $f_1 : \mathbb{R}^2 \to \mathbb{R}$ and $f_2 : \mathbb{R}^2 \to \mathbb{R}$.

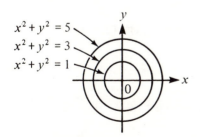

Contour map of $f_1 : \mathbb{R}^2 \to \mathbb{R}$

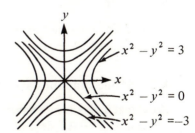

Contour map of $f_2 : \mathbb{R}^2 \to \mathbb{R}$

These contour maps can then be superimposed producing the pattern indicated in the diagram below on the left. The image of this pattern under the function $f: \mathbb{R}^2 \to \mathbb{R}^2$ is then the grid indicated in the diagram on the right.

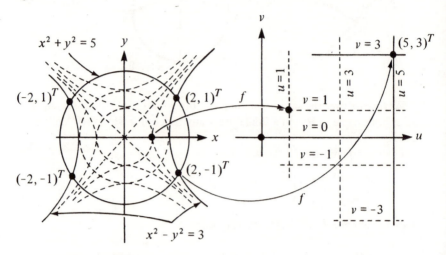

As the diagram indicates,

$$\binom{5}{3} = \binom{2^2 + (-1)^2}{2^2 - (-1)^2} = f(2, -1).$$

Thus the function f assigns the point $(5, 3)^T$ to the point $(2, -1)^T$. It is also true, however, that

$$\binom{5}{3} = f(2, -1) = f(2, 1) = f(-2, 1) = f(-2, -1).$$

Hence $(5, 3)^T$ is assigned to all four of the points $(2, 1)^T, (2, -1)^T,$ $(-2, 1)^T$ and $(-2, -1)^T$.

4.3   Affine functions and flats

Recall that if M is an $m \times n$ matrix and $\mathbf{c}$ is an $m \times 1$ vector, then the function $A: \mathbb{R}^n \to \mathbb{R}^m$ defined by

$$\mathbf{y} = A(\mathbf{x}) = M\mathbf{x} + \mathbf{c}$$

is called an *affine* function. As we know from §1.31, the graph of an affine function is a *flat*. If this flat passes through the point $(\boldsymbol{\xi}, \boldsymbol{\eta})^T$, then its equation may be rewritten in the form

$$\mathbf{y} - \boldsymbol{\eta} = M(\mathbf{x} - \boldsymbol{\xi}).$$

(i) $\underline{m = 1; n = 1}$. In this case M is the 1×1 matrix m and the equation takes the form

$$y - \eta = m(x - \xi).$$

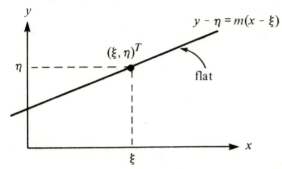

(ii) $\underline{m = 1; n = 2}$. In this case M is the 1×2 matrix

$$M = (l, m)$$

and the equation takes the form

$$y - \eta = (l, m)\begin{pmatrix} x_1 - \xi_1 \\ x_2 - \xi_2 \end{pmatrix}$$

i.e.

$$y - \eta = l(x_1 - \xi_1) + m(x_2 - \xi_2).$$

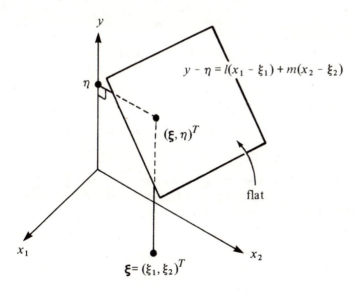

(iii) $\underline{m = 2; n = 1}$. In this case M is the 2×1 matrix

$$M = \begin{pmatrix} l \\ m \end{pmatrix}$$

and the equation takes the form

$$\begin{pmatrix} y_1 - \eta_1 \\ y_2 - \eta_2 \end{pmatrix} = \begin{pmatrix} l \\ m \end{pmatrix}(x - \xi)$$

i.e.

$$\left. \begin{aligned} y_1 - \eta_1 &= l(x - \xi) \\ y_2 - \eta_2 &= m(x - \xi). \end{aligned} \right\}$$

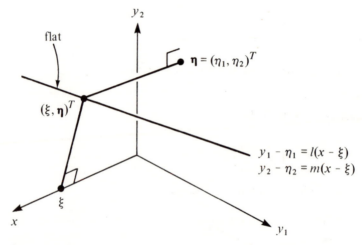

(Thus the flat is the line of intersection of the two planes $y_1 - \eta_1 = l(x - \xi)$ and $y_2 - \eta_2 = m(x - \xi)$.)

4.4 Derivatives of vector functions

A function $f: \mathbb{R}^n \to \mathbb{R}^m$ is *differentiable* at the point ξ if and only if there is a flat
$$y - \eta = M(x - \xi)$$
which is *tangent* to $y = f(x)$ where $x = \xi$. We must then have that $\eta = f(\xi)$.

If $f: \mathbb{R}^n \to \mathbb{R}^m$ is differentiable at ξ, we define its *derivative* $Df(\xi) = f'(\xi)$ to be the *matrix* M. We can then assert that the tangent to the graph $y = f(x)$ at the point $(\xi, f(\xi))^T$ has equation

$$y = f(\xi) + f'(\xi)(x - \xi).$$

(i) $f: \mathbb{R}^1 \to \mathbb{R}^1$. In this case, the derivative $f'(\xi)$ is a 1×1 matrix — i.e. a number.

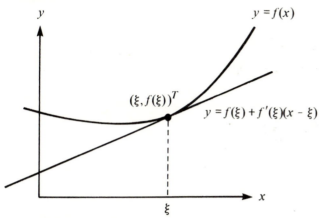

(ii) $f: \mathbb{R}^2 \to \mathbb{R}^1$. In this case, the derivative is a 1×2 matrix — i.e.

$$f'(\boldsymbol{\xi}) = (l, m)$$

and the tangent equation

$$y = f(\boldsymbol{\xi}) + f'(\boldsymbol{\xi})(\mathbf{x} - \boldsymbol{\xi})$$

assumes the form

$$y = f(\boldsymbol{\xi}) + (l, m)\begin{pmatrix} x_1 - \xi_1 \\ x_2 - \xi_2 \end{pmatrix}.$$

(iii) $f : \mathbb{R}^1 \to \mathbb{R}^2$. In this case, the derivative is a 2×1 matrix – i.e.

$$f'(\xi) = \begin{pmatrix} l \\ m \end{pmatrix}$$

and the tangent equation

$$y = f(\xi) + f'(\xi)(x - \xi)$$

assumes the form

$$y = f(\xi) + \begin{pmatrix} l \\ m \end{pmatrix}(x - \xi).$$

Writing out the equation $y = f(x)$ in terms of its component functions, we obtain that

$$
(1) \quad
\begin{cases}
\text{(i)} & y_1 = f_1(x_1, x_2, \ldots, x_n) \\
\text{(ii)} & y_2 = f_2(x_1, x_2, \ldots, x_n) \\
& \quad \vdots \\
\text{(m)} & y_m = f_m(x_1, x_2, \ldots, x_n).
\end{cases}
$$

Proceeding in a similar way with the equation $y - \boldsymbol{\eta} = M(x - \boldsymbol{\xi})$, we obtain

$$
\begin{pmatrix} y_1 - \eta_1 \\ y_2 - \eta_2 \\ \vdots \\ y_m - \eta_m \end{pmatrix}
=
\begin{pmatrix} l_{11} & l_{12} & \cdots & l_{1n} \\ l_{21} & l_{22} & \cdots & l_{2n} \\ \vdots & & & \\ l_{m1} & l_{m2} & \cdots & l_{mn} \end{pmatrix}
\begin{pmatrix} x_1 - \xi_1 \\ x_2 - \xi_2 \\ \vdots \\ x_n - \xi_n \end{pmatrix}
$$

and hence that

$$
(2) \begin{cases}
\text{(i)} & y_1 - \eta_1 = l_{11}(x_1 - \xi_1) + l_{12}(x_2 - \xi_2) + \ldots + l_{1n}(x_n - \xi_n) \\
\text{(ii)} & y_2 - \eta_2 = l_{21}(x_1 - \xi_1) + l_{22}(x_2 - \xi_2) + \ldots + l_{2n}(x_n - \xi_n) \\
& \vdots \\
\text{(m)} & y_m - \eta_m = l_{m1}(x_1 - \xi_1) + l_{m2}(x_2 - \xi_2) + \ldots + l_{mn}(x_n - \xi_n).
\end{cases}
$$

If $\mathbf{y} - \boldsymbol{\eta} = M(\mathbf{x} - \boldsymbol{\xi})$ is tangent to $\mathbf{y} = f(\mathbf{x})$, then the hyperplane (2(i)) is tangent to the 'hypersurface' (1(i)). Similarly (2(ii)) is tangent to (1(ii)) and so on.

But this observation means that we can use the remarks of §2.21 to evaluate the entries of the matrix M. We obtain that

$$
l_{11} = \frac{\partial f_1}{\partial x_1}, \quad l_{12} = \frac{\partial f_1}{\partial x_2}, \ldots, \quad l_{1n} = \frac{\partial f_1}{\partial x_n}
$$

$$
l_{21} = \frac{\partial f_2}{\partial x_1}, \quad l_{22} = \frac{\partial f_2}{\partial x_2}, \ldots, \quad l_{2n} = \frac{\partial f_2}{\partial x_n}
$$

$$
\vdots
$$

$$
l_{m1} = \frac{\partial f_m}{\partial x_1}, \quad l_{m2} = \frac{\partial f_m}{\partial x_2}, \ldots, l_{mn} = \frac{\partial f_m}{\partial x_n}
$$

where these partial derivatives are evaluated at the point $\boldsymbol{\xi}$. Thus, if $f : \mathbb{R}^m \to \mathbb{R}^n$ is differentiable at the point $\boldsymbol{\xi}$, then

$$
Df(\boldsymbol{\xi}) = f'(\boldsymbol{\xi}) = \begin{pmatrix}
\dfrac{\partial f_1}{\partial x_1} & \dfrac{\partial f_1}{\partial x_2} & \cdots & \dfrac{\partial f_1}{\partial x_n} \\[2mm]
\dfrac{\partial f_2}{\partial x_1} & \dfrac{\partial f_2}{\partial x_2} & \cdots & \dfrac{\partial f_2}{\partial x_n} \\[2mm]
\vdots & & & \\[2mm]
\dfrac{\partial f_m}{\partial x_1} & \dfrac{\partial f_m}{\partial x_2} & \cdots & \dfrac{\partial f_m}{\partial x_n}
\end{pmatrix}
$$

If $\mathbf{y} = f(\mathbf{x})$, we shall also find it very convenient to use the notation

$$
\frac{d\mathbf{y}}{d\mathbf{x}} = \begin{pmatrix}
\dfrac{\partial y_1}{\partial x_1} & \dfrac{\partial y_1}{\partial x_2} & \cdots & \dfrac{\partial y_1}{\partial x_n} \\[2mm]
\dfrac{\partial y_2}{\partial x_1} & \dfrac{\partial y_2}{\partial x_2} & \cdots & \dfrac{\partial y_n}{\partial x_n} \\[2mm]
\vdots & & & \\[2mm]
\dfrac{\partial y_m}{\partial x_1} & \dfrac{\partial y_m}{\partial x_2} & \cdots & \dfrac{\partial y_m}{\partial x_n}
\end{pmatrix}
$$

Example 4.5.

Consider the function $f: \mathbb{R}^3 \to \mathbb{R}^1$ defined by

$$v = f(\mathbf{u}) = xy^2 + yz^3 + zx^4,$$

where

$$\mathbf{u} = \begin{pmatrix} x \\ y \\ z \end{pmatrix}.$$

We have that

$$\frac{dv}{d\mathbf{u}} = \left(\frac{\partial v}{\partial x}, \frac{\partial v}{\partial y}, \frac{\partial v}{\partial z} \right) = (y^2 + 4zx^3, 2xy + z^3, 3yz^2 + x^4).$$

It follows that the tangent hyperplane to the 'hypersurface' $v = xy^2 + yz^3 + zx^4$ at $(x, y, z, v)^T = (1, 1, 1, 3)^T$ has equation

$$v - 3 = (5, 3, 4) \begin{pmatrix} x - 1 \\ y - 1 \\ z - 1 \end{pmatrix} = 5(x-1) + 3(y-1) + 4(z-1)$$

Example 4.6.

Consider the function $f: \mathbb{R}^1 \to \mathbb{R}^3$ defined by

$$\begin{pmatrix} x \\ y \\ z \end{pmatrix} = \mathbf{u} = f(t) = \begin{pmatrix} t \\ t^2 \\ t^3 \end{pmatrix}.$$

The function is therefore determined by the equations

$$\left. \begin{array}{l} x = t \\ y = t^2 \\ z = t^3. \end{array} \right\}$$

Thus

$$\frac{d\mathbf{u}}{dt} = \begin{pmatrix} \dfrac{\partial x}{\partial t} \\[2mm] \dfrac{\partial y}{\partial t} \\[2mm] \dfrac{\partial z}{\partial t} \end{pmatrix} = \begin{pmatrix} 1 \\ 2t \\ 3t^2 \end{pmatrix}.$$

The tangent at $(t, x, y, z)^T = (2, 2, 4, 8)^T$ is the line

$$\begin{pmatrix} x-2 \\ y-4 \\ z-8 \end{pmatrix} = \begin{pmatrix} 1 \\ 4 \\ 12 \end{pmatrix}(t-2)$$

i.e.

$$\left. \begin{array}{l} x - 2 = (t-2) \\ y - 4 = 4(t-2) \\ z - 8 = 12(t-2) \end{array} \right\}$$

which we can also write as

$$t - 2 = x - 2 = \frac{y-4}{4} = \frac{z-8}{12}.$$

Example 4.7.

Consider the function $f: \mathbb{R}^3 \to \mathbb{R}^2$ defined by

$$\begin{pmatrix} s \\ t \end{pmatrix} = \mathbf{v} = f(\mathbf{u}) = \begin{pmatrix} xy \\ xyz \end{pmatrix}$$

where

$$\mathbf{u} = \begin{pmatrix} x \\ y \\ z \end{pmatrix}.$$

The function is determined by the equations

$$s = xy$$

$$t = xyz.$$

Thus

$$\frac{d\mathbf{v}}{d\mathbf{u}} = \begin{pmatrix} \dfrac{\partial s}{\partial x} & \dfrac{\partial s}{\partial y} & \dfrac{\partial s}{\partial z} \\[2mm] \dfrac{\partial t}{\partial x} & \dfrac{\partial t}{\partial y} & \dfrac{\partial t}{\partial z} \end{pmatrix} = \begin{pmatrix} y & x & 0 \\ yz & xz & xy \end{pmatrix}$$

The tangent at $(x, y, z, s, t)^T = (1, 2, 3, 2, 6)^T$ is therefore the flat

$$\begin{pmatrix} s-2 \\ t-6 \end{pmatrix} = \begin{pmatrix} 1 & 2 & 0 \\ 6 & 3 & 2 \end{pmatrix}\begin{pmatrix} x-1 \\ y-2 \\ z-3 \end{pmatrix}.$$

4.8   Gradients and derivatives†

Suppose that $f: \mathbb{R}^n \to \mathbb{R}^1$ is a real-valued function. Then

$$
\nabla f = \begin{pmatrix} \dfrac{\partial f}{\partial x_1} \\[6pt] \dfrac{\partial f}{\partial x_2} \\[6pt] \vdots \\[6pt] \dfrac{\partial f}{\partial x_n} \end{pmatrix} ; \quad Df = \left(\dfrac{\partial f}{\partial x_1}, \dfrac{\partial f}{\partial x_2}, \dots, \dfrac{\partial f}{\partial x_n} \right).
$$

The *gradient* ∇f is therefore just the *transpose* of the *derivative* Df – i.e.

$$
\nabla f = (Df)^T.
$$

Our study of derivatives in this chapter is therefore just an extension of the work of chapter 2.

Recall in particular from chapter 2 that ∇f is normal to the contour $f(x) = \eta$ at the point $\boldsymbol{\xi}$ provided $\eta = f(\boldsymbol{\xi})$. This means that the hyperplane

$$
\frac{\partial f}{\partial x_1}(x_1 - \xi_1) + \frac{\partial f}{\partial x_2}(x_2 - \xi_2) + \dots + \frac{\partial f}{\partial x_n}(x_n - \xi_n) = 0
$$

is tangent to $f(x_1, x_2, \dots, x_n) = \eta$ (where the partial derivatives are calculated at the point $\boldsymbol{\xi}$).

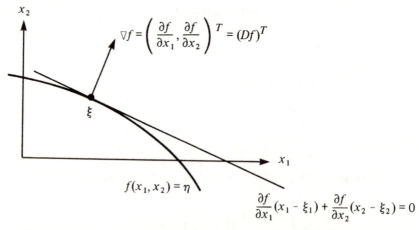

This observation is sometimes useful in discussing the derivative of a vector function $f: \mathbb{R}^n \to \mathbb{R}^m$. The affine function $A: \mathbb{R}^n \to \mathbb{R}^m$ defined by

$$y = A(x) = f(\xi) + f'(\xi)(x - \xi)$$

is tangent to $f: \mathbb{R}^n \to \mathbb{R}^m$ at the point (ξ, η) provided $\eta = f(\xi)$. It follows that the contours of this affine function are tangent to the corresponding contours of the function f at the point .

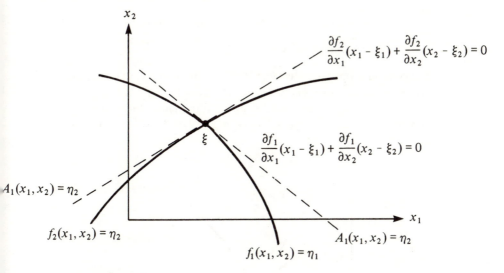

Example 4.9.†

Consider the function $f: \mathbb{R}^2 \to \mathbb{R}^2$ of example 4.2(ii). This is defined by the equations

$$u = f_1(x, y) = x^2 + y^2$$
$$v = f_2(x, y) = x^2 - y^2.$$

Since

$$\begin{pmatrix} \dfrac{\partial u}{\partial x} & \dfrac{\partial u}{\partial y} \\ \dfrac{\partial v}{\partial x} & \dfrac{\partial v}{\partial y} \end{pmatrix} = \begin{pmatrix} 2x & 2y \\ 2x & -2y \end{pmatrix},$$

we have that

$$f'(2, 1) = \begin{pmatrix} 4 & 2 \\ 4 & -2 \end{pmatrix}.$$

Also,

$$f(2, 1) = \begin{pmatrix} 5 \\ 3 \end{pmatrix}.$$

Hence the equation of the tangent to the function where $(x, y)^T = (2, 1)^T$ is

$$\begin{pmatrix} u \\ v \end{pmatrix} = A(x, y) = \begin{pmatrix} 5 \\ 3 \end{pmatrix} + \begin{pmatrix} 4 & 2 \\ 4 & -2 \end{pmatrix} \begin{pmatrix} x-2 \\ y-1 \end{pmatrix}$$

i.e.

$$\left. \begin{aligned} u &= A_1(x, y) = 5 + 4(x-2) + 2(y-1) \\ v &= A_2(x, y) = 3 + 4(x-2) - 2(y-1). \end{aligned} \right\}$$

The contour $A_1(x, y) = 5$ is tangent to the contour $f_1(x, y) = 5$ at the point $(x, y)^T = (2, 1)^T$ – i.e. $4(x-2) + 2(y-1) = 0$ is tangent to $x^2 + y^2 = 5$ at $(2, 1)^T$. Also, the contour $A_2(x, y) = 3$ is tangent to the contour $f_2(x, y) = 3$ at $(x, y)^T = (2, 1)^T$ – i.e. $4(x-2) - 2(y-1) = 0$ is tangent to $x^2 - y^2 = 3$ at $(2, 1)^T$.

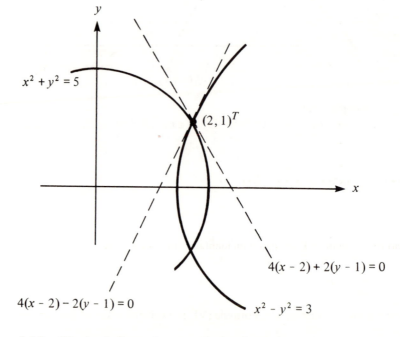

4.10   Manipulation of vector derivatives

In §2.3 we gave a list of rules for manipulating derivatives. It is a very satisfactory fact that these rules remain valid for vector derivatives *provided* that they make sense in this broader context. We list below some of the more useful versions of these rules:

$$\text{(I)} \quad \frac{d}{dx}(A\mathbf{y} + B\mathbf{z}) = A\frac{d\mathbf{y}}{dx} + B\frac{d\mathbf{z}}{dx}$$

(where A and B are constant matrices).

$$\text{(II)} \quad \frac{d}{dx}\left(\mathbf{y}^T\mathbf{z}\right) = \mathbf{z}^T\frac{d\mathbf{y}}{dx} + \mathbf{y}^T\frac{d\mathbf{z}}{dx}$$

$$\text{(III)} \quad \frac{d\mathbf{z}}{dx} = \frac{d\mathbf{z}}{d\mathbf{y}}\frac{d\mathbf{y}}{dx}$$

$$\text{(IV)} \quad \frac{d\mathbf{x}}{d\mathbf{y}} = \left(\frac{d\mathbf{y}}{d\mathbf{x}}\right)^{-1}$$

Notice the necessity of introducing the transpose signs in rule (II). The object $\mathbf{y}^T\mathbf{z}$ is then a scalar provided $\mathbf{y}$ and $\mathbf{z}$ are both $m \times 1$ column vectors. Notice also that the product in rule (IV) is a *matrix* product — i.e. the two derivatives are matrices and they are multiplied using matrix multiplications.

Also useful are the following formulae:

$$\text{(V)} \quad \frac{d}{dx}(A\mathbf{x} + \mathbf{b}) = A.$$

$$\text{(VI)} \quad \frac{d}{dx}\left(\mathbf{x}^T A\mathbf{x}\right) = \mathbf{x}^T\left(A^T + A\right).$$

The first of these is a consequence of the fact that an affine function is obviously tangent to itself. Note incidentally the special case

$$\frac{d\mathbf{x}}{d\mathbf{x}} = I,$$

where I is the identity matrix. Formula (VI) is a generalization of the familiar result that

$$\frac{d}{dx}(ax^2) = 2ax.$$

This is more apparent in the case when A is a *symmetric* matrix (i.e. $A^T = A$). In this special case

$$\frac{d}{d\mathbf{x}}(\mathbf{x}^T A \mathbf{x}) = 2\mathbf{x}^T A.$$

In order for (VI) to make sense, it is necessary that $\mathbf{x}$ be an $n \times 1$ column vector and that A be an $n \times n$ square matrix. In this case $\mathbf{x}^T A \mathbf{x}$ is a scalar.

Example 4.11.
It is instructive to see how rule (VI) may be deduced from rules (II) and (V). We have that

$$\frac{d}{d\mathbf{x}}\{\mathbf{x}^T A \mathbf{x}\} = \frac{d}{d\mathbf{x}}\{\mathbf{x}^T (A\mathbf{x})\}$$

$$= (A\mathbf{x})^T \frac{d\mathbf{x}}{d\mathbf{x}} + \mathbf{x}^T \frac{d}{d\mathbf{x}}(A\mathbf{x})$$

$$= \mathbf{x}^T A^T I + \mathbf{x}^T A$$

$$= \mathbf{x}^T (A^T + A).$$

Example 4.12.
Suppose that

$$z = \mathbf{u}^T A \mathbf{u} = (x, y)\begin{pmatrix} 1 & 2 \\ 2 & 4 \end{pmatrix}\begin{pmatrix} x \\ y \end{pmatrix} = x^2 + 4xy + 4y^2.$$

Then

$$\frac{\partial z}{\partial x} = 2x + 4y; \quad \frac{\partial z}{\partial y} = 4x + 8y.$$

Thus

$$\frac{dz}{d\mathbf{u}} = \left(\frac{\partial z}{\partial x}, \frac{\partial z}{\partial y}\right) = (2x + 4y, 4x + 8y).$$

Alternatively, using rule (VI),

$$\frac{dz}{d\mathbf{u}} = \frac{d}{d\mathbf{u}}(\mathbf{u}^T A \mathbf{u}) = 2\mathbf{u}^T A$$

$$= 2(x, y)\begin{pmatrix} 1 & 2 \\ 2 & 4 \end{pmatrix}$$

$$= (2x + 4y, 4x + 8y).$$

Example 4.13.
In 'least squares analysis' one is given an *inconsistent* system $A\mathbf{x} = \mathbf{b}$ of linear equations and asked to find the vector (or vectors) $\mathbf{x}$ which minimizes $\|A\mathbf{x} - \mathbf{b}\|$. Consider

$$y = \|A\mathbf{x} - \mathbf{b}\|^2 = (A\mathbf{x} - \mathbf{b})^T(A\mathbf{x} - \mathbf{b}) = \mathbf{u}^T\mathbf{u},$$

where $\mathbf{u} = A\mathbf{x} - \mathbf{b}$. Then, by rule (III),

$$\frac{dy}{d\mathbf{x}} = \frac{dy}{d\mathbf{u}}\frac{d\mathbf{u}}{d\mathbf{x}} = (2\mathbf{u}^T I)A = 2(A\mathbf{x} - \mathbf{b})^T A.$$

For the stationary points, we require that

$$\frac{\partial y}{\partial x_1} = \frac{\partial y}{\partial x_2} = \frac{\partial y}{\partial x_3} = \dots = \frac{\partial y}{\partial x_n} = 0$$

i.e.

$$\frac{dy}{d\mathbf{x}} = \left(\frac{\partial y}{\partial x_1}, \frac{\partial y}{\partial x_2}, \dots, \frac{\partial y}{\partial x_n}\right) = (0, 0, \dots, 0) = \mathbf{0}.$$

We therefore have to solve

$$2(A\mathbf{x} - \mathbf{b})^T A = \mathbf{0}$$
$$A^T(A\mathbf{x} - \mathbf{b}) = \mathbf{0}$$
$$A^T A\mathbf{x} = A^T\mathbf{b}.$$

Since a minimum certainly does exist, *these* equations must be solvable. If the square matrix $A^T A$ happens to be non-singular, there is a unique solution given by

$$\mathbf{x} = (A^T A)^{-1}A^T\mathbf{b}.$$

4.14 Chain rule

Suppose that $f: \mathbb{R}^n \to \mathbb{R}$ and that $g: \mathbb{R} \to \mathbb{R}^n$. We shall write

$$y = f(\mathbf{x}) \quad \text{and} \quad \mathbf{x} = g(t).$$

Then

$$\frac{dy}{d\mathbf{x}} = \left(\frac{\partial y}{\partial x_1}, \frac{\partial y}{\partial x_2}, \dots, \frac{\partial y}{\partial x_n}\right)$$

$$\frac{d\mathbf{x}}{dt} = \begin{pmatrix} \dfrac{dx_1}{dt} \\ \dfrac{dx_2}{dt} \\ \vdots \\ \dfrac{dx_n}{dt} \end{pmatrix}.$$

Let $h : \mathbb{R} \to \mathbb{R}$ be the function defined by $y = h(t) = f(g(t))$. Then rule (III) asserts that

$$\frac{dy}{dt} = \frac{dy}{d\mathbf{x}} \frac{d\mathbf{x}}{dt}$$

$$= \left(\frac{\partial y}{\partial x_1}, \frac{\partial y}{\partial x_2}, \ldots, \frac{\partial y}{\partial x_n} \right) \begin{pmatrix} \dfrac{dx_1}{dt} \\[2mm] \dfrac{dx_2}{dt} \\ \cdot \\ \cdot \\ \cdot \\ \dfrac{dx_n}{dt} \end{pmatrix} .$$

Thus

$$\frac{dy}{dt} = \frac{\partial y}{\partial x_1} \frac{dx_1}{dt} + \frac{\partial y}{\partial x_2} \frac{dx_2}{dt} + \ldots + \frac{\partial y}{\partial x_n} \frac{dx_n}{dt}.$$

This is the chain rule which we met previously in §2.17.

More generally, suppose that $f : \mathbb{R}^m \to \mathbb{R}$ and $g : \mathbb{R}^n \to \mathbb{R}^m$. We may then write

$$y = f(\mathbf{x}) \quad \text{and} \quad \mathbf{x} = g(\mathbf{u}).$$

We have that

$$\frac{dy}{d\mathbf{x}} = \left(\frac{\partial y}{\partial x_1}, \frac{\partial y}{\partial x_2}, \ldots, \frac{\partial y}{\partial x_m} \right)$$

$$\frac{d\mathbf{x}}{d\mathbf{u}} = \begin{pmatrix} \dfrac{\partial x_1}{\partial u_1} & \dfrac{\partial x_1}{\partial u_2} & , \ldots, & \dfrac{\partial x_1}{\partial u_n} \\[2mm] \dfrac{\partial x_2}{\partial u_1} & \dfrac{\partial x_2}{\partial u_2} & , \ldots, & \dfrac{\partial x_2}{\partial u_n} \\ \cdot \\ \cdot \\ \dfrac{\partial x_m}{\partial u_1} & \dfrac{\partial x_m}{\partial u_2} & , \ldots, & \dfrac{\partial x_m}{\partial u_n} \end{pmatrix}$$

Thus, by rule (III),

$$\left(\frac{\partial y}{\partial u_1}, \frac{\partial y}{\partial u_2}, \ldots, \frac{\partial y}{\partial u_n} \right) = \frac{dy}{d\mathbf{u}} = \frac{dy}{d\mathbf{x}} \frac{d\mathbf{x}}{d\mathbf{u}}$$

$$= \left(\frac{\partial y}{\partial x_1}, \ldots, \frac{\partial y}{\partial x_m} \right) \begin{pmatrix} \dfrac{\partial x_1}{\partial u_1}, & \ldots, & \dfrac{\partial x_1}{\partial u_n} \\ \cdot & & \\ \cdot & & \\ \cdot & & \\ \dfrac{\partial x_m}{\partial u_1}, & \ldots, & \dfrac{\partial x_m}{\partial u_n} \end{pmatrix}$$

Writing this out in full,

$$\frac{\partial y}{\partial u_1} = \frac{\partial y}{\partial x_1} \frac{\partial x_1}{\partial u_1} + \frac{\partial y}{\partial x_2} \frac{\partial x_2}{\partial u_1} + \ldots + \frac{\partial y}{\partial x_m} \frac{\partial x_m}{\partial u_1}$$

$$\frac{\partial y}{\partial u_2} = \frac{\partial y}{\partial x_1} \frac{\partial x_1}{\partial u_2} + \frac{\partial y}{\partial x_2} \frac{\partial x_2}{\partial u_2} + \ldots + \frac{\partial y}{\partial x_m} \frac{\partial x_m}{\partial u_2}$$

$$\vdots$$

$$\frac{\partial y}{\partial u_n} = \frac{\partial y}{\partial x_1} \frac{\partial x_1}{\partial u_n} + \frac{\partial y}{\partial x_2} \frac{\partial x_2}{\partial u_n} + \ldots + \frac{\partial y}{\partial x_m} \frac{\partial x_m}{\partial u_n}.$$

Example 4.15.

Suppose that $z = xy^2$ where

$$x = \tfrac{1}{2}(s^2 - t^2)$$
$$y = st.$$

Then

$$\frac{\partial z}{\partial s} = \frac{\partial z}{\partial x} \frac{\partial x}{\partial s} + \frac{\partial z}{\partial y} \frac{\partial y}{\partial s}$$

In these partial derivatives it is, as always, important to be quite clear about what variables are being held constant. This is made clear in the equation below:

$$\left(\frac{\partial z}{\partial s} \right)_t = \left(\frac{\partial z}{\partial x} \right)_y \left(\frac{\partial x}{\partial s} \right)_t + \left(\frac{\partial z}{\partial y} \right)_x \left(\frac{\partial y}{\partial s} \right)_t.$$

We obtain that

$$\left(\frac{\partial z}{\partial s} \right)_t = y^2 s + 2xyt$$

$$= s^3 t^2 + (s^2 - t^2)st^2$$

$$= 2s^3 t^2 - st^4.$$

Similarly,

$$\left(\frac{\partial z}{\partial t}\right)_s = \left(\frac{\partial z}{\partial x}\right)_y \left(\frac{\partial x}{\partial t}\right)_s + \left(\frac{\partial z}{\partial y}\right)_x \left(\frac{\partial y}{\partial t}\right)_s$$

$$= -y^2 t + 2yxs$$

$$= -s^2 t^3 + (s^2 - t^2)s^2 t$$

$$= -2s^2 t^3 + s^4 t.$$

4.16   Stationary points

We have seen in example 4.13 that, if $y = f(\mathbf{x})$, then the stationary points of a function $f : \mathbb{R}^n \to \mathbb{R}$ occur when

$$\frac{dy}{d\mathbf{x}} = \mathbf{0}.$$

A similar very satisfactory shorthand notation is available for constrained optimization problems. Suppose that $f : \mathbb{R}^n \to \mathbb{R}$ and that $g : \mathbb{R}^n \to \mathbb{R}^m$. The problem is to find the stationary points of $y = f(\mathbf{x})$ subject to the constraint $g(\mathbf{x}) = \mathbf{0}$.

The Lagrangian for this problem is

$$L = f(\mathbf{x}) + \boldsymbol{\lambda}^T g(\mathbf{x}) = f(\mathbf{x})^T + g(\mathbf{x})^T \boldsymbol{\lambda}$$

(where $\boldsymbol{\lambda}$ is an $m \times 1$ column vector) and, as always, we require the stationary points of L. These are found by solving the simultaneous equations

$$\begin{cases} \dfrac{\partial L}{\partial \mathbf{x}} = \mathbf{0} - \text{i.e.} \, f'(\mathbf{x}) + \boldsymbol{\lambda}^T g'(\mathbf{x}) = \mathbf{0} \\[2mm] \dfrac{\partial L}{\partial \boldsymbol{\lambda}} = \mathbf{0} - \text{i.e.} \, g(\mathbf{x}) = \mathbf{0}. \end{cases}$$

Example 4.17.

Suppose that A is a non-singular $n \times n$ symmetric matrix and that $\mathbf{m}$ is a $1 \times n$ row vector. Prove that a maximum or minimum value of

$$y = \mathbf{mx}.$$

subject to the constraint $\mathbf{x}^T A \mathbf{x} = 1$ must satisfy

$$y^2 = \mathbf{m} A^{-1} \mathbf{m}^T.$$

We form the Lagrangian

$$L = \mathbf{mx} + \lambda(\mathbf{x}^T A \mathbf{x} - 1) = 0.$$

The stationary points are found by solving

(1) $\quad \dfrac{\partial L}{\partial \mathbf{x}} = \mathbf{m} + 2\lambda \mathbf{x}^T A = \mathbf{0}$

(2) $\quad \dfrac{\partial L}{\partial \lambda} = \mathbf{x}^T A \mathbf{x} - 1 = 0.$

Multiplying (1) through by A^{-1} we obtain that

$$\mathbf{m} A^{-1} + 2\lambda \mathbf{x}^T = \mathbf{0}$$

$$\mathbf{x}^T = -\frac{1}{2\lambda} \mathbf{m} A^{-1}$$

$$\mathbf{x} = -\frac{1}{2\lambda} (\mathbf{m} A^{-1})^T$$

$$= -\frac{1}{2\lambda} A^{-1} \mathbf{m}^T \quad \text{(recall that } A^T = A\text{).}$$

Substituting in (2) yields that

$$\left(-\frac{1}{2\lambda} \mathbf{m} A^{-1}\right) A \left(-\frac{1}{2\lambda} A^{-1} \mathbf{m}^T\right) = 1$$

$$(2\lambda)^2 = \mathbf{m} A^{-1} \mathbf{m}^T.$$

But, multiplying (1) through by $\mathbf{x}$,

$$\mathbf{m}\mathbf{x} + 2\lambda \mathbf{x}^T A \mathbf{x} = 0$$

$$y + 2\lambda = 0$$

$$y^2 = (2\lambda)^2 = \mathbf{m} A^{-1} \mathbf{m}^T$$

as required.

4.18 Second derivatives

The second derivative of a function $f : \mathbb{R} \to \mathbb{R}$ at the point ξ is denoted by $f''(\xi)$ or by $D^2 f(\xi)$. This is simply the derivative at the point ξ of the derivative of f. If $y = f(x)$, we also use the notation

$$\frac{d^2 y}{dx^2} = \frac{d}{dx}\left(\frac{dy}{dx}\right).$$

Higher order derivatives can also be defined in a similar way. We denote the nth order derivative of $f : \mathbb{R} \to \mathbb{R}$ at the point ξ by $f^{(n)}(\xi)$ or by $D^n f(\xi)$. Alternatively, we may write

$$\frac{d^n y}{dx^n} = \frac{d}{dx}\left(\frac{d^{n-1}y}{dx^{n-1}}\right).$$

Example 4.19. ◫

Let $y = \log(1 + x)$. Then

$$\frac{dy}{dx} = \frac{1}{1+x}$$

$$\frac{d^2 y}{dx^2} = -\frac{1}{(1+x)^2}$$

$$\frac{d^3 y}{dx^3} = \frac{2}{(1+x)^3}$$

and, in general,

$$\frac{d^n y}{dx^n} = \frac{(-1)^{n-1}(n-1)!}{(1+x)^n}.$$

We next discuss the second order derivative of a function $f:\mathbb{R}^n \to \mathbb{R}$. We shall *not* attempt to discuss higher order derivatives of such functions. *Nor* shall we discuss the second derivatives of functions $f:\mathbb{R}^n \to \mathbb{R}^m$ when $m > 1$. A study of these latter cases would take us too far afield.

Consider a function $f:\mathbb{R}^2 \to \mathbb{R}$ and write

$$z = f(x, y).$$

From such a function one can form four second order *partial* derivatives. These are defined by

$$\frac{\partial^2 z}{\partial x^2} = \frac{\partial}{\partial x}\left(\frac{\partial z}{\partial x}\right) \qquad \frac{\partial^2 z}{\partial y \partial x} = \frac{\partial}{\partial y}\left(\frac{\partial z}{\partial x}\right)$$

$$\frac{\partial^2 z}{\partial x \partial y} = \frac{\partial}{\partial x}\left(\frac{\partial z}{\partial y}\right) \qquad \frac{\partial^2 z}{\partial y^2} = \frac{\partial}{\partial y}\left(\frac{\partial z}{\partial y}\right).$$

The second order partial derivatives of a function $f:\mathbb{R}^n \to \mathbb{R}$ are defined in a similar way. Thus, for example,

$$\frac{\partial^2 f}{\partial x_2 \partial x_5} = \frac{\partial}{\partial x_2}\left(\frac{\partial f}{\partial x_5}\right).$$

Example 4.20.

Suppose that $z = x^2 y + xy^3$. Then

$$\frac{\partial z}{\partial x} = 2xy + y^3 \, ; \quad \frac{\partial z}{\partial y} = x^2 + 3xy^2 \, .$$

Hence

$$\frac{\partial^2 z}{\partial x^2} = \frac{\partial}{\partial x}(2xy + y^3) = 2y$$

$$\frac{\partial^2 z}{\partial y \partial x} = \frac{\partial}{\partial y}(2xy + y^3) = 2x + 3y^2$$

$$\frac{\partial^2 z}{\partial x \partial y} = \frac{\partial}{\partial x}(x^2 + 3xy^2) = 2x + 3y^2$$

$$\frac{\partial^2 z}{\partial y^2} = \frac{\partial}{\partial y}(x^2 + 3xy^2) = 6xy.$$

Note in the above example that

$$\frac{\partial^2 z}{\partial y \partial x} = \frac{\partial^2 z}{\partial x \partial y}.$$

This is a special case of the general rule for a function $f : \mathbb{R}^n \to \mathbb{R}$:

$$\boxed{\frac{\partial^2 f}{\partial x_i \partial x_j} = \frac{\partial^2 f}{\partial x_j \partial x_i}}$$

(There are exceptions to this rule, but the exceptional functions are rather peculiar and we therefore exclude these exceptional functions from our study.)

Having defined the second order *partial* derivatives of $f : \mathbb{R}^n \to \mathbb{R}$, we can now consider the second order *derivative* of f.

We have that

$$\frac{df}{d\mathbf{x}} = \left(\frac{\partial f}{\partial x_1}, \frac{\partial f}{\partial x_2}, \dots, \frac{\partial f}{\partial x_n} \right)$$

is a $1 \times n$ row vector. Its transpose is therefore a column vector and so we may define

$$\left(\frac{d^2 f}{d\mathbf{x}^2} \right)^T = \frac{d}{d\mathbf{x}} \left(\frac{df}{d\mathbf{x}} \right)^T = \begin{pmatrix} \dfrac{\partial^2 f}{\partial x_1^2} & \dfrac{\partial^2 f}{\partial x_2 \partial x_1} & \cdots & \dfrac{\partial^2 f}{\partial x_n \partial x_1} \\ \dfrac{\partial^2 f}{\partial x_1 \partial x_2} & \dfrac{\partial^2 f}{\partial x_2^2} & \cdots & \dfrac{\partial^2 f}{\partial x_n \partial x_2} \\ \vdots & & & \\ \dfrac{\partial^2 f}{\partial x_1 \partial x_n} & \dfrac{\partial^2 f}{\partial x_2 \partial x_n} & \cdots & \dfrac{\partial^2 f}{\partial x_n^2} \end{pmatrix}$$

The second derivative of a real-valued function $f: \mathbb{R}^n \to \mathbb{R}$ is therefore an $n \times n$ *matrix* whose entries are the second order partial derivatives of f. Observe that this matrix is *symmetric*.

The geometric interpretation is that $\mathbf{y} = f'(\boldsymbol{\xi})^T + f''(\boldsymbol{\xi})(\mathbf{x} - \boldsymbol{\xi})$ is tangent to $\mathbf{y} = f'(\mathbf{x})^T$ at the point where $\mathbf{x} = \boldsymbol{\xi}$.

Example 4.21.
Let $f: \mathbb{R}^2 \to \mathbb{R}$ be defined by

$$z = f(x, y) = x^2 y + xy^3.$$

The second derivative is

$$\begin{pmatrix} \dfrac{\partial^2 z}{\partial x^2} & \dfrac{\partial^2 z}{\partial x \partial y} \\[2mm] \dfrac{\partial^2 z}{\partial y \partial x} & \dfrac{\partial^2 z}{\partial y^2} \end{pmatrix} = \begin{pmatrix} 2y & 2x + 3y^2 \\[2mm] 2x + 3y^2 & 6xy \end{pmatrix}$$

4.22 Taylor series

The following power series expansions will be familiar

$$e^x = 1 + x + \frac{x^2}{2!} + \frac{x^3}{3!} + \ldots$$

$$\cos x = 1 - \frac{x^2}{2!} + \frac{x^4}{4!} - \frac{x^6}{6!} + \ldots$$

$$\sin x = x - \frac{x^3}{3!} + \frac{x^5}{5!} - \frac{x^7}{7!} + \ldots$$

$$\log(1 + x) = x - \frac{x^2}{2} + \frac{x^3}{3} - \frac{x^4}{4} + \ldots (-1 < x \leqslant 1)$$

$$(1 + x)^\alpha = 1 + \alpha x + \frac{\alpha(\alpha - 1)}{2!} x^2 + \ldots (-1 < x < 1)$$

These are all examples of Taylor series expansions of a function $f: \mathbb{R} \to \mathbb{R}$ about the point 0. In general, the Taylor series expansion of a function $f: \mathbb{R} \to \mathbb{R}$ about the point ξ is given by

$$f(x) = f(\xi) + \frac{(x - \xi)}{1!}f'(\xi) + \frac{(x - \xi)^2}{2!}f''(\xi) + \frac{(x - \xi)^3}{3!}f'''(\xi) + \dots$$

This formula is valid for some set of values of x including the point ξ. Sometimes this set contains only the point ξ (in which case the formula is pretty useless). On other occasions the formula is valid for *all* values of x. Usually, however, the formula is valid for values of x close to ξ and invalid for values of x not close to ξ.

Example 4.23.[H]
Suppose that $f(x) = \log(1 + x)$. As we know from example 4.19

$$f(0) = 0, \quad f'(0) = 1, \quad f''(0) = -1, \quad f'''(0) = 2, \dots.$$

The Taylor expansion of $\log(1 + x)$ about the point 0 is therefore

$$\log(1 + x) = 0 + \frac{x}{1!} + \frac{x^2(-1)}{2!} + \frac{x^3(2)}{3!} + \frac{x^4(-6)}{4!} + \dots$$

$$= x - \frac{x^2}{2} + \frac{x^3}{3} - \frac{x^4}{4} + \dots.$$

The range of validity of the formula is $-1 < x \leq 1$. (Those with pocket calculators may find it instructive to calculate the sum of the first n terms of the expansion when $x = -1$, $x = 1$ and $x = 1.01$.)

We now turn to the Taylor series expansion about the point $\boldsymbol{\xi}$ of a real-valued function $f: \mathbb{R}^n \to \mathbb{R}$.

Let $\mathbf{X}$ and $\boldsymbol{\xi}$ be fixed vectors in $\mathbb{R}^n$ and write $\mathbf{x} = \boldsymbol{\xi} + t(\mathbf{X} - \boldsymbol{\xi})$. Define $F: \mathbb{R} \to \mathbb{R}$ by

$$F(t) = f(\mathbf{x}) = f(\boldsymbol{\xi} + t(\mathbf{X} - \boldsymbol{\xi})).$$

Using the chain rule, we obtain that

$$\frac{dF}{dt} = \frac{df}{d\mathbf{x}}\frac{d\mathbf{x}}{dt} = \frac{df}{d\mathbf{x}}(\mathbf{X} - \boldsymbol{\xi}) = (\mathbf{X} - \boldsymbol{\xi})^T \left(\frac{df}{d\mathbf{x}}\right)^T$$

(for the last step one needs to take note of the fact that we are dealing with a scalar quantity and so this is equal to its transpose)

$$\frac{d^2 F}{dt^2} = \frac{d}{dt}\left(\frac{dF}{dt}\right) = (\mathbf{X}-\boldsymbol{\xi})^T \frac{d}{dt}\left(\frac{df}{dx}\right)^T$$

$$= (\mathbf{X}-\boldsymbol{\xi})^T \frac{d}{dx}\left(\frac{df}{dx}\right)^T \frac{dx}{dt}$$

$$= (\mathbf{X}-\boldsymbol{\xi})^T \left(\frac{d^2 f}{dx^2}\right)^T (\mathbf{X}-\boldsymbol{\xi})$$

$$= (\mathbf{X}-\boldsymbol{\xi})^T \frac{d^2 f}{dx^2}(\mathbf{X}-\boldsymbol{\xi}).$$

But

$$f(\mathbf{X}) = F(1) = F(0) + \frac{1}{1!}F'(0) + \frac{1}{2!}F''(0) + \dots.$$

It follows that

$$f(\mathbf{X}) = f(\boldsymbol{\xi}) + \frac{1}{1!}f'(\boldsymbol{\xi})(\mathbf{X} - \boldsymbol{\xi}) + \frac{1}{2!}(\mathbf{X} - \boldsymbol{\xi})^T f''(\boldsymbol{\xi})(\mathbf{X} - \boldsymbol{\xi}) + \dots$$

Example 4.24.
Suppose that $f: \mathbb{R}^2 \to \mathbb{R}$. Then

$$f'(\xi, \eta) = \left(\frac{\partial f}{\partial x}, \frac{\partial f}{\partial y}\right) = (f_x, f_y)$$

$$f''(\xi, \eta) = \begin{pmatrix} \dfrac{\partial^2 f}{\partial x^2} & \dfrac{\partial^2 f}{\partial x \partial y} \\[2mm] \dfrac{\partial^2 f}{\partial y \partial x} & \dfrac{\partial^2 f}{\partial y^2} \end{pmatrix} = \begin{pmatrix} f_{xx} & f_{xy} \\ f_{yx} & f_{yy} \end{pmatrix}$$

provided the partial derivatives are evaluated at the point $(\xi, \eta)^T$.
The Taylor expansion of f about $(\xi, \eta)^T$ is

$$f(\xi + h, \eta + k) = f + (f_x, f_y)\begin{pmatrix} h \\ k \end{pmatrix} + \frac{1}{2}(h, k)\begin{pmatrix} f_{xx} & f_{xy} \\ f_{yx} & f_{yy} \end{pmatrix}\begin{pmatrix} h \\ k \end{pmatrix} + \dots$$

$$= f + \{hf_x + kf_y\} + \frac{1}{2}\{h^2 f_{xx} + 2hk f_{xy} + k^2 f_{yy}\} + \dots.$$

Exercises 4.25

1. The equations

$$u = f_1(x,y) = \tfrac{1}{2}(x^2 - y^2) \left.\right\}$$
$$v = f_2(x,y) = xy$$

define a function $f : \mathbb{R}^2 \to \mathbb{R}^2$ whose component functions are f_1 and f_2.
Sketch the contours $f_1(x,y) = 6$ and $f_2(x,y) = 8$ on the same diagram.
Indicate on the diagram the values of $(x,y)^T$ for which

$$f(x,y) = (6,8)^T.$$

2.* The equations

$$u = f_1(x,y) = x^2 + y^2 \left.\right\}$$
$$v = f_2(x,y) = x + y$$

define a function $f : \mathbb{R}^2 \to \mathbb{R}^2$ whose component functions are f_1 and f_2.
Sketch the contours $f_1(x,y) = 5, f_1(x,y) = 8, f_2(x,y) = 3$ and
$f_2(x,y) = 4$ on the same diagram. Indicate on the diagram the set of $(x,y)^T$
which correspond to values of $(u,v)^T$ for which $5 \leqslant u \leqslant 8$ and $3 \leqslant v \leqslant 4$.

3. Let S be the set of vectors $(x,y,z)^T$ in $\mathbb{R}^3$ for which $y \neq 0$ and $z \neq 0$.
Find the derivative at the point $(x,y,z)^T = (1,2,3)^T$ of the function
$f : S \to \mathbb{R}^2$ defined by the equations

$$u = \frac{x}{y} \left.\right\}$$
$$v = \frac{y}{z}.$$

Write down the equations for the tangent flat at the point where $(x,y,z)^T = (1,2,3)^T$.

4.* Find the derivative at the given point of the functions defined by the
following sets of equations. Write down the equations for the corresponding
tangent flats in each case.

(i) $y = x_1^2 + 2x_1 x_2 + x_2^3$ at $(x_1, x_2)^T = (1,2)^T$

(ii) $y_1 = \cos(x_1 x_2) \left.\right\}$ at $(x_1, x_2)^T = (\pi, 1)^T$

$$y_2 = \sin \frac{x_1}{x_2}$$

(iii) $r = \{x^2 + y^2\}^{1/2} \left.\right\}$ at $(x,y)^T = (1,0)^T$

$$\theta = \arctan \frac{y}{x}$$

(iv) $u = x$ at $(x, y)^T = (1, 1)^T$

    $v = x + y$

    $w = xy$

(v) $u = 1$ at $t = 0$

    $v = t$

    $w = t^2$

(vi) $u = xy$ at $(x, y, z)^T = (1, 1, 1)^T$.

    $v = yz$

5. Show that $z = f(x^2 + y^2)$ satisfies the partial differential equation

$$y \frac{\partial z}{\partial x} = x \frac{\partial z}{\partial y}$$

for any differentiable function $f: \mathbb{R} \to \mathbb{R}$.

6.* Show that $z = f(y \log x)$ satisfies the partial differential equation

$$x \frac{\partial z}{\partial x} + y \frac{\partial z}{\partial y} = xy \frac{\partial^2 z}{\partial x \partial y} - x^2 \log x \frac{\partial^2 z}{\partial x^2}$$

for any differentiable function $f: \mathbb{R} \to \mathbb{R}$.

7. Find a system of linear equations satisfied by the stationary points of

$$y = (A\mathbf{x} + \mathbf{a})^T (B\mathbf{x} + \mathbf{b})$$

where A and B are $m \times n$ matrices and $\mathbf{a}$ and $\mathbf{b}$ are $m \times 1$ column vectors.

8.* Let A be a symmetric $n \times n$ matrix. Show that, if $\boldsymbol{\xi}$ is a stationary point of

$$y = \mathbf{x}^T A \mathbf{x}$$

subject to the constraint $\mathbf{x}^T \mathbf{x} = 1$, then

$$A\boldsymbol{\xi} = \lambda \boldsymbol{\xi}$$

for some scalar λ. Show also that λ is the value of y when $\mathbf{x} = \boldsymbol{\xi}$.

9.† Let B be an $m \times n$ matrix and let $\mathbf{c}$ be an $m \times 1$ column vector. Show that there is a unique stationary point for

$$y = \mathbf{x}^T \mathbf{x}$$

subject to the constraint

$$B\mathbf{x} = \mathbf{c}$$

provided that the matrix BB^T is non-singular. Find this stationary point.

10.*† Let A be a symmetric, non-singular $n \times n$ matrix. Let B be an $m \times n$ matrix and let c be an $m \times 1$ column vector. Show that the unique stationary point for

$$y = x^T A x$$

subject to the constraint

$$Bx = c$$

is given by

$$x = A^{-1} B^T (BA^{-1} B^T)^{-1} c$$

provided that $BA^{-1} B^T$ is non-singular.

11. Find the second derivative of the real-valued function $f: \mathbb{R}^2 \to \mathbb{R}$ defined by
$$z = f(x, y) = xe^{-y}.$$

Write down the first three terms of the Taylor expansion about the point $(x, y)^T = (1, 0)^T$.

12.* Let S be the set of all vectors $(x, y, z)^T$ in $\mathbb{R}^3$ for which $x > 0, y > 0$ and $z > 0$. Find the second derivative of the real-valued function $f: S \to \mathbb{R}$ defined by

$$u = f(x, y, z) = \log(x + yz^{-1}).$$

Write down the first three terms of the Taylor expansion about the point $(x, y, z)^T = (1, 1, 1)^T$.

SOME APPLICATIONS (OPTIONAL)

4.26   Least squares analysis

Suppose that a process has a single input and results in a single output. The quantity u of output will then be a function of the quantity t of input – i.e. $u = f(t)$. Suppose that we have a theory that f is an affine function – i.e.

$$u = f(t) = mt + c$$

and we would like to know the values of the two constants m and c. The obvious course of action is to run an experiment and obtain a table of corresponding values of u and t. Suppose that the table obtained is that given below:

t	1	2	3	4	5
u	-0.4	0.8	2.6	3.2	5.4

The most notable feature about this data is that it is not consistent with the hypothesis that $u = mx + c$ – i.e. the system of equations

$$
\begin{aligned}
-0.4 &= m1 + c \\
0.8 &= m2 + c \\
2.6 &= m3 + c \\
3.2 &= m4 + c \\
5.4 &= m5 + c
\end{aligned}
$$

is *inconsistent*. This system can be written $\mathbf{u} = T\mathbf{x}$, where

$$
\mathbf{u} = \begin{pmatrix} -0.4 \\ 0.8 \\ 2.6 \\ 3.2 \\ 5.4 \end{pmatrix}; \quad
T = \begin{pmatrix} 1 & 1 \\ 2 & 1 \\ 3 & 1 \\ 4 & 1 \\ 5 & 1 \end{pmatrix}; \quad
\text{and} \quad \mathbf{x} = \begin{pmatrix} m \\ c \end{pmatrix}.
$$

But the fact that $\mathbf{u} = T\mathbf{x}$ is inconsistent does not necessarily imply that the theory that $u = mt + c$ is false. Perhaps the data contains errors. If the theory is correct and the errors are not too large, the data can then be used to *estimate m* and *c*.

The problem is then to find the straight line $y = mx + c$ which 'fits the data best'. One way of interpreting this is to seek the values of m and c which minimize

$$y = r_1^2 + r_2^2 + r_3^2 + r_4^2 + r_5^2.$$

This is called the 'least squares' estimation method for obvious reasons.

Observe that

$$r_k^2 = (u_k - mt_k - c)^2$$

and so

$$y = \|\mathbf{u} - T\mathbf{x}\|^2.$$

The least squares method is therefore to find the value of $\mathbf{x} = (m, c)^T$ which minimizes $\|\mathbf{u} - T\mathbf{x}\|$. We studied this problem in example 4.13.

4.27 Kuhn–Tucker conditions

Suppose that $f : \mathbb{R}^n \to \mathbb{R}$ and $g : \mathbb{R}^n \to \mathbb{R}^m$ are functions. We have studied the problem of maximizing or minimizing

$$f(\mathbf{x})$$

subject to the constraint

$$g(\mathbf{x}) = \mathbf{0}.$$

One forms the Lagrangian

$$L(\mathbf{x}, \mathbf{y}) = f(\mathbf{x}) + \mathbf{y}^T g(\mathbf{x})$$

and examines the solutions of the *Lagrangian conditions*

(i) $\quad \dfrac{\partial L}{\partial \mathbf{x}} = f'(\mathbf{x}) + \mathbf{y}^T g'(\mathbf{x}) = \mathbf{0}$

(ii) $\quad \dfrac{\partial L}{\partial \mathbf{y}} = g(\mathbf{x})^T = \mathbf{0}$ – i.e. $g(\mathbf{x}) = \mathbf{0}.$

A more general problem is to maximize or minimize

$$f(\mathbf{x})$$

subject to the constraint

$$g(\mathbf{x}) \geqslant \mathbf{0}.$$

(Recall that $g(\mathbf{x}) \geqslant 0$ means that $g_1(\mathbf{x}) \geqslant 0, g_2(\mathbf{x}) \geqslant 0, \ldots,$ and $g_m(\mathbf{x}) \geqslant 0$.) The Lagrangian conditions then have to be replaced by the *Kuhn–Tucker conditions*. In the case of a *maximum*, these are the following:

$$
\begin{cases}
\text{(i)} & f'(\mathbf{x}) + \mathbf{y}^T g'(\mathbf{x}) = \mathbf{0} \\
\text{(ii)} & g(\mathbf{x}) \geqslant \mathbf{0} \\
\text{(iii)} & \mathbf{y}^T g(\mathbf{x}) = 0 \\
\text{(iv)} & \mathbf{y} \geqslant \mathbf{0}.
\end{cases}
$$

In the case of a *minimum*, (iv) is replaced by $\mathbf{y} \leqslant \mathbf{0}$.

The diagram below illustrates a possible configuration in the case $n = 2$ and $m = 3$ when $f(\mathbf{x})$ achieves a maximum (subject to the constraint $g(\mathbf{x}) \geqslant 0$) at the point $\mathbf{x} = \tilde{\mathbf{x}}$.

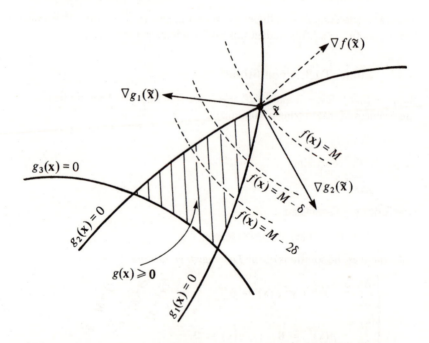

(Since $g_3(\tilde{\mathbf{x}}) > 0$, items (ii), (iii) and (iv) of the Kuhn–Tucker conditions imply that $y_3 = 0$. Item (i) then asserts that $\nabla f = -y_1 \nabla g_1 - y_2 \nabla g_2$. Since $y_1 \geqslant 0$ and $y_2 \geqslant 0$ by (iv), this guarantees that $\langle \nabla f, \mathbf{u} \rangle \leqslant 0$ for each direction $\mathbf{u}$ which points into the region defined by $g(\mathbf{x}) \geqslant 0$.)

As an example, consider the case $g(\mathbf{x}) = \mathbf{x}$. Then $g'(\mathbf{x}) = I$ and the Kuhn–Tucker conditions become

$$\begin{cases} f'(\mathbf{x}) + \mathbf{y}^T = \mathbf{0} \\ \mathbf{y}^T\mathbf{x} = \mathbf{0} \\ \mathbf{x} \geq \mathbf{0}, \mathbf{y} \geq \mathbf{0} \end{cases}$$

These imply that

$$\begin{cases} f'(\mathbf{x}) \leq \mathbf{0} \quad -\text{i.e.} \quad \nabla f(\mathbf{x}) \leq \mathbf{0} \\ \mathbf{x} \geq \mathbf{0} \qquad\qquad\qquad \mathbf{x} \geq \mathbf{0} \\ f'(\mathbf{x})\mathbf{x} = \mathbf{0} \qquad \langle \nabla f(\mathbf{x}), \mathbf{x} \rangle = 0. \end{cases}$$

The diagrams below illustrate the various possibilities in the case $f: \mathbb{R}^2 \to \mathbb{R}$:

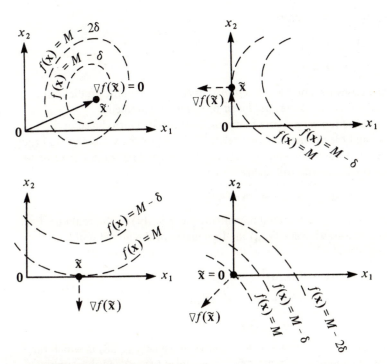

4.28 Linear programming

We considered linear programming in §1.36. Our example concerned a housewife who sought to *maximize* profit

$$\mathbf{p}^T\mathbf{x}$$

subject to the constraints

$$Ax \leqslant b$$
$$x \geqslant 0.$$

Here the vector **b** represents the housewife's stock of ingredients and $A\mathbf{x}$ represents the input required to produce the output **x**.

In §1.37, we elaborated on this theme by introducing a bakery combine which sought to persuade the housewife not to sell the products of her baking by quoting a price vector **q** for her ingredients. The idea of the combine was to make it as attractive to her to sell her ingredients to the combine at these prices as to bake them and sell the results of her baking. We decided that the combine should choose **q** so as to *minimize*

$$\mathbf{b}^T\mathbf{q}$$

subject to the constraints

$$A^T\mathbf{q} \geqslant \mathbf{p}$$
$$\mathbf{q} \geqslant 0.$$

The housewife's problem is a *primal* linear programming problem and the combine's problem is the *dual* linear programming problem.

In this section we propose to discuss how the idea introduced in this chapter and the previous chapter relate to these problems. Observe, to begin with, that, if the housewife produces **x** and the combine quotes price vector **q**, then the housewife will realize an amount

$$L(\mathbf{x}, \mathbf{q}) = \mathbf{p}^T\mathbf{x} + \mathbf{q}^T(\mathbf{b} - A\mathbf{x})$$

assuming she sells her unused ingredients (i.e. $\mathbf{b} - A\mathbf{x}$) to the combine. For obvious reasons we call $L(\mathbf{x}, \mathbf{q})$ the Lagrangian of the primal problem. Note that it is also the Lagrangian of the dual problem because

$$L(\mathbf{x}, \mathbf{q}) = L(\mathbf{x}, \mathbf{q})^T$$
$$= \mathbf{x}^T\mathbf{p} + (\mathbf{b}^T - \mathbf{x}^TA^T)\mathbf{q}$$
$$= \mathbf{b}^T\mathbf{q} + \mathbf{x}^T(\mathbf{p} - A^T\mathbf{q}).$$

We can regard the situation as a species of 'game' (§1.38) in which the housewife chooses **x** in an attempt to maximize $L(\mathbf{x}, \mathbf{q})$ and the combine chooses **q** in an attempt to minimize $L(\mathbf{x}, \mathbf{q})$. A solution $(\tilde{\mathbf{x}}, \tilde{\mathbf{q}})$ to this game has the property that $\mathbf{x} = \tilde{\mathbf{x}}$ *maximizes* $L(\mathbf{x}, \tilde{\mathbf{q}})$ subject to the constraint $\mathbf{x} \geqslant 0$ and $\mathbf{q} = \tilde{\mathbf{q}}$ *minimizes* $L(\tilde{\mathbf{x}}, \mathbf{q})$ subject to the constraint $\mathbf{q} \geqslant 0$. (We noted in §1.38 that such a pair $(\tilde{\mathbf{x}}, \tilde{\mathbf{q}})$ is called a Nash equilibrium for the game.)

We considered the problem of maximizing $f(\mathbf{x})$ subject to the constraint

$\mathbf{x} \geqslant \mathbf{0}$ at the end of §4.27. Applying this work in the case when $f(\mathbf{x}) = L(\mathbf{x}, \tilde{\mathbf{q}})$, we obtain that

(i) $\mathbf{p}^T - \tilde{\mathbf{q}}^T A \leqslant \mathbf{0}$ – i.e. $A^T \tilde{\mathbf{q}} \geqslant \mathbf{p}$

(ii) $\tilde{\mathbf{x}} \geqslant \mathbf{0}$

(iii) $\mathbf{p}^T \tilde{\mathbf{x}} - \tilde{\mathbf{q}}^T A \tilde{\mathbf{x}} = 0$ – i.e. $\mathbf{p}^T \tilde{\mathbf{x}} = \tilde{\mathbf{q}}^T A \tilde{\mathbf{x}}$.

Similarly, since $\mathbf{q} = \tilde{\mathbf{q}}$ minimizes $L(\tilde{\mathbf{x}}, \mathbf{q})$ subject to the constraint $\mathbf{q} \geqslant \mathbf{0}$

(iv) $\mathbf{b}^T - \tilde{\mathbf{x}}^T A^T \geqslant \mathbf{0}$ – i.e. $A\tilde{\mathbf{x}} \leqslant \mathbf{b}$

(v) $\tilde{\mathbf{q}} \geqslant \mathbf{0}$

(vi) $\mathbf{b}^T \tilde{\mathbf{q}} - \tilde{\mathbf{x}}^T A^T \tilde{\mathbf{q}} = 0$ – i.e. $\mathbf{b}^T \tilde{\mathbf{q}} = \tilde{\mathbf{q}}^T A \tilde{\mathbf{x}}$.

Suppose that $\mathbf{x} \geqslant \mathbf{0}$ and $A\mathbf{x} \leqslant \mathbf{b}$. Then

$$\begin{aligned} \mathbf{p}^T \mathbf{x} \leqslant \mathbf{p}^T \mathbf{x} + \tilde{\mathbf{q}}^T (\mathbf{b} - A\mathbf{x}) &= L(\mathbf{x}, \tilde{\mathbf{q}}) \\ &\leqslant \mathbf{p}^T \tilde{\mathbf{x}} + \tilde{\mathbf{q}}^T (\mathbf{b} - A\tilde{\mathbf{x}}) \\ &= \mathbf{p}^T \tilde{\mathbf{x}} \qquad \text{(by vi).} \end{aligned}$$

Since $A\tilde{\mathbf{x}} \leqslant \mathbf{b}$ (by iv), it follows that $\mathbf{x} = \tilde{\mathbf{x}}$ solves our primal linear programming problem. A similar argument shows that $\mathbf{q} = \tilde{\mathbf{q}}$ solves the dual linear programming problem.

The most significant feature of this analysis is that provided by (iii) and (vi) which show that

$$\mathbf{p}^T \tilde{\mathbf{x}} = \mathbf{b}^T \tilde{\mathbf{q}}.$$

This result is called the *duality theorem* of linear programming. Stated in full, it asserts that if the primal and the dual programs both have solutions, then the maximum value attained in the primal problem is equal to the minimum value attained in the dual problem.

5

Maxima and minima

A familiar technique in classifying the stationary points of a real-valued function of one real variable is to examine the sign of the second derivative. The same technique works in the case of a real-valued function of a vector variable except that, in this case, the second derivative is a matrix and instead of asking whether this second derivative is positive or negative we ask instead whether it is 'positive definite' or 'negative definite'. The chapter begins with several sections on linear algebra which explain these notions rather more extensively than is strictly necessary since it is possible for most purposes to manage just by remembering statements (i) and (ii) of §5.6. Note, however, that although more linear algebra is supplied than is strictly necessary to use the technique, what is supplied still remains only a survey of the appropriate ideas and is not an adequate introduction for those to whom these ideas are entirely new.

5.1 Change of basis[□]

In this and the next section we shall review some necessary material from linear algebra.

A set of n vectors $\mathbf{b}_1, \mathbf{b}_2, \ldots, \mathbf{b}_n$ is a *basis* for $\mathbb{R}^n$ if each vector $\mathbf{x}$ in $\mathbb{R}^n$ can be expressed uniquely in the form

$$\mathbf{x} = X_1\mathbf{b}_1 + X_2\mathbf{b}_2 + \ldots + X_n\mathbf{b}_n.$$

The entries in the column vector

$$\mathbf{X} = \begin{pmatrix} X_1 \\ X_2 \\ \vdots \\ X_n \end{pmatrix}$$

are then called the *co-ordinates* of **x** with respect to the basis $\mathbf{b}_1, \mathbf{b}_2, \ldots, \mathbf{b}_n$.

Any set of n linearly independent vectors in $\mathbb{R}^n$ is a basis for $\mathbb{R}^n$. The same criterion can be expressed in terms of the matrix

$$B = \begin{pmatrix} | & | & & | \\ \mathbf{b}_1 & \mathbf{b}_2 & \ldots & \mathbf{b}_n \\ | & | & & | \end{pmatrix} = \begin{pmatrix} b_{11} & b_{12} & \ldots b_{1n} \\ b_{21} & b_{22} & \ldots b_{2n} \\ \vdots & & \\ b_{n1} & b_{n2} & \ldots b_{nn} \end{pmatrix}.$$

To say that B is *non-singular* is equivalent to any one of the following statements:

(i) B^{-1} exists.

(ii) $\det (B) \neq 0$.

(iii) B has rank n.

(iv) The columns of B are linearly independent.

(v) The rows of B are linearly independent.

Thus the columns (or rows) of an $n \times n$ matrix B are a basis for $\mathbb{R}^n$ if and only if B is non-singular.

The *natural basis* for $\mathbb{R}^n$ is the set of vectors

$$\mathbf{a}_1 = \begin{pmatrix} 1 \\ 0 \\ 0 \\ \vdots \\ 0 \end{pmatrix} ; \quad \mathbf{a}_2 = \begin{pmatrix} 0 \\ 1 \\ 0 \\ \vdots \\ 0 \end{pmatrix} ; \ldots ; \quad \mathbf{a}_n = \begin{pmatrix} 0 \\ 0 \\ \vdots \\ 0 \\ 1 \end{pmatrix}.$$

The reason for calling these vectors the natural basis is that

$$\mathbf{x} = \begin{pmatrix} x_1 \\ x_2 \\ \vdots \\ x_n \end{pmatrix} = x_1 \mathbf{a}_1 + x_2 \mathbf{a}_2 + \ldots + x_n \mathbf{a}_n.$$

Sometimes it is convenient to change the basis from the natural basis $\mathbf{a}_1, \mathbf{a}_2, \ldots, \mathbf{a}_n$ to a new basis $\mathbf{b}_1, \mathbf{b}_2, \ldots, \mathbf{b}_n$. One then needs to be able to express the column **X** of new co-ordinates in terms of the column **x** of old co-ordinates. This turns out to be quite easy. We have that

$$BX = \begin{pmatrix} b_{11} & b_{12} \dots b_{1n} \\ b_{21} & b_{22} \dots b_{2n} \\ \vdots \\ b_{n1} & b_{n2} \dots b_{nn} \end{pmatrix} \begin{pmatrix} X_1 \\ X_2 \\ \vdots \\ X_n \end{pmatrix} = \begin{pmatrix} b_{11}X_1 + b_{12}X_2 + \dots + b_{1n}X_n \\ b_{21}X_1 + b_{22}X_2 + \dots + b_{2n}X_n \\ \vdots \\ b_{n1}X_1 + b_{n2}X_2 + \dots + b_{nn}X_n \end{pmatrix}$$

$$= X_1 \begin{pmatrix} b_{11} \\ b_{21} \\ \vdots \\ b_{n1} \end{pmatrix} + X_2 \begin{pmatrix} b_{12} \\ b_{22} \\ \vdots \\ b_{n2} \end{pmatrix} + \dots + X_n \begin{pmatrix} b_{1n} \\ b_{2n} \\ \vdots \\ b_{nn} \end{pmatrix}$$

$$= X_1 \mathbf{b}_1 + X_2 \mathbf{b}_2 + \dots + X_n \mathbf{b}_n = \mathbf{x}.$$

It follows that

$$\mathbf{x} = B\mathbf{X}.$$

i.e.

$$\mathbf{X} = B^{-1}\mathbf{x}.$$

Example 5.2.[□]
In $\mathbb{R}^2$ the basis is changed from the natural basis

$$\mathbf{a}_1 = \begin{pmatrix} 1 \\ 0 \end{pmatrix}; \quad \mathbf{a}_2 = \begin{pmatrix} 0 \\ 1 \end{pmatrix}$$

to the new basis

$$\mathbf{b}_1 = \begin{pmatrix} 1 \\ 1 \end{pmatrix}; \quad \mathbf{b}_2 = \begin{pmatrix} 2 \\ 1 \end{pmatrix}.$$

The problem is to express the new co-ordinates in terms of the old co-ordinates. We have that

$$B = \begin{pmatrix} 1 & 2 \\ 1 & 1 \end{pmatrix}; \quad B^{-1} = \begin{pmatrix} -1 & 2 \\ 1 & -1 \end{pmatrix}$$

and hence

$$\begin{pmatrix} X_1 \\ X_2 \end{pmatrix} = \begin{pmatrix} -1 & 2 \\ 1 & -1 \end{pmatrix} \begin{pmatrix} x_1 \\ x_2 \end{pmatrix}.$$

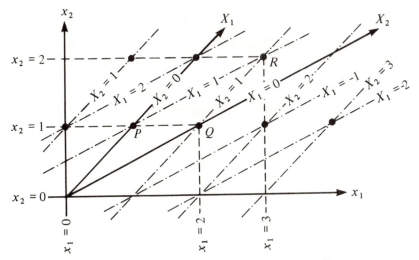

Observe that the old co-ordinates for the point R are $(3, 2)^T$ but the new co-ordinates are $(1, 1)^T$.

The natural basis is *orthonormal*. This means that its constituent vectors are mutually orthogonal and of unit length. If the same is true of the *new* basis, then the problem of changing bases simplifies. We have that the vectors $\mathbf{b}_1, \mathbf{b}_2, \ldots, \mathbf{b}_n$ are orthonormal if and only if

$$B^T B = \begin{pmatrix} \text{---} \mathbf{b}_1^T \text{---} \\ \text{---} \mathbf{b}_2^T \text{---} \\ \text{---} \mathbf{b}_n^T \text{---} \end{pmatrix} \begin{pmatrix} | & | & & | \\ \mathbf{b}_1 & \mathbf{b}_2 & \ldots & \mathbf{b}_n \\ | & | & & | \end{pmatrix}$$

$$= \begin{pmatrix} \mathbf{b}_1^T\mathbf{b}_1 & \mathbf{b}_1^T\mathbf{b}_2 \ldots \mathbf{b}_1^T\mathbf{b}_n \\ \mathbf{b}_2^T\mathbf{b}_1 & \mathbf{b}_2^T\mathbf{b}_2 \ldots \mathbf{b}_n^T\mathbf{b}_1 \\ \vdots \\ \mathbf{b}_2^T\mathbf{b}_n & \mathbf{b}_n^T\mathbf{b}_2 \ldots \mathbf{b}_n^T\mathbf{b}_n \end{pmatrix}$$

$$= \begin{pmatrix} \|\mathbf{b}_1\|^2 & \langle \mathbf{b}_1, \mathbf{b}_2 \rangle \ldots \langle \mathbf{b}_1, \mathbf{b}_n \rangle \\ \langle \mathbf{b}_2, \mathbf{b}_1 \rangle & \|\mathbf{b}_2\|^2 \ldots \langle \mathbf{b}_2, \mathbf{b}_n \rangle \\ \vdots \\ \langle \mathbf{b}_n, \mathbf{b}_1 \rangle & \langle \mathbf{b}_n, \mathbf{b}_2 \rangle \ldots \|\mathbf{b}_n\|^2 \end{pmatrix} = \begin{pmatrix} 1 & 0 \ldots 0 \\ 0 & 1 \ldots 0 \\ \vdots \\ 0 & 0 \ldots 1 \end{pmatrix} = I.$$

A matrix B satisfying $B^T B = I$ is called *orthogonal*. As we have seen, a matrix B is orthogonal if and only if its columns are orthonormal. It is useful to note that an orthogonal matrix is non-singular and that its inverse is simply its transpose — i.e.

$$B^{-1} = B^T.$$

This observation can save a lot of calculation.

Example 5.3.□

Suppose that the basis in $\mathbb{R}^2$ is changed from the natural basis

$$\mathbf{a}_1 = \begin{pmatrix} 1 \\ 0 \end{pmatrix}; \quad \mathbf{a}_2 = \begin{pmatrix} 0 \\ 1 \end{pmatrix}$$

to the basis

$$\mathbf{b}_1 = \begin{pmatrix} \cos \theta \\ \sin \theta \end{pmatrix}; \quad \mathbf{b}_2 = \begin{pmatrix} -\sin \theta \\ \cos \theta \end{pmatrix}.$$

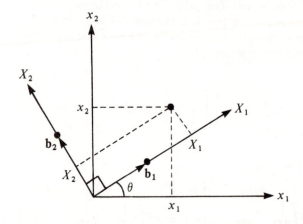

We have that

$$B = \begin{pmatrix} \cos \theta & -\sin \theta \\ \sin \theta & \cos \theta \end{pmatrix}$$

and so

$$B^{-1} = B^T = \begin{pmatrix} \cos \theta & \sin \theta \\ -\sin \theta & \cos \theta \end{pmatrix}.$$

Thus

$$\begin{pmatrix} X_1 \\ X_2 \end{pmatrix} = \begin{pmatrix} \cos \theta & \sin \theta \\ -\sin \theta & \cos \theta \end{pmatrix} \begin{pmatrix} x_1 \\ x_2 \end{pmatrix}.$$

5.4 Quadratic forms[II]

If A is an $n \times n$ matrix, the equation

$$\det (A - \lambda I) = 0$$

is called the characteristic equation of A. This equation is a polynomial of degree n in λ and hence has n roots. These roots are called the *eigenvalues* of the matrix A. We shall denote these eigenvalues by $\lambda_1, \lambda_2, \lambda_3, \ldots, \lambda_n$.

In general, the eigenvalues of A will be complex numbers. However, we shall be interested only in the case when A is *symmetric*. In this case the eigenvalues are always *real* numbers.

If λ is an eigenvalue of the matrix A, then there are vectors $\mathbf{e} \neq \mathbf{0}$ such that

$$A\mathbf{e} = \lambda\mathbf{e}.$$

Such a vector $\mathbf{e}$ is called an *eigenvector* corresponding to λ. We shall restrict our attention to eigenvectors of unit length — i.e. satisfying $\| \mathbf{e} \| = 1$.

If A is symmetric and its eigenvalues $\lambda_1, \lambda_2, \ldots, \lambda_n$ are *distinct*, then the corresponding unit eigenvectors $\mathbf{e}_1, \mathbf{e}_2, \ldots, \mathbf{e}_n$ are mutually orthogonal and so form an orthonormal set. Let E be the matrix

$$E = \begin{pmatrix} | & | & & | \\ \mathbf{e}_1 & \mathbf{e}_2 & \ldots & \mathbf{e}_n \\ | & | & & | \end{pmatrix}.$$

Then E is an *orthogonal* matrix — i.e. $E^T E = I$. Also,

$$E^T A E = E^T \begin{pmatrix} | & | & & | \\ A\mathbf{e}_1 & A\mathbf{e}_2 & \ldots & A\mathbf{e}_n \\ | & | & & | \end{pmatrix} = E^T \begin{pmatrix} | & | & & | \\ \lambda_1\mathbf{e}_1 & \lambda_2\mathbf{e}_2 & \ldots & \lambda_n\mathbf{e}_n \\ | & | & & | \end{pmatrix}$$

$$= \begin{pmatrix} \lambda_1\mathbf{e}_1^T\mathbf{e}_1 & \lambda_2\mathbf{e}_1^T\mathbf{e}_2 & \ldots & \lambda_n\mathbf{e}_1^T\mathbf{e}_n \\ \lambda_1\mathbf{e}_2^T\mathbf{e}_1 & \lambda_2\mathbf{e}_2^T\mathbf{e}_2 & \ldots & \lambda_n\mathbf{e}_2^T\mathbf{e}_n \\ \vdots & & & \vdots \\ \lambda_1\mathbf{e}_n^T\mathbf{e}_1 & \lambda_2\mathbf{e}_n^T\mathbf{e}_2 & \ldots & \lambda_n\mathbf{e}_n^T\mathbf{e}_n \end{pmatrix} = \begin{pmatrix} \lambda_1 & 0 & \ldots & 0 \\ 0 & \lambda_2 & \ldots & 0 \\ \vdots & & & \vdots \\ 0 & 0 & \ldots & \lambda_n \end{pmatrix}$$

and thus

$$E^T AE = D$$

where D is the 'diagonal' matrix which has the eigenvalues of A on its 'main diagonal' and zeros elsewhere.

(Note that, if the eigenvalues of A are not distinct – i.e. the equation $\det(A - \lambda I) = 0$ has multiple roots – then the discussion above remains valid except that several of the eigenvectors $e_1, e_2, \ldots, e_n$ will be derived from the *same* eigenvalue.)

A *quadratic form* is an expression

$$z = x^T Ax$$

where x is an $n \times 1$ column vector and A is an $n \times n$ *symmetric* matrix.

Given such a quadratic form, it is natural to introduce the change of variable

$$x = Ey$$

(or, equivalently, $y = E^T x$). Then

$$z = x^T Ax = (Ey)^T AEy = y^T (E^T AE)y = y^T Dy.$$

It follows that

$$z = (y_1, y_2, \ldots, y_n) \begin{pmatrix} \lambda_1 & 0 & \ldots & 0 \\ 0 & \lambda_2 & \ldots & 0 \\ & \cdot & & \\ & \cdot & & \\ 0 & 0 & \ldots & \lambda_n \end{pmatrix} \begin{pmatrix} y_1 \\ y_2 \\ \cdot \\ \cdot \\ y_n \end{pmatrix}$$

i.e.

$$z = \lambda_1 y_1^2 + \lambda_2 y_2^2 + \ldots + \lambda_n y_n^2.$$

The change of variable $y = E^T x$ therefore reduces the quadratic form to something reasonably simple. This change of variable, of course, amounts to changing from the natural basis to the basis $e_1, e_2, \ldots, e_n$.

Example 5.5.□

Consider the quadratic form

$$z = 5x_1^2 - 6x_1 x_2 + 5x_2^2.$$

We write this in the form

$$z = (x_1, x_2) \begin{pmatrix} 5 & -3 \\ -3 & 5 \end{pmatrix} \begin{pmatrix} x_1 \\ x_2 \end{pmatrix}.$$

The characteristic equation of the matrix is

$$\left| \begin{pmatrix} 5 & -3 \\ -3 & 5 \end{pmatrix} - \lambda \begin{pmatrix} 1 & 0 \\ 0 & 1 \end{pmatrix} \right| = \begin{vmatrix} 5-\lambda & -3 \\ -3 & 5-\lambda \end{vmatrix} = 0$$

i.e.

$$(5-\lambda)^2 - 9 = 0$$

$$5 - \lambda = \pm 3$$

$$\lambda = 2 \quad \underline{\text{OR}} \quad \lambda = 8.$$

The eigenvalues of the matrix are therefore $\lambda_1 = 2$ and $\lambda_2 = 8$.

We can therefore change the variables from $(x_1, x_2)^T$ to $(y_1, y_2)^T$ in such a way that the quadratic form reduces to

$$z = 2y_1^2 + 8y_2^2.$$

This makes it obvious that z has a minimum at the origin and that the contour $5x_1^2 - 6x_1x_2 + 5x_2^2 = c$ is an ellipse:

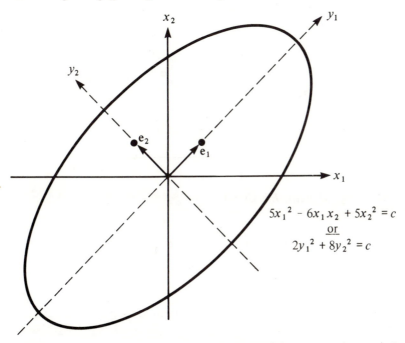

The new y_1 and y_2 axes point in the directions of the eigenvectors e_1 and e_2. We can find e_1 by solving

$$\begin{pmatrix} 5 & -3 \\ -3 & 5 \end{pmatrix} \begin{pmatrix} e \\ f \end{pmatrix} = 2 \begin{pmatrix} e \\ f \end{pmatrix} - \text{i.e.} \quad \left. \begin{matrix} 5e - 3f = 2e \\ -3e + 5f = 2f. \end{matrix} \right\}$$

Both equations yield $e = f$. We want an eigenvector of unit length and so we take $e = f = 1/\sqrt{2}$. Thus

$$\mathbf{e}_1 = \frac{1}{\sqrt{2}}\begin{pmatrix} 1 \\ 1 \end{pmatrix}.$$

We find $\mathbf{e}_2$ by solving

$$\begin{pmatrix} 5 & -3 \\ -3 & 5 \end{pmatrix}\begin{pmatrix} e \\ f \end{pmatrix} = 8\begin{pmatrix} e \\ f \end{pmatrix} - \text{i.e.} \quad \left.\begin{array}{r} 5e - 3f = 8e \\ -3e + 5f = 8f. \end{array}\right\}$$

Both equations yield $e = -f$. Again we want a unit eigenvector and we we take $e = -f = -1/\sqrt{2}$. Thus

$$\mathbf{e}_2 = \frac{1}{\sqrt{2}}\begin{pmatrix} -1 \\ 1 \end{pmatrix}.$$

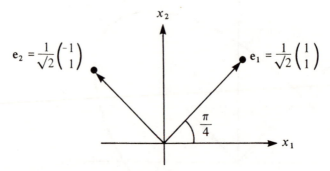

The orthogonal matrix E is given by

$$E = \begin{pmatrix} | & | \\ \mathbf{e}_1 & \mathbf{e}_2 \\ | & | \end{pmatrix} = \begin{pmatrix} \dfrac{1}{\sqrt{2}} & -\dfrac{1}{\sqrt{2}} \\[2mm] \dfrac{1}{\sqrt{2}} & \dfrac{1}{\sqrt{2}} \end{pmatrix}.$$

Thus the new co-ordinates $(y_1, y_2)^T$ are related to the old co-ordinates $(x_1, x_2)^T$ by the formulae

$$\begin{pmatrix} x_1 \\ x_2 \end{pmatrix} = \begin{pmatrix} \dfrac{1}{\sqrt{2}} & -\dfrac{1}{\sqrt{2}} \\[2mm] \dfrac{1}{\sqrt{2}} & \dfrac{1}{\sqrt{2}} \end{pmatrix}\begin{pmatrix} y_1 \\ y_2 \end{pmatrix}; \quad \begin{pmatrix} y_1 \\ y_2 \end{pmatrix} = \begin{pmatrix} \dfrac{1}{\sqrt{2}} & \dfrac{1}{\sqrt{2}} \\[2mm] -\dfrac{1}{\sqrt{2}} & \dfrac{1}{\sqrt{2}} \end{pmatrix}\begin{pmatrix} x_1 \\ x_2 \end{pmatrix}.$$

Finally, it is of some interest to note that

$$E = \begin{pmatrix} \dfrac{1}{\sqrt{2}} & -\dfrac{1}{\sqrt{2}} \\ \dfrac{1}{\sqrt{2}} & \dfrac{1}{\sqrt{2}} \end{pmatrix} = \begin{pmatrix} \cos\theta & -\sin\theta \\ \sin\theta & \cos\theta \end{pmatrix}$$

where $\theta = \pi/4$. The new axes are therefore obtained from the old axes by a rotation through $45°$ (see example 5.3).

5.6   Positive and negative definite

A symmetric matrix is called *positive definite* if *all* its eigenvalues are positive. A symmetric matrix is called *negative definite* if all its eigenvalues are negative.

But, as explained in §5.4,

$$\mathbf{x}^T A \mathbf{x} = \lambda_1 y_1^2 + \lambda_2 y_2^2 + \ldots + \lambda_n y_n^2$$

and hence A is positive definite if and only if

$$z = \mathbf{x}^T A \mathbf{x} > 0$$

unless $\mathbf{x} = \mathbf{0}$. Similarly, A is negative definite if and only if

$$z = \mathbf{x}^T A \mathbf{x} < 0$$

unless $\mathbf{x} = \mathbf{0}$.

It may be helpful to consider the diagrams for the two-dimensional case drawn below:

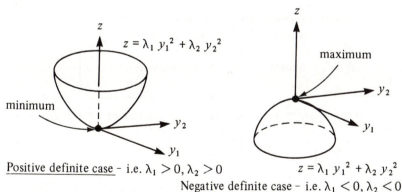

Positive definite case – i.e. $\lambda_1 > 0, \lambda_2 > 0$

$z = \lambda_1 y_1{}^2 + \lambda_2 y_2{}^2$

Negative definite case – i.e. $\lambda_1 < 0, \lambda_2 < 0$

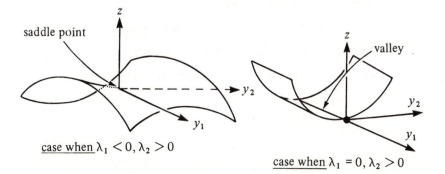

case when $\lambda_1 < 0, \lambda_2 > 0$

case when $\lambda_1 = 0, \lambda_2 > 0$

Fortunately, it is not necessary to calculate the eigenvalues of A to determine whether or not A is positive definite or negative definite. The *principal minors* of a symmetric matrix

$$A = \begin{pmatrix} a_{11} & a_{12} \ldots a_{1n} \\ a_{21} & a_{22} \ldots a_{2n} \\ \phantom{a_{11}} \vdots \\ a_{n1} & a_{n2} \ldots a_{nn} \end{pmatrix}$$

are the determinants

$$a_{11}, \quad \begin{vmatrix} a_{11} & a_{12} \\ a_{21} & a_{22} \end{vmatrix}, \quad \begin{vmatrix} a_{11} & a_{12} & a_{13} \\ a_{21} & a_{22} & a_{23} \\ a_{31} & a_{32} & a_{33} \end{vmatrix}, \ldots, \quad \begin{vmatrix} a_{11} & a_{12} \ldots a_{1n} \\ a_{21} & a_{22} \ldots a_{2n} \\ \phantom{a_{11}} \vdots \\ a_{n1} & a_{n2} \ldots a_{nn} \end{vmatrix}$$

We have that

> (I) A symmetric matrix is positive definite
> if and only if *all* of its principal minors
> are *positive*.
>   (II) A symmetric matrix is negative definite
> if and only if all of its principal minors
> of *even* order are *positive* and all its
> principal minors of *odd* order are *negative*.

This result can be proved quite easily for the case of a 2 × 2 symmetric matrix

$$A = \begin{pmatrix} \alpha & \gamma \\ \gamma & \beta \end{pmatrix}.$$

The characteristic equation is

$$\begin{vmatrix} \alpha - \lambda & \gamma \\ \gamma & \beta - \lambda \end{vmatrix} = (\alpha - \lambda)(\beta - \lambda) - \gamma^2 \\ = \lambda^2 - \lambda(\alpha + \beta) + (\alpha\beta - \gamma^2).$$

If the eigenvalues are λ_1 and λ_2, we therefore have that

$$(\lambda - \lambda_1)(\lambda - \lambda_2) = \lambda^2 - \lambda(\alpha + \beta) + (\alpha\beta - \gamma^2)$$

and so

$$\begin{cases} \lambda_1 + \lambda_2 = \alpha + \beta \\ \lambda_1 \lambda_2 = \alpha\beta - \gamma^2 = \begin{vmatrix} \alpha & \gamma \\ \gamma & \beta \end{vmatrix} = \det(A). \end{cases}$$

These results, incidentally, are special cases of more general results. For any $n \times n$ symmetric matrix A,

$$\lambda_1 + \lambda_2 + \ldots + \lambda_n = a_{11} + a_{22} + \ldots + a_{nn}$$

$$\lambda_1 \lambda_2 \ldots \lambda_n = \det(A).$$

Returning to the 2×2 case, we observe that λ_1 and λ_2 have the same sign if and only if $\lambda_1 \lambda_2 > 0$ – i.e.

$$\det(A) = \alpha\beta - \gamma^2 = \lambda_1 \lambda_2 > 0.$$

But $\alpha\beta > \gamma^2$ implies that α and β have the same sign. If $\lambda_1 > 0$ and $\lambda_2 > 0$, then $\alpha + \beta = \lambda_1 + \lambda_2 > 0$ and hence $\alpha > 0$. Similarly, if $\lambda_1 < 0$ and $\lambda_2 < 0$, then $\alpha + \beta = \lambda_1 + \lambda_2 < 0$ and hence $\alpha < 0$.

Example 5.7

(i) Consider the matrix

$$A = \begin{pmatrix} 5 & -3 \\ -3 & 5 \end{pmatrix}.$$

We know that this matrix has eigenvalues $\lambda_1 = 2$ and $\lambda_2 = 8$ and hence is positive definite (see example 5.5). The matrix can therefore be used to check our criterion (I) concerning the principal minors of a matrix. Observe that

(a) $\begin{vmatrix} 5 & -3 \\ -3 & 5 \end{vmatrix} = 25 - 9 = 16 > 0$

(b) $5 > 0$.

Thus criterion (I) is satisfied.

(ii) Consider the matrix

$$B = \begin{pmatrix} 1 & 6 \\ 6 & 4 \end{pmatrix}.$$

Calculating the principal minors, we obtain that

(a) $\begin{vmatrix} 1 & 6 \\ 6 & 4 \end{vmatrix} = 4 - 36 = -32 < 0$

(b) $1 > 0.$

Thus the matrix is neither positive definite nor negative definite because neither criterion (I) nor criterion (II) is satisfied. (Note that (b) is unnecessary for this conclusion.)

(iii) Consider the matrix

$$C = \begin{pmatrix} -2 & 1 \\ 1 & -2 \end{pmatrix}.$$

Calculating the principal minors, we obtain that

(a) $\begin{vmatrix} -2 & 1 \\ 1 & -2 \end{vmatrix} = 4 - 1 = 3 > 0$

(b) $-2 < 0.$

Hence criterion (II) is satisfied and so C is negative definite.

Since $\lambda_1 \lambda_2 \ldots \lambda_n = \det(A)$, it follows that, if $\det(A) = 0$, then at least one of the eigenvalues of A is zero. In view of this result, it is tempting to suppose that if some of the principal minors of A are zero and the rest are positive, then the matrix A will necessarily have non-negative eigenvalues. But this is FALSE as the next example shows. When some of the principal minors are zero, criteria (I) and (II) have nothing to say about the signs of the eigenvalues of the matrix A.

Example 5.8.
Consider the matrix

$$A = \begin{pmatrix} 1 & 1 & 0 \\ 1 & 1 & 0 \\ 0 & 0 & t \end{pmatrix}.$$

The principal minors are all non-negative because

(a) $\begin{vmatrix} 1 & 1 & 0 \\ 1 & 1 & 0 \\ 0 & 0 & t \end{vmatrix} = t - t = 0$

(b) $\begin{vmatrix} 1 & 1 \\ 1 & 1 \end{vmatrix} = 1 - 1 = 0$

(c) $1 > 0$.

But the characteristic equation of A is

$$\begin{vmatrix} 1-\lambda & 1 & 0 \\ 1 & 1-\lambda & 0 \\ 0 & 0 & t-\lambda \end{vmatrix} = (t-\lambda)\{(1-\lambda)^2 - 1\} = 0$$

and hence the eigenvalues are $\lambda_1 = t, \lambda_2 = 0, \lambda_3 = 2$. But t may be positive or it may be negative. The values of the principal minors are therefore not much use in this case.

5.9 Maxima and minima

We have seen that the stationary points for a function $f: \mathbb{R}^n \to \mathbb{R}$ are those at which all the partial derivatives are zero. Thus ξ is a stationary point if and only if

$$f'(\xi) = 0.$$

How does one determine whether a stationary point ξ is a local maximum, a local minimum or a saddle point?

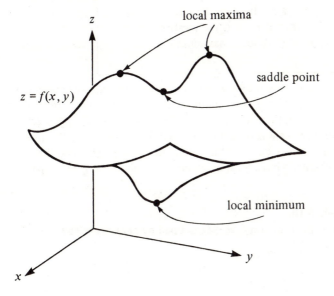

Taylor's theorem asserts that

$$f(\mathbf{x}) = f(\boldsymbol{\xi}) + \frac{1}{1!}f'(\boldsymbol{\xi})(\mathbf{x} - \boldsymbol{\xi}) + \frac{1}{2!}(\mathbf{x} - \boldsymbol{\xi})^T f''(\boldsymbol{\xi})(\mathbf{x} - \boldsymbol{\xi}) + \ldots.$$

At a stationary point, $f'(\boldsymbol{\xi}) = \mathbf{0}$ and so

$$f(\mathbf{x}) - f(\boldsymbol{\xi}) = \frac{1}{2!}(\mathbf{x} - \boldsymbol{\xi})^T f''(\boldsymbol{\xi})(\mathbf{x} - \boldsymbol{\xi}) + \ldots.$$

Provided that $\mathbf{x}$ is sufficiently close to $\boldsymbol{\xi}$, the third and higher order terms in this expression are negligible compared with

$$\frac{1}{2!}(\mathbf{x} - \boldsymbol{\xi})^T f''(\boldsymbol{\xi})(\mathbf{x} - \boldsymbol{\xi})$$

(unless this term happens to be zero). Recall that $f''(\boldsymbol{\xi})$ is an $n \times n$ symmetric matrix. If we write $\mathbf{u} = \mathbf{x} - \boldsymbol{\xi}$, we therefore have to deal with a quadratic form $\frac{1}{2}\mathbf{u}^T f''(\boldsymbol{\xi})\mathbf{u}$.

If this quadratic form is always positive (unless $\mathbf{u} = \mathbf{0}$), then $f(\mathbf{x}) > f(\boldsymbol{\xi})$ for all $\mathbf{x}$ close to $\boldsymbol{\xi}$ (except $\mathbf{x} = \boldsymbol{\xi}$). If the quadratic form is always negative (unless $\mathbf{u} = \mathbf{0}$), then $f(\mathbf{x}) < f(\boldsymbol{\xi})$ for all $\mathbf{x}$ close to $\boldsymbol{\xi}$ (except $\mathbf{x} = \boldsymbol{\xi}$).

Summarizing these results, we obtain that:

> (I) If $f'(\boldsymbol{\xi}) = \mathbf{0}$ and $f''(\boldsymbol{\xi})$ is positive definite, then $\boldsymbol{\xi}$ is a local minimum.
> (II) If $f'(\boldsymbol{\xi}) = \mathbf{0}$ and $f''(\boldsymbol{\xi})$ is negative definite, then $\boldsymbol{\xi}$ is a local maximum.

Note that a stationary point $\boldsymbol{\xi}$ may well be a local maximum or minimum even though $f''(\boldsymbol{\xi})$ is neither positive definite nor negative definite. In particular, if $\det f''(\boldsymbol{\xi}) = 0$, then at least one of the eigenvalues of $f''(\boldsymbol{\xi})$ is zero. Hence $\mathbf{u}^T f''(\boldsymbol{\xi})\mathbf{u}$ is *zero* along some line through the origin. Along this line the third and higher order terms of the Taylor are *not* negligible and the question as to whether $\boldsymbol{\xi}$ has a local maximum or minimum at $\boldsymbol{\xi}$ will depend on their nature.

Example 5.10.
Consider the function $f: \mathbb{R} \to \mathbb{R}$ defined by $f(x) = x^4$. Then

$$f'(0) = 4 \cdot 0^3 = 0$$

$$f''(0) = 12 \cdot 0^2 = 0.$$

Thus 0 is a stationary point but the 1×1 matrix $f''(0)$ is neither positive definite nor negative definite. However, the function has a minimum at $x = 0$.

If $f''(\xi)$ has some positive eigenvalues and some negative eigenvalues, then the function has a *saddle point* at ξ. A useful criterion is the following:

> (III) If $f'(\xi) = 0$ and $\det f''(\xi) \neq 0$ but $f''(\xi)$ is neither positive definite nor negative definite, then ξ is a saddle point.

In the case of a function $f : \mathbb{R} \to \mathbb{R}$ the above rules for classifying a stationary point are very easy to apply. If $f''(\xi) > 0$, then ξ is a local minimum (rule I). If $f''(\xi) < 0$, then ξ is a local maximum (rule II). Note that $f''(\xi) = 0$ does *not* necessarily mean that ξ is a point of inflexion (see example 5.10).

In the case of a function $f : \mathbb{R}^2 \to \mathbb{R}$, it is best to begin the process of classifying a stationary point by calculating

$$\Delta = \det f''(\xi) = \begin{vmatrix} f_{xx} & f_{xy} \\ f_{xy} & f_{yy} \end{vmatrix} = f_{xx} f_{yy} - f_{xy}^2.$$

There are then three possibilities:

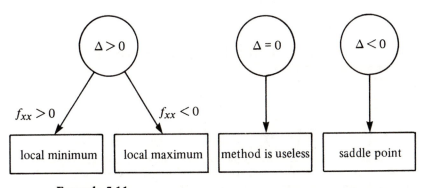

Example 5.11.

In example 3.5, we considered the function $f : \mathbb{R} \to \mathbb{R}$ defined by $f(x) = x^3 - 3x^2 + 2x$ and discovered two stationary points $x = \{1 - (1/\sqrt{3})\}$ and $x = \{1 + (1/\sqrt{3})\}$. We have that

$$f'(x) = 3x^2 - 6x + 2$$

$$f''(x) = 6x - 6 = 6(x - 1).$$

Thus

$$f''\left(1 - \frac{1}{\sqrt{3}}\right) = -\frac{6}{\sqrt{3}} < 0 \\ f''\left(1 + \frac{1}{\sqrt{3}}\right) = \frac{6}{\sqrt{3}} > 0$$

and so $\{1 - (1/\sqrt{3})\}$ is a local maximum and $\{1 + (1/\sqrt{3})\}$ is a local minimum.

Example 5.12.
Consider the function $f : \mathbb{R}^2 \to \mathbb{R}$ defined by

$$f(x, y) = x^3 - 3xy^2 + y^4.$$

The stationary points are found by solving

$$\frac{\partial f}{\partial x} = 3x^2 - 3y^2 = 0 \\ \frac{\partial f}{\partial y} = -6xy + 4y^3 = 0.$$

We obtain that

$$(x = y \quad \text{OR} \quad x = -y)$$

and

$$(y = 0 \quad \text{OR} \quad y^2 = \tfrac{3}{2}x).$$

The stationary points are therefore $(0, 0)^T$, $(\tfrac{3}{2}, \tfrac{3}{2})^T$ and $(\tfrac{3}{2}, \tfrac{-3}{2})^T$.
    The second derivative is

$$f''(x, y) = \begin{pmatrix} \dfrac{\partial^2 f}{\partial x^2} & \dfrac{\partial^2 f}{\partial x \partial y} \\ \dfrac{\partial^2 f}{\partial x \partial y} & \dfrac{\partial^2 f}{\partial y^2} \end{pmatrix} = \begin{pmatrix} 6x & -6y \\ -6y & -6x + 12y^2 \end{pmatrix}$$

and so the principal minors are δ and Δ where

$$\Delta = \det f''(x, y) = \begin{vmatrix} 6x & -6y \\ -6y & -6x + 12y^2 \end{vmatrix} \\ = 36(-x^2 + 2xy^2 - y^2)$$

and

$$\delta = f_{xx}(x, y) = 6x.$$

(i) At the point $(\frac{3}{2}, \frac{3}{2})^T$, we have that

$$\Delta = 36(-\frac{9}{4} + 2 \cdot \frac{3}{2} \cdot \frac{9}{4} - \frac{9}{4}) = 81 > 0$$

$$\delta = 6 \cdot \frac{3}{2} = 9 > 0$$

and hence this point is a local minimum.

(ii) At the point $(\frac{3}{2}, -\frac{3}{2})^T$, we have that

$$\Delta = 81 > 0$$

$$\delta = 9 > 0$$

as before. Hence this point is also a local minimum.

(iii) At the point $(0, 0)^T$, $\Delta = 0$ and $\delta = 0$. Hence the method is useless. Observe, however, that

$$z = f(0, y) = y^4$$

$$z = f(x, 0) = x^3.$$

Thus the function certainly does not have a local maximum or minimum at $(0, 0)^T$.

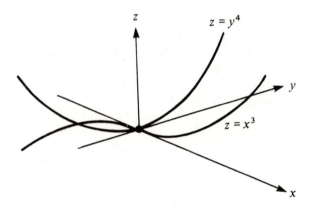

Example 5.13.
Consider the function $f: \mathbb{R}^3 \to \mathbb{R}$ defined by

$$f(x, y, z) = xy + yz + zx.$$

The stationary points are found by solving the equations

$$\frac{\partial f}{\partial x} = y + z = 0; \quad \frac{\partial f}{\partial y} = x + z = 0; \quad \frac{\partial f}{\partial z} = y + x = 0$$

and hence $(0, 0, 0)^T$ is the only stationary point. The second derivative is

$$\begin{pmatrix} f_{xx} & f_{xy} & f_{xz} \\ f_{yx} & f_{yy} & f_{yz} \\ f_{zx} & f_{zy} & f_{zz} \end{pmatrix} = \begin{pmatrix} 0 & 1 & 1 \\ 1 & 0 & 1 \\ 1 & 1 & 0 \end{pmatrix}.$$

The principal minors are

$$\begin{vmatrix} 0 & 1 & 1 \\ 1 & 0 & 1 \\ 1 & 1 & 0 \end{vmatrix} = 2 < 0; \quad \begin{vmatrix} 0 & 1 \\ 1 & 0 \end{vmatrix} = -1 < 0; \quad 0 = 0.$$

Since $\det f''(0, 0, 0) \neq 0$ and $f''(0, 0, 0)$ is neither positive definite nor negative definite, $(0, 0, 0)^T$ is a saddle point.

5.14† Convex and concave functions

The discussion in this chapter so far has centred on the identification of *local* maxima and minima. But, as noted in §3.2, the identification of local maxima and minima is usually only a step on the way to finding *global* maxima or minima for an optimization problem. However, in the case of a *concave* function, a local maximum is automatically a global maximum. Similarly, in the case of a *convex* function, a local minimum is automatically a global minimum. This observation is of some importance since concave and convex functions appear quite frequently in applications.

A *convex set* in $\mathbb{R}^n$ is a set S with the property that the line segment joining any two points of S lies entirely inside S.

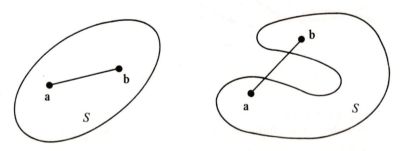

Convex set in $\mathbb{R}^2$ Non-convex set in $\mathbb{R}^2$

A *convex function* $f : \mathbb{R}^n \to \mathbb{R}$ is a function with the property that the set of points in $\mathbb{R}^{n+1}$ which lie *above* the graph $y = f(\mathbf{x})$ (see §2.1) is convex. A *concave function* $f : \mathbb{R}^n \to \mathbb{R}$ is a function with the property that the set of points which lie *below* the graph $y = f(\mathbf{x})$ is convex.

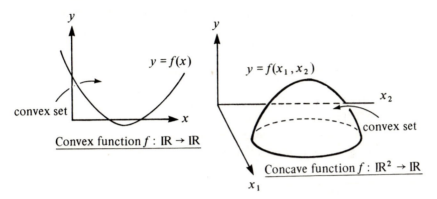

The diagrams above make it clear why a local minimum for a convex function is necessarily a global minimum and why a local maximum for a concave function is necessarily a global maximum.

If a function $f: \mathbb{R}^n \to \mathbb{R}$ is twice differentiable, it is possible to determine whether or not it is convex or concave by examining the second derivative. We consider the case of a concave function. As the diagram below indicates in the case $n = 1$, the tangent hyperplanes to the graph of a concave function $f: \mathbb{R}^n \to \mathbb{R}$ all lie on or above the graph.

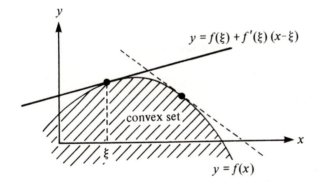

As we know from §4.4, the equation of the tangent hyperplane where $\mathbf{x} = \boldsymbol{\xi}$ is given by

$$y = f(\boldsymbol{\xi}) + f'(\boldsymbol{\xi})(\mathbf{x} - \boldsymbol{\xi}).$$

To say that this lies on or above the graph $y = f(\mathbf{x})$ is the same as asserting that

$$f(\mathbf{x}) \leqslant f(\boldsymbol{\xi}) + f'(\boldsymbol{\xi})(\mathbf{x} - \boldsymbol{\xi})$$

for all values of $\mathbf{x}$. Applying Taylor's theorem as in §5.9, we are led to the conclusion that $f''(\boldsymbol{\xi})$ must be non-positive definite for all $\boldsymbol{\xi}$. Similarly, the second derivative of a convex function must always be non-negative definite. The general result is stated below.

(I) A function f is convex if and only if $f''(\boldsymbol{\xi})$ is non-negative definite for all $\boldsymbol{\xi}$.

(II) A function f is concave if and only if $f''(\boldsymbol{\xi})$ is non-positive definite for all $\boldsymbol{\xi}$.

(In terms of the eigenvalues, non-negative definite means that $\lambda_1 \geqslant 0, \lambda_2 \geqslant 0, \ldots, \lambda_n \geqslant 0$. Similarly, non-positive definite means that $\lambda_1 \leqslant 0, \lambda_2 \leqslant 0, \ldots, \lambda_n \leqslant 0$. Sometimes the terms 'positive semi-definite' and 'negative semi-definite' are used instead.)

Examples 5.15

(i) Consider the function $f: \mathbb{R} \to \mathbb{R}$ defined by $f(x) = -x^4$. We have that $f'(x) = -4x^3$ and $f''(x) = -12x^2$. Since $f''(x) \leqslant 0$ for all x, it follows that this function is concave.

(ii) Consider the function $f: \mathbb{R} \to \mathbb{R}$ defined by $f(x) = e^x$. We have that $f'(x) = e^x$ and $f''(x) = e^x$. Since $f''(x) \geqslant 0$ for all x, it follows that this function is convex.

(iii) Consider the function $f: \mathbb{R}^2 \to \mathbb{R}$ defined by

$$f(x_1, x_2) = 5x_1^2 - 6x_1x_2 + 5x_2^2.$$

The second derivative is

$$f''(x_1, x_2) = \begin{pmatrix} 10 & -6 \\ -6 & 10 \end{pmatrix}$$

which is positive definite (example 5.7(i)) and hence non-negative definite. The function is therefore convex.

(iv) Consider the function $f: \mathbb{R}^2 \to \mathbb{R}$ defined by

$$f(x_1, x_2) = 2x_1x_2 - x_1^2 - x_2^2.$$

The second derivative is

$$f''(x_1, x_2) = \begin{pmatrix} -2 & 2 \\ 2 & -2 \end{pmatrix}.$$

The eigenvalues are found by solving $(-2-\lambda)^2 = 4$ for which the roots are $\lambda_1 = -4$ and $\lambda_2 = 0$. Thus the second derivative is always non-positive definite and so the function is concave.

Exercises 5.16

1. Classify the stationary points of the function $f: \mathbb{R} \to \mathbb{R}$ defined by

$$f(x) = e^{2x} + e^{-3x}$$

using the second derivative (see exercise 3.12(1)).

2.* Classify the stationary points of the functions $f: \mathbb{R} \to \mathbb{R}$ defined by

(i) $\quad f(x) = \dfrac{x^2 + x - 1}{x^2 + 1}$ (ii) $\quad f(x) = x^3 e^{-x^2}$

using the second derivative (see exercise 3.12(2)).

3. Classify the stationary points of the function $f: \mathbb{R}^2 \to \mathbb{R}$ defined by

$$f(x, y) = x^2 y + y^3 x - xy$$

using the second derivative (see example 3.8)

4.* Classify the stationary points of the functions $f: \mathbb{R}^2 \to \mathbb{R}$ defined by

(i) $f(x, y) = x^2 y + y^3 x - xy^2$
(ii) $f(x, y) = e^{x+y}(x^2 - 2xy + 3y^2)$
(iii) $f(x, y) = y^3 + 3x^2 y - 3x^2 - 3y^2 + 2$
(iv) $f(x, y) = 8x^2 y - x^3 y - 5y^3 x$
(v) $f(x, y, z) = x^2 y + y^2 z + z^2 x$.

using the second derivative (see exercise 3.12(6)).

5.† Find the stationary points of the function $f: \mathbb{R}^3 \to \mathbb{R}$ defined by

$$f(x, y, z) = 5x^2 + 5y^2 + 9z^2 - 6xz - 12yz$$

and determine their nature.

6.*† Find the stationary points of the function $f: \mathbb{R}^6 \to \mathbb{R}$ defined by

$$f(u, v, w, x, y, z) = (u^3 - 3uv^2 + v^4) + (w^3 - 3wx^2 + x^4)$$
$$+ (y^3 - 3yz^2 + z^4)$$

and determine their nature (see example 5.12).

7.† By an orthogonal transformation of the variables, reduce $z = xy$ to the form $z = \lambda_1 X^2 + \lambda_2 Y^2$. Draw a diagram indicating the curve $xy = 1$ together with the new X and Y axes. Discuss the behaviour of the function $f: \mathbb{R}^2 \to \mathbb{R}$ defined by $f(x, y) = xy$ at the point $(0, 0)^T$.

8.*† By an orthogonal transformation of the variables, reduce $u = xy + yz + zx$ to the form $u = \lambda_1 X^2 + \lambda_2 Y^2 + \lambda_3 Z^2$. Use your result to discuss the behavior of the function $f: \mathbb{R}^3 \to \mathbb{R}$ defined by $f(x, y, z) = xy + yz + zx$ at the point $(0, 0, 0)^T$.

9.[†] Show that the function $f: \mathbb{R}^3 \to \mathbb{R}$ of question 5 is convex. What is a global minimum for this function?

10.*[†] Sketch the graphs of the functions considered in examples 5.15.

SOME APPLICATIONS (OPTIONAL)

5.17   Saddle points

It is easy to see why maxima and minima are important but applications for saddle points are perhaps not so evident.

We therefore give an application from game theory. In a two-person zero-sum game, one is given a payoff function

$$L(\mathbf{p}, \mathbf{q})$$

which the first player seeks to maximize by choosing an appropriate $\mathbf{p}$ from his set P of feasible strategies, while the second player simultaneously seeks to minimize $L(\mathbf{p}, \mathbf{q})$ by choosing an appropriate $\mathbf{q}$ from his set Q of feasible strategies.

In §1.38, we considered the case in which

$$L(\mathbf{p}, \mathbf{q}) = \mathbf{p}^T A \mathbf{q}$$

where A is an $m \times n$ matrix. The $m \times 1$ vector

$$\mathbf{p} = (p_1, p_2, \dots, p_m)^T$$

in this case represents a 'mixed' strategy in which p_k is the probability with which the kth 'pure' strategy is to be played. Thus P is the set of all vectors $\mathbf{p}$ for which $p_1 \geqslant 0, p_2 \geqslant 0, \dots, p_m \geqslant 0$ and $p_1 + p_2 + \dots + p_m = 1$. Similarly, Q is the set of all vectors $\mathbf{q}$ for which $q_1 \geqslant 0, q_2 \geqslant 0, \dots, q_n \geqslant 0$ and $q_1 + q_2 + \dots + q_n = 1$.

In §4.28, we considered the zero-sum game whose payoff function

$$L(\mathbf{x}, \mathbf{q}) = \mathbf{p}^T \mathbf{x} + \mathbf{q}^T (\mathbf{b} - A\mathbf{x})$$

is the Lagrangian of the linear programming problem

$$\max \mathbf{p}^T \mathbf{x}$$

subject to

$$Ax \leqslant b$$
$$x \geqslant 0.$$

(Here **p** is a *constant* vector representing the prices at which the commodity bundle **x** can be sold.) The housewife sought to choose $\mathbf{x} \geqslant \mathbf{0}$ to maximise $L(\mathbf{x}, \mathbf{q})$ and the combine to choose $\mathbf{q} \geqslant 0$ so as to minimise $L(\mathbf{x}, \mathbf{q})$.

A pair $(\tilde{\mathbf{p}}, \tilde{\mathbf{q}})$ of strategies for a zero-sum game with payoff function $L(\mathbf{p}, \mathbf{q})$ is a Nash equilibrium if and only if

$$L(\mathbf{p}, \tilde{\mathbf{q}}) \leqslant L(\tilde{\mathbf{p}}, \tilde{\mathbf{q}}) \leqslant L(\tilde{\mathbf{p}}, \mathbf{q})$$

for all **p** in the set P and for all **q** in the set Q. If $(\tilde{\mathbf{p}}, \tilde{\mathbf{q}})$ is a Nash equilibrium, then the choice $\mathbf{p} = \tilde{\mathbf{p}}$ is an optimal response by the first player to the choice of $\mathbf{q} = \tilde{\mathbf{q}}$ by the second player. Simultaneously, the choice $\mathbf{q} = \tilde{\mathbf{q}}$ by the second player is an optimal response by the second player to the choice $\mathbf{p} = \tilde{\mathbf{p}}$ by the first player. Any solution to a two-person zero-sum game must clearly satisfy this criterion.

If $(\tilde{\mathbf{p}}, \tilde{\mathbf{q}})$ is a Nash equilibrium, then the function $L(\mathbf{p}, \tilde{\mathbf{q}})$ achieves a maximum at $\mathbf{p} = \tilde{\mathbf{p}}$ (subject to the constraint that **p** lies in P). Similarly, the function $L(\tilde{\mathbf{p}}, \mathbf{q})$ achieves a minimum at $\mathbf{q} = \tilde{\mathbf{q}}$ (subject to the constraint that **q** lies in Q). If $\tilde{\mathbf{p}}$ is an interior point of P and $\tilde{\mathbf{q}}$ is an interior point of Q, then

$$\frac{\partial L}{\partial \mathbf{p}}(\tilde{\mathbf{p}}, \tilde{\mathbf{q}}) = \mathbf{0}; \quad \frac{\partial L}{\partial \mathbf{q}}(\tilde{\mathbf{p}}, \tilde{\mathbf{q}}) = \mathbf{0}$$

and hence $(\tilde{\mathbf{p}}, \tilde{\mathbf{q}})$ is a *stationary point* of L.

However, $(\tilde{\mathbf{p}}, \tilde{\mathbf{q}})$ is neither a local maximum nor a local minimum for L. It is, in fact, a *saddle point* for L.

5.18 Constrained optimization

Suppose that $f: \mathbb{R}^n \to \mathbb{R}$ and that $g: \mathbb{R}^n \to \mathbb{R}^m$. The stationary points of $f(\mathbf{x})$ subject to the constraint $g(\mathbf{x}) = \mathbf{0}$ are found by calculating the stationary points of the Lagrangian

$$L(\mathbf{x}, \mathbf{y}) = f(\mathbf{x}) + \mathbf{y}^T g(\mathbf{x}).$$

But how does one classify these constrained stationary points?

The condition for a local maximum is that the final $n - m$ principal minors of the matrix below should alternate in sign with the sign of the first being $(-1)^{m+1}$

$$
\begin{pmatrix}
0 & \cdots & 0 & \dfrac{\partial g_1}{\partial x_1} & \cdots & \dfrac{\partial g_1}{\partial x_n} \\[2ex]
\vdots & & \vdots & \vdots & & \\[1ex]
0 & & 0 & \dfrac{\partial g_m}{\partial x_1} & \cdots & \dfrac{\partial g_m}{\partial x_n} \\[2ex]
\dfrac{\partial g_1}{\partial x_1} & & \dfrac{\partial g_m}{\partial x_1} & \dfrac{\partial^2 L}{\partial x_1^2} & \cdots & \dfrac{\partial^2 L}{\partial x_1 \partial x_n} \\[2ex]
\vdots & & & \vdots & & \\[1ex]
\dfrac{\partial g_1}{\partial x_n} & \cdots & \dfrac{\partial g_m}{\partial x_n} & \dfrac{\partial^2 L}{\partial x_n \partial x_1} & \cdots & \dfrac{\partial^2 L}{\partial x_n^2}
\end{pmatrix}
$$

We shall not even begin to discuss the reasons why this criterion is valid.

6

Inverse functions

People often find inverse functions confusing. Some of this confusion is due to the fact that they are usually discussed in a highly cavalier fashion when first introduced. School children, for example, are taught to 'solve' equations like $y = \sin x$ by writing $x = \sin^{-1} y$ and to think of the latter equation as defining a 'multi-valued function'. But the standard definition of a function (given in §2.1) is insistent that functions are 'single-valued'. That confusion should result is therefore not surprising.

In this chapter we begin with a fairly careful account of what an inverse function is and show how the idea applies in the case of the familiar exponential, logarithm and trigonometric functions. Although this discussion is essentially self-contained, it is not intended to introduce these functions to those who have not met them before. Its purpose is to clarify a number of points which are commonly found somewhat puzzling. This applies in particular to the 'inverse trigonometric functions'. It is suggested that even those who feel they know the exponential, logarithm and trigonometric functions quite well should check on their facility with these functions by attempting exercise 6.9 before proceeding further.

The chapter then moves on to a discussion of local inverse functions in the vector case and to the application of this work to general co-ordinate systems. The ideas are explained at some length but it would be inappropriate to spend a lot of time on these sections if difficulties are being encountered elsewhere. For most purposes it is enough to have some understanding of why things go wrong when the Jacobian is zero and to be able to change the variables in differential operators as described in §6.19.

6.1 Intervals

A set I of real numbers is an interval if, whenever it contains two real numbers, it also contains all real numbers between them. The set $\mathbb{R}$ of all real numbers is an example of an interval. Some further examples, together with the notation used to denote them, are given in the table below.

Notation	Description
(a, b)	the set of all x such that $a < x < b$
$[a, b]$	the set of all x such that $a \leqslant x \leqslant b$
(a, ∞)	the set of all x such that $x > a$
$[a, \infty)$	the set of all x such that $x \geqslant a$

6.2   Inverse functions

Let X and Y be sets and suppose that $f : X \to Y$ is a function. This means that, for each x in the set X, there is a unique y in the set Y such that $y = f(x)$. If $g : Y \to X$ is another function and has the property that

$$y = f(x) \quad \text{if and only if} \quad x = g(y),$$

then we call g the *inverse function* to f.

Observe that $x = g(y)$ is what we obtain by solving the equation $y = f(x)$ for x in terms of y. However, in the general case, the equation $y = f(x)$ may have no solutions at all or else may have many solutions. For example, in the diagram of a function $f : \mathbb{R} \to \mathbb{R}$ drawn below, the equation $y_1 = f(x)$ has no solutions, while the equation $y_2 = f(x)$ has five solutions (namely x_1, x_2, x_3, x_4 and x_5).

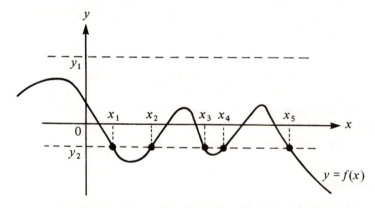

But in order that $g : Y \to X$ be a function it is necessary that, for *each* y in the set Y, there is a *unique* x in the set X such that $x = g(y)$. Thus, for $f : X \to Y$ to admit an inverse function, it is necessary that, for *each* y in the set Y, the equation $y = f(x)$ has a *unique* solution x in the set X.

The diagrams below illustrate two functions $f: I \to J$ which have this property and which therefore admit inverse functions.

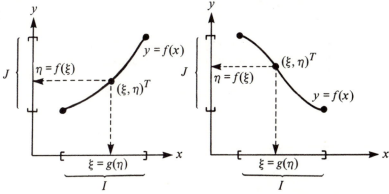

If $f: X \to Y$ has the inverse function $g: Y \to X$, then $y = f(x)$ if and only if $x = g(y)$. It follows that, for each x in the set X and for each y in the set Y,

$$f(g(y)) = f(x) = y$$
$$g(f(x)) = g(y) = x.$$

Example 6.3.

Consider the function $f: \mathbb{R} \to \mathbb{R}$ defined by $y = f(x) = x^2$. This function has *no* inverse function $g: \mathbb{R} \to \mathbb{R}$. Observe that the equation

$$-1 = x^2$$

has *no* (real) solution, while the equation

$$1 = x^2$$

has *two* solutions.

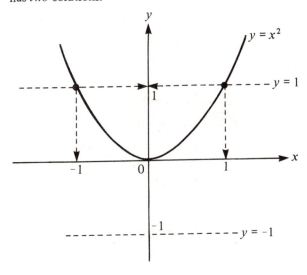

Example 6.4. ◻

Let n be a natural number (i.e. 1, 2, 3, etc.). Consider the function $f: [0, \infty) \to [0, \infty)$ defined by

$$y = f(x) = x^n.$$

Note that we are here deliberately excluding consideration of negative real numbers and restricting attention to *non-negative* values of x and y.

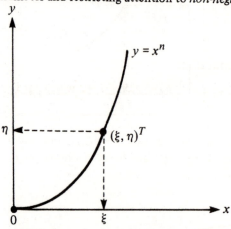

For each $y \geqslant 0$, the equation $y = x^n$ has a unique solution $x \geqslant 0$ and hence the function $f: [0, \infty) \to [0, \infty)$ has an inverse $g: [0, \infty) \to [0, \infty)$. We use the notation

$$x = g(y) = y^{1/n}.$$

Thus, if $x \geqslant 0$ and $y \geqslant 0$,

$$x = y^{1/n} \quad \text{if and only if} \quad y = x^n.$$

It is instructive to observe that the graph of $x = y^{1/n}$ is just the same as the graph of $y = x^n$ but looked at from a different viewpoint. The graph of $x = y^{1/n}$ is what a man would see if he drew the graph of $y = x^n$ on a piece of glass and lay on his side behind the glass to look at it.

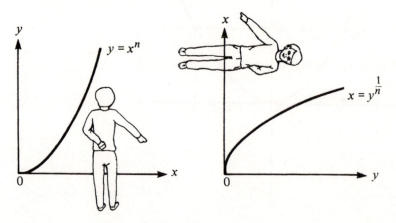

The same principle, of course, applies in respect of the graph of *any* inverse function.

If m is any integer (i.e. $0, \pm 1, \pm 2$, etc.) and $x \geqslant 0$, we define $x^{m/n}$ by

$$x^{m/n} = (x^m)^{1/n}.$$

Note in particular that

$$(x^n)^{1/n} = x.$$

6.5 Derivatives of inverse functions

The diagrams below illustrate a differentiable function $f: I \to J$ which has a differentiable inverse function $g: J \to I$.

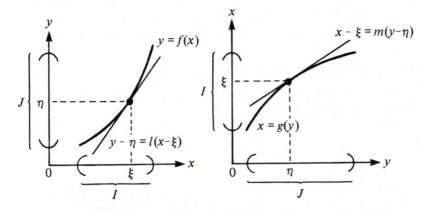

We know that the second diagram is really the same as the first but looked at from a different viewpoint. In particular, the lines $y - \eta = l(x - \xi)$ and $x - \xi = m(y - \eta)$ are really the *same* lines. Thus $m = l^{-1}$. But $l = f'(\xi)$ and $m = g'(\eta)$. This explains the formula

$$\frac{dx}{dy} = \left(\frac{dy}{dx}\right)^{-1}$$

(which we have already met as formula (V) of §2.3).

Example 6.6.
Let $x = y^{1/n}$. Then $y = x^n$. Hence

$$Dy^{1/n} = \frac{dx}{dy} = \left(\frac{dy}{dx}\right)^{-1} = \frac{1}{nx^{n-1}} = \frac{1}{n}(y^{1/n})^{1-n} = \frac{1}{n}y^{1/n-1}.$$

(Furthermore, if $z = y^{m/n}$ and $t = y^m$. Then $z = t^{1/n}$ and so

$$Dy^{m/n} = \frac{dz}{dy} = \frac{dz}{dt}\frac{dt}{dy} = \frac{1}{n}t^{1/n-1}my^{m-1} = \frac{m}{n}y^{m/n-1}.$$

This justifies the formula $Dy^{\alpha} = \alpha y^{\alpha-1}$ in the case when α is a rational number (e.g. $\frac{1}{2}, 3, -\frac{7}{5}$) and $y > 0$.)

It is quite apparent from the formula for the derivative of an inverse function that special thought has to be given as to what happens in the case when the derivative of f is zero. We take this point up in §6.12. But, for the moment, we shall consider only the case in which the derivative of $f : I \to J$ is *always* positive or *always* negative (except possibly at the endpoints of the interval I). In this case the existence of a differentiable inverse function $g : J \to I$ is *guaranteed* provided that the interval J is properly chosen.

In the next two sections we shall illustrate how this result applies in the case of some familiar special functions. This will simultaneously provide an opportunity for some useful revision of the properties of these functions.

6.7   Exponential and logarithm functions[□]

Given a real number $a > 0$, we can plot a graph of $y = a^r$ for each rational number r. (A rational number is a fraction – e.g. $\frac{1}{2}, 3, -\frac{7}{5}$.)

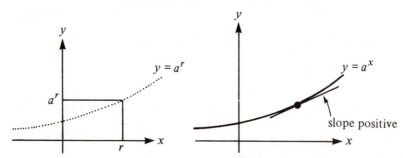

For irrational numbers x (e.g. $\sqrt{2}, \pi, e$), we assign values to a^x in such a way as to 'fill up the holes' in the graph of $y = a^r$. The function $f : \mathbb{R} \to (0, \infty)$ then defined by

$$y = f(x) = a^x$$

is differentiable for all x and its derivative is always positive.

We may deduce the existence of an inverse function $g : (0, \infty) \to \mathbb{R}$. This inverse function is used to define the *logarithm to base a*. We have that

$$x = \log_a y = g(y) \quad (y > 0).$$

Thus $x = \log_a y$ if and only if $y = a^x$. (Note that $\log_a y$ is defined only for *positive* values of y because a^x takes only positive values.)

Let $y_1 = a^{x_1}$ and $y_2 = a^{x_2}$. Then

$$\begin{cases} \log_a(y_1 y_2) = \log_a(a^{x_1} a^{x_2}) = \log_a(a^{x_1+x_2}) = x_1 + x_2 \\ \qquad\qquad = \log_a y_1 + \log_a y_2 \\ \log_a(y_1^b) = \log_a(a^{x_1})^b = \log_a(a^{x_1 b}) = x_1 b = b \log_a y_1. \end{cases}$$

These formulae make it possible to carry out quite complicated calculations fairly easily even without an electronic calculator provided a book of logarithm tables is available. These contain tables of values of $x = \log_{10} y$ and of $y = \text{antilog}_{10} x = 10^x$. To calculate $y_1 y_2$, one simply uses the fact that

$$y_1 y_2 = \text{antilog}_{10}(\log_{10} y_1 + \log_{10} y_2) = 10^{x_1+x_2}.$$

The number $e \doteq 2.718$ has the special property that the slope of the tangent to $y = e^x$ when $x = 0$ is equal to one. The function $f: \mathbb{R} \to (0, \infty)$ defined by $y = f(x) = e^x$ is called the *exponential function*. Its inverse function $g: (0, \infty) \to \mathbb{R}$ is the logarithm to base e. This is called the *natural logarithm*. In mathematical texts it is usual to write $\log_e y = \log y$ since little occasion arises for the use of bases other than e. We shall follow this practice and use $\log y$ strictly for the natural logarithm (see §2.3). An alternative notation is $\log_e y = \ln y$.

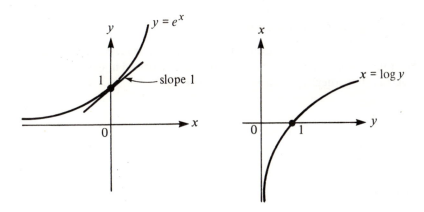

Observe that

$$\frac{d}{dx}(e^x) = \lim_{h \to 0}\left(\frac{e^{x+h} - e^x}{h}\right) = e^x \lim_{h \to 0}\left(\frac{e^h - e^0}{h}\right).$$

But the second limit is the slope of the exponential function at $x = 0$ and e was chosen to make this equal to one. It follows that

$$\frac{d}{dx}(e^x) = e^x$$

Recall that $y = e^x$ if and only if $x = \log y$. It follows that

$$\frac{d}{dy}(\log y) = \frac{dx}{dy} = \left(\frac{dy}{dx}\right)^{-1} = \frac{1}{e^x} = \frac{1}{y}$$

i.e.

$$\frac{d}{dy}(\log y) = \frac{1}{y} \quad (y > 0)$$

Note finally that all powers can be expressed in terms of the exponential and natural logarithm functions. We have that

$$e^{b \log a} = e^{\log(a^b)} = a^b \quad (a > 0).$$

6.8   Trigonometric functions[¤]

In the triangle below, the angle x is to be understood as measured in *radians*. What this means in terms of sectors of a circle is indicated in the second diagram. (In particular, π radians $= 180$ degrees.)

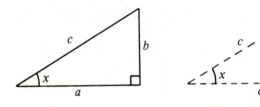

For $0 \leqslant x < \pi/2$, the *sine* and *cosine* functions are defined by

$$\sin x = \frac{b}{c}; \quad \cos x = \frac{a}{c}.$$

The graphs below indicate how $\sin x$ and $\cos x$ are defined for other values of x. Observe that $\sin x = \cos(x - \pi/2)$.

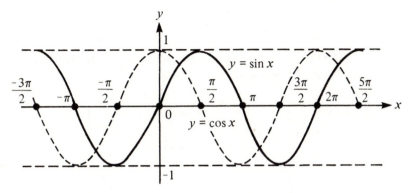

Both the sine and the cosine function are periodic with period 2π (i.e. $\sin(x + 2\pi) = \sin x$ and $\cos(x + 2\pi) = \cos x$). This simply reflects the fact that there are 2π radians in a full circle.

It is important to remember the formulae

$$\frac{d}{dx}(\sin x) = \cos x$$
$$\frac{d}{dx}(\cos x) = -\sin x$$

A justification for these formulae may be based on the diagram below:

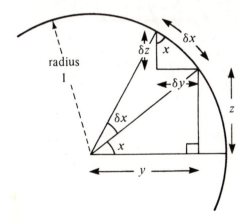

From the larger right-angled triangle, we have that $y = \cos x$ and $z = \sin x$. From the smaller right-angled 'triangle',

$$\cos x \doteqdot \frac{\delta z}{\delta x} \quad , \quad \sin x \doteqdot -\frac{\delta y}{\delta x}.$$

The definitions for the other trigonometric functions may be expressed in terms of the sine and cosine functions as below:

$$\begin{cases} \tan x = \dfrac{\sin x}{\cos x} & \left(= \dfrac{b}{a} \quad \text{when} \quad 0 \leqslant x < \dfrac{\pi}{2} \right) \\[3mm] \sec x = \dfrac{1}{\cos x} \\[3mm] \operatorname{cosec} x = \dfrac{1}{\sin x} \\[3mm] \cot x = \dfrac{1}{\tan x}. \end{cases}$$

Note that the objects on the left are *not* defined when the denominators on the right are zero. The *tangent* function is the most important of these further functions. Its graph is illustrated below:

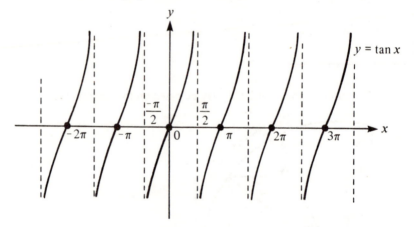

Using the formula for differentiating a quotient, we obtain that

$$\boxed{\dfrac{d}{dx}(\tan x) = \sec^2 x}$$

We can now discuss the so-called 'inverse trigonometric functions'. The first point that needs to be made is that the function $f: \mathbb{R} \to \mathbb{R}$ defined by

$y = f(x) = \sin x$ does *not* have an inverse function. In particular, the equation

$$2 = \sin x$$

has no solutions at all, while the equation

$$0 = \sin x$$

has an infinite number of solutions (i.e. $x = n\pi$ where $n = 0, \pm 1, \pm 2, \ldots$).
Similar remarks apply to the other trigonometric functions.

However, if we restrict our attention to values of x satisfying
$-\pi/2 \leqslant x \leqslant \pi/2$ and values of y satisfying $-1 \leqslant y \leqslant 1$, then the equation

$$y = \sin x$$

does always have a unique solution. Thus the function
$F: [-\pi/2, \pi/2] \to [-1, 1]$ defined by $y = F(x) = \sin x$ admits an inverse
function. We call this inverse function the *arcsine* function. Observe that

$$x = \arcsin y \quad \text{if and only if} \quad y = \sin x$$

provided that $-\pi/2 \leqslant x \leqslant \pi/2$ and $-1 \leqslant y \leqslant 1$.

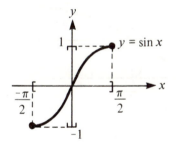 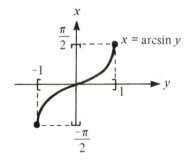

Similarly, the equation

$$y = \cos x$$

always has a unique solution provided that we restrict our attention to values
of x satisfying $0 \leqslant x \leqslant \pi$ and values of y satisfying $-1 \leqslant y \leqslant 1$. The *arc-cosine* function may therefore be defined by writing

$$x = \arccos y \quad \text{if and only if} \quad y = \cos x$$

provided that $0 \leqslant x \leqslant \pi$ and $-1 \leqslant y \leqslant 1$.

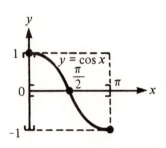

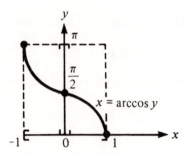

Finally, observe that the equation

$$y = \tan x$$

always has a unique solution provided that we restrict our attention to values of x satisfying $-\pi/2 < x < \pi/2$. The *arctangent* function may therefore be defined by

$$x = \arctan y \quad \text{if and only if} \quad y = \tan x$$

provided that $-\pi/2 < x < \pi/2$.

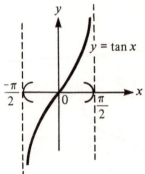

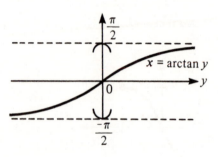

The derivatives of the arcsine, arccosine and arctangent functions are important in integration theory. We begin with the arcsine function.

If $-1 < y < 1$, we have that

$$\frac{d}{dy}(\arcsin y) = \frac{dx}{dy} = \left(\frac{dy}{dx}\right)^{-1} = \frac{1}{\cos x} = \frac{1}{\sqrt{(1-\sin^2 x)}} = \frac{1}{\sqrt{(1-y^2)}}.$$

(Here we have used the fact that $x = \arcsin y$ if and only if $y = \sin x$ provided $-1 \le y \le 1$ and $-\pi/2 \le x \le \pi/2$. The fact that $\cos^2 x + \sin^2 x = 1$ will be familiar (apply Pythagoras's theorem to the right-angled triangle at the beginning of §6.8). In solving $\cos^2 x + \sin^2 x = 1$ for $\cos x$, we take $\sqrt{(1-\sin^2 x)}$ rather than $-\sqrt{(1-\sin^2 x)}$ because we know that the graph $x = \arcsin y$ has positive slope.)

Similarly, if $-1 < y < 1$,

$$\frac{d}{dy}(\arccos y) = \frac{dx}{dy} = \left(\frac{dy}{dx}\right)^{-1} = \frac{1}{-\sin x} = -\frac{1}{\sqrt{(1 - \cos^2 x)}}$$

$$= -\frac{1}{\sqrt{(1 - y^2)}}.$$

Finally, for all values of y,

$$\frac{d}{dy}(\arctan y) = \frac{dx}{dy} = \left(\frac{dy}{dx}\right)^{-1} = \frac{1}{\sec^2 x} = \frac{1}{1 + \tan^2 x} = \frac{1}{1 + y^2}.$$

(Note that the formula $\sec^2 x = 1 + \tan^2 x$ follows from the fact that $\cos^2 x + \sin^2 x = 1$.)

The table below summarizes these results:

$$\frac{d}{dy}(\arcsin y) = \frac{1}{\sqrt{(1-y^2)}} \quad (-1 < y < 1)$$

$$\frac{d}{dy}(\arccos y) = -\frac{1}{\sqrt{(1-y^2)}} \quad (-1 < y < 1)$$

$$\frac{d}{dy}(\arctan y) = \frac{1}{1 + y^2}$$

Exercises 6.9□

1.□ Differentiate the following expressions:

(i) $\log(\log x)$ $(x > 1)$ (ii) $(\log(x + 1))^2$ $(x > -1)$

(iii) $e^{(\log x)^{1/2}}$ $(x > 1)$ (iv) 2^x.

2.*□ Differentiate the following expressions:

(i) $\operatorname{cosec} x$ $(x \neq n\pi)$ (ii) $\cot(\frac{1}{2}x)$ $(x \neq 2n\pi)$

(iii) $\log(\sin x)$ $(0 < x < \pi)$ (iv) $\log(\cot \frac{1}{2}x)$ $(0 < x < \pi)$.

3.□ Sketch the graphs of the functions defined by $\sinh x = \frac{1}{2}(e^x - e^{-x})$ and $\cosh x = \frac{1}{2}(e^x + e^{-x})$. Prove that

(i) $\frac{d}{dx}(\sinh x) = \cosh x$ (ii) $\frac{d}{dx}(\cosh x) = \sinh x$.

Explain why the equation $y = \sinh x$ always has a unique solution for x in terms of y. We write $x = \sinh^{-1} y$ if and only if $y = \sinh x$. Prove that $\cosh^2 x - \sinh^2 x = 1$ and hence establish the formula

$$\frac{d}{dy}\{\sinh^{-1} y\} = \frac{1}{\sqrt{(1+y^2)}}.$$

It is not possible to argue in a similar way from the equation $y = \cosh x$. Explain why not.

4.*$^\sqcap$ Sketch the graph of the function defined by

$$y = \tanh x = \frac{\sinh x}{\cosh x}.$$

Calculate the derivative of this function and explain how $\tanh^{-1} y$ is defined for $-1 < y < 1$. Prove that

$$\frac{d}{dy}(\tanh^{-1} y) = \frac{1}{1-y^2} \quad (-1 < y < 1).$$

5.$^\sqcap$ Differentiate $\log \{y + \sqrt{(y^2 + 1)}\}$ and discuss the relevance of your result to question 3.

6.*$^\sqcap$ Differentiate $\frac{1}{2} \log \{(1 + y)/(1 - y)\}$ and discuss the relevance of your result to question 4.

7.$^\sqcap$ Explain why the Taylor series of e^x about the point 0 is

$$1 + x + \frac{x^2}{2!} + \frac{x^3}{3!} + \frac{x^4}{4!} + \ldots.$$

Assuming that e^x is equal to its Taylor series expansion, prove that, for any fixed n, $\quad x^{n+1} e^{-x} < (n + 1)! \quad (x > 0).$

Deduce that $x^n e^{-x} \to 0$ as $x \to \infty$.

8.*$^\sqcap$ Deduce from the formula $\sin x = \cos \{-x + \frac{1}{2}\pi\}$ that

$$\arccos y + \arcsin y = \frac{\pi}{2} \quad (-1 \le y \le 1).$$

6.10   Local inverses

We have seen some examples of functions $f : \mathbb{R} \to \mathbb{R}$ which do not have an inverse function $g : \mathbb{R} \to \mathbb{R}$ because the equation $y = f(x)$ fails to have a solution for x in terms of y or else has too many solutions. In many cases, however, one can get round this problem by restricting the sets of values of x and y which one takes into account. This leads to the notion of a 'local inverse'.

We say that $g: J \to I$ is a *local inverse* for $f: \mathbb{R} \to \mathbb{R}$ if it is true that

$$x = g(y) \quad \text{if and only if} \quad y = f(x)$$

provided that x is in the interval I and y is in the interval J. The existence of the local inverse therefore requires that, for each y in the interval J, the equation $y = f(x)$ has a unique solution in the interval I.

If ξ is an interior point of the interval I and $\eta = f(\xi)$ is an interior point of the interval J, we say that $g: J \to I$ is a local inverse for f *at the point* ξ.

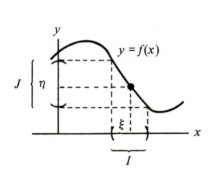
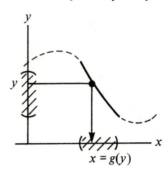

Examples 6.11

(i) The function $f: \mathbb{R} \to \mathbb{R}$ defined by $y = f(x) = x^2$ has no inverse function (see example 6.2). But it does have a *local* inverse at $x = 1$. This is the function $g_1: [0, \infty) \to [0, \infty)$ defined by $x = g_1(y) = \sqrt{y}$.

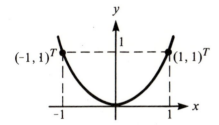
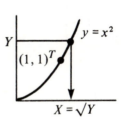

It also has a *local* inverse at $x = -1$. This is the function $g_2: [0, \infty) \to (-\infty, 0]$ defined by $x = g_2(y) = -\sqrt{y}$.

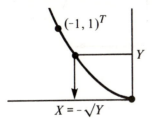

(ii) The function $f: \mathbb{R} \to \mathbb{R}$ defined by $y = f(x) = \sin x$ has no inverse function (see §6.8). But it does have a *local* inverse at $x = \pi/6$. This is the function $g: [-1, 1] \to [-\pi/2, \pi/2]$ defined by

$$x = g(y) = \arcsin y$$

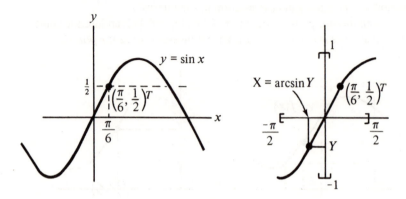

6.12   Critical points

For a function $f: \mathbb{R} \to \mathbb{R}$ with a continuous derivative, the existence of a local inverse at $x = \xi$ is guaranteed if $f'(\xi) \neq 0$. What is more, the local inverse is then differentiable at $\eta = f(\xi)$ and the derivative may be evaluated using the formula

$$\frac{dx}{dy} = \left(\frac{dy}{dx}\right)^{-1}$$

(see §6.5).

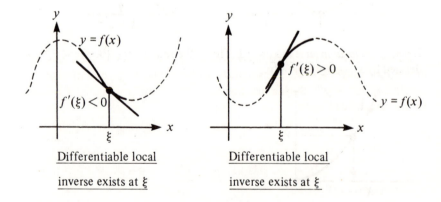

Differentiable local   Differentiable local

inverse exists at ξ inverse exists at ξ

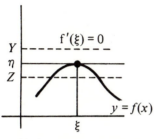

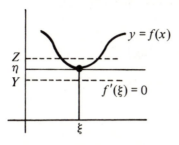

No local inverse exists at ξ   No local inverse exists at ξ

Note in the lower pair of diagrams that the equation $Y = f(x)$ has *no* solutions while $Z = f(x)$ has *two* solutions.

The fact that $f'(\xi) = 0$ does not always mean that no local inverse exists at ξ. However, if a local inverse does exist, its tangent at the point $\eta = f(\xi)$ will be vertical and hence the local inverse is not differentiable.

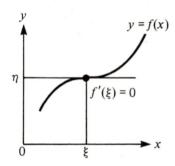

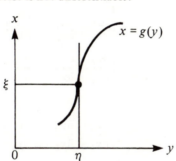

Tangent to the local inverse

at η is vertical

So far we have carried out a fairly exhaustive examination of local inverses of *real* functions $f: \mathbb{R} \to \mathbb{R}$. The situation is very similar in the case of vector functions $f: \mathbb{R}^n \to \mathbb{R}^m$. Such a function can have a local inverse only if $n = m$. In this case the derivative

$$\frac{d\mathbf{y}}{d\mathbf{x}}$$

is a square matrix. If this matrix is *non-singular* at $\mathbf{x} = \boldsymbol{\xi}$, then $f: \mathbb{R}^n \to \mathbb{R}^n$ has a differentiable local inverse at the point $\boldsymbol{\xi}$. The derivative of the local inverse at $\boldsymbol{\eta} = f(\boldsymbol{\xi})$ is given by the formula

$$(1) \qquad \frac{d\mathbf{x}}{d\mathbf{y}} = \left(\frac{d\mathbf{y}}{d\mathbf{x}}\right)^{-1}.$$

If the matrix

$$\frac{d\mathbf{y}}{d\mathbf{x}}$$

is singular at $\mathbf{x} = \boldsymbol{\xi}$, we say that $\boldsymbol{\xi}$ is a *critical point* for the function $f: \mathbb{R}^n \to \mathbb{R}^n$. At a critical point, the function either has no local inverse or else the local inverse is not differentiable. Note in particular that formula (1) makes no sense at a critical point.

6.13   Jacobians

A square matrix is singular if and only if its determinant is zero. It follows that the critical points for a function $f: \mathbb{R}^n \to \mathbb{R}^n$ occur where

$$\det\left(\frac{d\mathbf{y}}{d\mathbf{x}}\right) = \begin{vmatrix} \dfrac{\partial y_1}{\partial x_1} & \dfrac{\partial y_1}{\partial x_2} & \cdots & \dfrac{\partial y_1}{\partial x_n} \\[2ex] \dfrac{\partial y_2}{\partial x_1} & \dfrac{\partial y_2}{\partial x_2} & \cdots & \dfrac{\partial y_2}{\partial x_n} \\[1ex] \vdots & & & \\[1ex] \dfrac{\partial y_n}{\partial x_1} & \dfrac{\partial y_n}{\partial x_2} & \cdots & \dfrac{\partial y_n}{\partial x_n} \end{vmatrix} = 0.$$

We call this determinant the *Jacobian* of the function $f: \mathbb{R}^n \to \mathbb{R}^n$. The critical points occur where the Jacobian vanishes.

Example 6.14.
Consider the function $f: \mathbb{R}^2 \to \mathbb{R}^2$ defined by

$$\begin{pmatrix} u \\ v \end{pmatrix} = f(x, y) = \begin{pmatrix} x^2 + y^2 \\ x^2 - y^2 \end{pmatrix}.$$

We have already studied this function to some extent in example 4.2(ii).

This function has no proper inverse. The diagram below indicates that the equations

$$5 = x^2 + y^2$$
$$3 = x^2 - y^2.$$

have *four* solutions (namely $(2, 1)^T, (2, -1)^T, (-2, 1)^T$ and $(-2, -1)^T$).

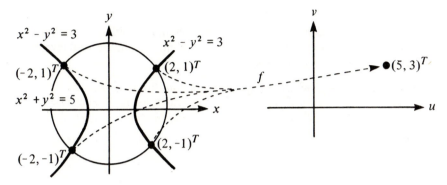

If we wish to find the points at which f has a *local* inverse, we must begin by calculating the derivative of f. This is given by

$$\begin{pmatrix} \dfrac{\partial u}{\partial x} & \dfrac{\partial u}{\partial y} \\ \dfrac{\partial v}{\partial x} & \dfrac{\partial v}{\partial y} \end{pmatrix} = \begin{pmatrix} 2x & 2y \\ 2x & -2y \end{pmatrix}.$$

The Jacobian is therefore

$$\begin{vmatrix} \dfrac{\partial u}{\partial x} & \dfrac{\partial u}{\partial y} \\ \dfrac{\partial v}{\partial x} & \dfrac{\partial v}{\partial y} \end{vmatrix} = \begin{vmatrix} 2x & 2y \\ 2x & -2y \end{vmatrix} = -8xy.$$

and so the critical points occur when $x = 0$ or $y = 0$.

At a non-critical point (i.e. where $x \neq 0$ and $y \neq 0$), a local inverse exists. We can compute the derivative of the local inverse by observing that

$$\begin{pmatrix} \dfrac{\partial x}{\partial u} & \dfrac{\partial x}{\partial v} \\ \dfrac{\partial y}{\partial u} & \dfrac{\partial y}{\partial v} \end{pmatrix} = \begin{pmatrix} \dfrac{\partial u}{\partial x} & \dfrac{\partial u}{\partial y} \\ \dfrac{\partial v}{\partial x} & \dfrac{\partial v}{\partial y} \end{pmatrix}^{-1} = \begin{pmatrix} 2x & 2y \\ 2x & -2y \end{pmatrix}^{-1}.$$

Hence

$$\begin{pmatrix} \dfrac{\partial x}{\partial u} & \dfrac{\partial x}{\partial v} \\[2mm] \dfrac{\partial y}{\partial u} & \dfrac{\partial y}{\partial v} \end{pmatrix} = -\frac{1}{8xy}\begin{pmatrix} -2y & -2y \\ -2x & 2x \end{pmatrix}.$$

In this case (although certainly not in general), it is fairly easy to find a formula for the local inverse at a particular point. To find the local inverse at $(2, -1)^T$, we begin with the equations

$$\left. \begin{aligned} u &= x^2 + y^2 \\ v &= x^2 - y^2 \end{aligned} \right\}$$

which have to be solved to give $(x, y)^T$ in terms of $(u, v)^T$. We have that

$$\left. \begin{aligned} x^2 &= \tfrac{1}{2}(u + v) \\ y^2 &= \tfrac{1}{2}(u - v). \end{aligned} \right\}$$

These equations have four solutions depending on the signs we take when extracting square roots. We want x to be positive and y to be negative and so we take the solution

$$\left. \begin{aligned} x &= \sqrt{\tfrac{1}{2}(u + v)} \\ y &= -\sqrt{\tfrac{1}{2}(u - v)}. \end{aligned} \right\}$$

The local inverse at $(2, -1)^T$ is therefore the function $g : J \to I$ defined by

$$\begin{pmatrix} x \\ y \end{pmatrix} = g(u, v) = \begin{pmatrix} \sqrt{\tfrac{1}{2}(u + v)} \\ -\sqrt{\tfrac{1}{2}(u - v)} \end{pmatrix}.$$

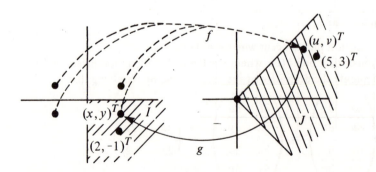

We have already calculated the derivative of g. Observe in particular that

$$g'(5,3) = -\frac{1}{8 \cdot 2(-1)} \begin{pmatrix} -2(-1) & -2(-1) \\ 2 \cdot 2 & 2 \cdot 2 \end{pmatrix} = \begin{pmatrix} \frac{1}{8} & \frac{1}{8} \\ -\frac{1}{4} & \frac{1}{4} \end{pmatrix}$$

It is also instructive to see what happens at a critical point. Consider, for example, the critical point $(x, y)^T = (2, 0)^T$.

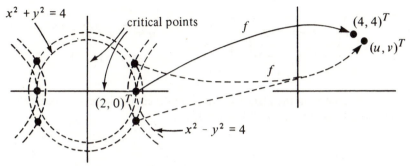

Notice that, however close $(u, v)^T$ is taken to $(4, 4)^T$ the equation $(u, v)^T = f(x, y)$ has *two* solutions close to $(2, 0)^T$. *No* local inverse can therefore exist at $(2, 0)^T$.

It is usually evident from a contour map that something peculiar is going on at a critical point. In this example, the peculiarity is that the contours *touch* at the critical points.

6.15 Co-ordinate systems

The use of the usual system of Cartesian co-ordinates in $\mathbb{R}^2$ amounts to locating points in the plane by reference to a rectangular grid as in the diagram below.

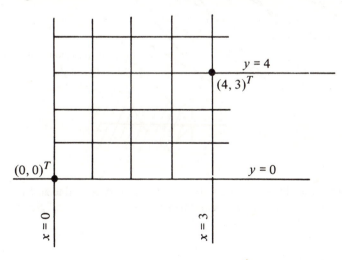

But it is not always convenient to use a rectangular grid. In example 5.2, we studied a different type of grid. This resulted from change from the natural basis $\{(1, 0)^T, (0, 1)^T\}$ to a new basis $\{(1, 1)^T, (2, 1)^T\}$. We found that the new co-ordinates $(u, v)^T$ of a point with respect to the new basis are related to the original Cartesian co-ordinates by the formula

$$\begin{pmatrix} u \\ v \end{pmatrix} = \begin{pmatrix} -1 & 2 \\ 1 & -1 \end{pmatrix}\begin{pmatrix} x \\ y \end{pmatrix} \quad - \text{ i.e.} \quad \left.\begin{array}{l} u = -x + 2y \\ v = x - y. \end{array}\right\}$$

To specify a point in the plane by giving its new co-ordinates is to locate the point with respect to the grid drawn below:

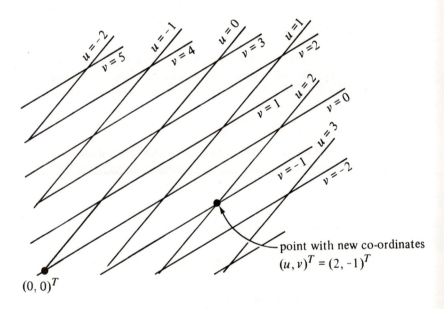

point with new co-ordinates
$(u, v)^T = (2, -1)^T$

$(0, 0)^T$

There is no reason why we should restrict ourselves to straight line grids. In many cases matters can be greatly simplified by using a 'curvilinear' grid as indicated below:

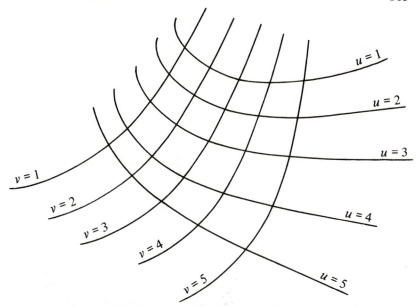

When using such 'curvilinear' co-ordinates, one needs of course to have a formula which expresses the new co-ordinates $(u, v)^T$ of a point in terms of its Cartesian co-ordinates $(x, y)^T$. Suppose that the appropriate formula is

$$u = \phi(x, y)$$
$$v = \psi(x, y).$$

Then the diagram below illustrates a point P_0 whose original Cartesian co-ordinates are $(x_0, y_0)^T$ and whose new 'curvilinear' co-ordinates are $(u_0, v_0)^T$.

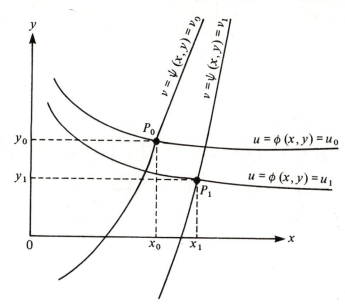

It is often helpful to introduce the function f defined by

$$\begin{pmatrix} u \\ v \end{pmatrix} = f(x, y) = \begin{pmatrix} \phi(x, y) \\ \psi(x, y) \end{pmatrix}.$$

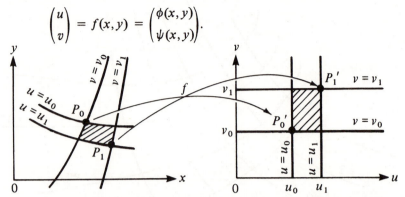

One can then not only think in terms of relabelling points P_0 and P_1 in the $(x, y)^T$ plane with new co-ordinates $(u, v)^T$: one can also simultaneously think in terms of transforming the points P_0 and P_1 into new points P_0' and P_1' in a new $(u, v)^T$ plane. In our diagram above, for example, the complicated shaded region in the $(x, y)^T$ plane is transformed into the simple shaded rectangular region in the $(u, v)^T$ plane.

It is *not* true that *any* function $f: \mathbb{R}^n \to \mathbb{R}^n$ can be used to define a new co-ordinate system in $\mathbb{R}^n$. In order that $f: \mathbb{R}^n \to \mathbb{R}^n$ can be used to define a new co-ordinate system, it is essential that f admits at least a local *inverse* function $g: J \to I$. The component functions of f can then be used to define new co-ordinates $(u_1, u_2, \ldots, u_n)^T$ for a point $(x_1, x_2, \ldots, x_n)^T$ in the set I by means of the formulae

$$\left. \begin{aligned} u_1 &= f_1(x_1, x_2, \ldots, x_n) \\ u_2 &= f_2(x_1, x_2, \ldots, x_n) \\ &\;\;\vdots \\ u_n &= f_n(x_1, x_2, \ldots, x_n). \end{aligned} \right\}$$

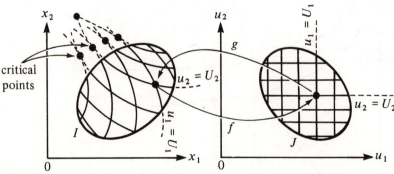

If the function f does *not* have an appropriate inverse, then it is useless for the purpose of introducing a new co-ordinate system because it will then assign the *same* new co-ordinates to two *different* points. It follows in particular that there is no point in trying to use a function $f: \mathbb{R}^n \to \mathbb{R}^n$ for the purpose of defining a new co-ordinate system in a set I if a critical point of f is an interior point of I. Thus, for each interior point $\mathbf{x}$ of the set I, we require that the Jacobian

$$\det f'(\mathbf{x}) \neq 0.$$

Example 6.16.
We have studied the function $f: \mathbb{R}^2 \to \mathbb{R}^2$ defined by

$$\left. \begin{aligned} u &= x^2 + y^2 \\ v &= x^2 - y^2 \end{aligned} \right\}$$

in example 4.2(ii) and in example 6.14. This function can be used to introduce a new co-ordinate system into any one of the four quadrants of the $(x, y)^T$ plane. In example 6.14, we focussed attention on the fourth quadrant but in the diagram below we have chosen to illustrate what happens in the first quadrant (i.e. the set I of those $(x, y)^T$ for which $x > 0$ and $y > 0$).

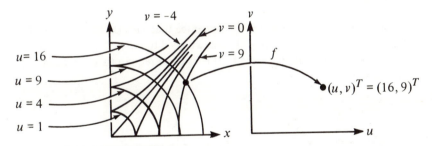

Notice that $(u, v)^T = (16, 9)^T$ corresponds to only *one* point in the first quadrant of the $(x, y)^T$ plane. In general, if $(u, v)^T$ are the new co-ordinates of a point in the first quadrant with old co-ordinates $(x, y)^T$, then

$$\left. \begin{aligned} x &= \sqrt{\tfrac{1}{2}(u + v)} \\ y &= \sqrt{\tfrac{1}{2}(u - v)}. \end{aligned} \right\}$$

6.17 Polar co-ordinates

Often one is given the old co-ordinates $(x, y)^T$ as functions of the new co-ordinates instead of the other way around. Consider, for example, the function $f: \mathbb{R}^2 \to \mathbb{R}^2$ defined by

$$x = r \cos \theta$$
$$y = r \sin \theta.$$

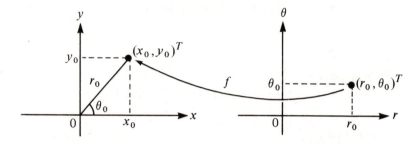

The important point to note here is the geometric interpretation of r and θ in the $(x, y)^T$ plane. This explains why r and θ are called the *polar co-ordinates* of the point $(x, y)^T$.

In general, the equations

$$x = r \cos \theta$$
$$y = r \sin \theta$$

have many solutions for r and θ. To guarantee that each $(x, y)^T$ is assigned a *unique* pair $(r, \theta)^T$ of polar co-ordinates we need to restrict the values of r and θ which we are willing to admit. It is usual to require that $r > 0$ and $-\pi < \theta \leqslant \pi$.

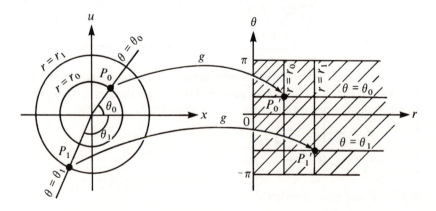

The Jacobian of f is given by

$$\det f'(r,\theta) = \begin{vmatrix} \dfrac{\partial x}{\partial r} & \dfrac{\partial x}{\partial \theta} \\[2mm] \dfrac{\partial y}{\partial r} & \dfrac{\partial y}{\partial \theta} \end{vmatrix}$$

$$= \begin{vmatrix} \cos\theta & -r\sin\theta \\ \sin\theta & r\cos\theta \end{vmatrix} = r(\cos^2\theta + \sin^2\theta) = r.$$

This is zero only when $r = 0$.

Note finally that, for $r > 0$ and $-\pi/2 < \theta < \pi/2$, the unique solution of the equations

$$\left. \begin{aligned} x &= r\cos\theta \\ y &= r\sin\theta \end{aligned} \right\}$$

is given by

$$\left. \begin{aligned} r &= \sqrt{(x^2 + y^2)} \\ \theta &= \arctan\frac{y}{x}. \end{aligned} \right\}$$

Spherical polar co-ordinates are sometimes useful in $\mathbb{R}^3$. The appropriate formulae are

$$\left. \begin{aligned} x &= r\cos\theta\sin\phi \\ y &= r\sin\theta\sin\phi \\ z &= r\cos\phi. \end{aligned} \right\}$$

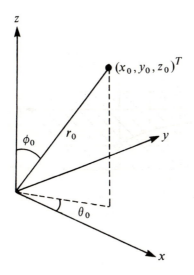

The geometric significance of these co-ordinates is indicated above. To guarantee a unique solution to the equations, one requires that $r > 0$, $-\pi < \theta \leqslant \pi$ and $0 \leqslant \phi < \pi$.

6.18   Differential operators

It is often the case that a change of variables is made to simplify a given integral or differential expression. Integrals will be left to a later chapter. In this section we shall restrict our attention to the question of how differential expressions transform when the co-ordinate system is changed.

The basic rule is that for differentiating a 'function of a function'. If $z = \Phi(\mathbf{u})$ and $\mathbf{u} = f(\mathbf{x})$ where $\mathbf{x}$ and $\mathbf{u}$ are $n \times 1$ column vectors, this rule takes the form

$$\frac{dz}{d\mathbf{x}} = \frac{dz}{d\mathbf{u}} \frac{d\mathbf{u}}{d\mathbf{x}}.$$

In the case when $n = 2$, this formula reduces to

$$\left(\frac{\partial z}{\partial x_1}, \frac{\partial z}{\partial x_2} \right) = \left(\frac{\partial z}{\partial u_1}, \frac{\partial z}{\partial u_2} \right) \begin{pmatrix} \dfrac{\partial u_1}{\partial x_1} & \dfrac{\partial u_1}{\partial x_2} \\[2mm] \dfrac{\partial u_2}{\partial x_1} & \dfrac{\partial u_2}{\partial x_2} \end{pmatrix}$$

i.e.

$$\left. \begin{aligned} \frac{\partial z}{\partial x_1} &= \frac{\partial u_1}{\partial x_1} \frac{\partial z}{\partial u_1} + \frac{\partial u_2}{\partial x_1} \frac{\partial z}{\partial u_2} \\[2mm] \frac{\partial z}{\partial x_2} &= \frac{\partial u_1}{\partial x_2} \frac{\partial z}{\partial u_1} + \frac{\partial u_2}{\partial x_2} \frac{\partial z}{\partial u_2}. \end{aligned} \right\}$$

These equations are valid regardless of the function $z = \Phi(\mathbf{u})$ to which they are applied. It is therefore common to rewrite them in *operator* notation as below:

$$\frac{\partial}{\partial x_1} = \frac{\partial u_1}{\partial x_1} \frac{\partial}{\partial u_1} + \frac{\partial u_2}{\partial x_1} \frac{\partial}{\partial u_2}$$

$$\frac{\partial}{\partial x_2} = \frac{\partial u_1}{\partial x_2} \frac{\partial}{\partial u_1} + \frac{\partial u_2}{\partial x_2} \frac{\partial}{\partial u_2}.$$

The meaning of these formulae is simply that, if the left-hand side is applied to z, then the result will be exactly the same as applying the right-hand side to z.

Example 6.19.
The formulae

$$u = x^2 + y^2$$
$$v = x^2 - y^2$$

are used to change the co-ordinate system. How does the partial differential equation

$$\frac{1}{x}\frac{\partial z}{\partial x} + \frac{1}{y}\frac{\partial z}{\partial y} = 1$$

transform?

We have that

$$\frac{\partial}{\partial x} = \frac{\partial u}{\partial x}\frac{\partial}{\partial u} + \frac{\partial v}{\partial x}\frac{\partial}{\partial v}$$

$$= 2x\frac{\partial}{\partial u} + 2x\frac{\partial}{\partial v}$$

$$\frac{\partial}{\partial y} = \frac{\partial u}{\partial y}\frac{\partial}{\partial u} + \frac{\partial v}{\partial y}\frac{\partial}{\partial v}$$

$$= 2y\frac{\partial}{\partial u} - 2y\frac{\partial}{\partial v}.$$

It follows that

$$\frac{1}{x}\frac{\partial z}{\partial x} + \frac{1}{y}\frac{\partial z}{\partial y} = \left(2\frac{\partial z}{\partial u} + 2\frac{\partial z}{\partial v}\right) + \left(2\frac{\partial z}{\partial u} - 2\frac{\partial z}{\partial v}\right)$$

$$= 4\frac{\partial z}{\partial u}.$$

The given partial differential equation therefore assumes the particularly simple form

$$\frac{\partial z}{\partial u} = \frac{1}{4}.$$

Matters are not quite so simple when, instead of the new co-ordinates being given as functions of the old co-ordinates, we are instead given the old co-ordinates as functions of the new co-ordinates — i.e. $\mathbf{x} = f(\mathbf{u})$. In principle, one can of course solve the equation $\mathbf{x} = f(\mathbf{u})$ to obtain $\mathbf{u} = g(\mathbf{x})$. But this is often quite difficult to do and even when it is not particularly difficult it is usually easier to use the formula

$$\frac{d\mathbf{u}}{d\mathbf{x}} = \left(\frac{d\mathbf{x}}{d\mathbf{u}}\right)^{-1}$$

instead.

Example 6.20.
The formula

$$\left.\begin{array}{l} x = u^2 + v^2 \\ y = u^2 - v^2 \end{array}\right\}$$

is used to change the co-ordinate system. How does the partial differential equation

$$\frac{\partial z}{\partial x} + \frac{\partial z}{\partial y} = 1$$

transform?

We have that

$$\frac{\partial}{\partial x} = \frac{\partial u}{\partial x}\frac{\partial}{\partial u} + \frac{\partial v}{\partial x}\frac{\partial}{\partial v}$$

$$\frac{\partial}{\partial y} = \frac{\partial u}{\partial y}\frac{\partial}{\partial u} + \frac{\partial v}{\partial y}\frac{\partial}{\partial v}.$$

Our problem is to calculate $\partial u/\partial x, \partial v/\partial x, \partial u/\partial y, \partial v/\partial y$. We use the formula

$$\begin{pmatrix} \dfrac{\partial u}{\partial x} & \dfrac{\partial u}{\partial y} \\ \dfrac{\partial v}{\partial x} & \dfrac{\partial v}{\partial y} \end{pmatrix} = \begin{pmatrix} \dfrac{\partial x}{\partial u} & \dfrac{\partial x}{\partial v} \\ \dfrac{\partial y}{\partial u} & \dfrac{\partial y}{\partial v} \end{pmatrix}^{-1} = \begin{pmatrix} 2u & 2v \\ 2u & -2v \end{pmatrix}^{-1}$$

$$= -\frac{1}{8uv}\begin{pmatrix} -2v & -2v \\ -2u & 2u \end{pmatrix} = \begin{pmatrix} \dfrac{1}{4u} & \dfrac{1}{4u} \\ \dfrac{1}{4v} & -\dfrac{1}{4v} \end{pmatrix}.$$

It follows that

$$\frac{\partial}{\partial x} = \frac{1}{4u}\frac{\partial}{\partial u} + \frac{1}{4v}\frac{\partial}{\partial v}$$

$$\frac{\partial}{\partial y} = \frac{1}{4u}\frac{\partial}{\partial u} - \frac{1}{4v}\frac{\partial}{\partial v}.$$

Thus

$$\frac{\partial z}{\partial x} + \frac{\partial z}{\partial y} = \left(\frac{1}{4u}\frac{\partial z}{\partial u} + \frac{1}{4v}\frac{\partial z}{\partial v}\right) + \left(\frac{1}{4u}\frac{\partial z}{\partial u} - \frac{1}{4v}\frac{\partial z}{\partial v}\right)$$

$$= \frac{1}{2u}\frac{\partial z}{\partial u}.$$

The given partial differential equation therefore assumes the form

$$\frac{\partial z}{\partial u} = 2u.$$

Example 6.21.
Repeat the previous example for the equation

$$\frac{\partial^2 z}{\partial x \partial y} = 0.$$

We have that

$$\frac{\partial^2 z}{\partial x \partial y} = \frac{\partial}{\partial x}\left(\frac{\partial z}{\partial y}\right) = \frac{\partial}{\partial x}\left(\frac{1}{4u}\frac{\partial z}{\partial u} - \frac{1}{4v}\frac{\partial z}{\partial v}\right)$$

$$= \left(\frac{1}{4u}\frac{\partial}{\partial u} + \frac{1}{4v}\frac{\partial}{\partial v}\right)\left(\frac{1}{4u}\frac{\partial z}{\partial u} - \frac{1}{4v}\frac{\partial z}{\partial v}\right)$$

$$= \frac{1}{4u}\frac{\partial}{\partial u}\left(\frac{1}{4u}\frac{\partial z}{\partial u} - \frac{1}{4v}\frac{\partial z}{\partial v}\right) + \frac{1}{4v}\frac{\partial}{\partial v}\left(\frac{1}{4u}\frac{\partial z}{\partial u} - \frac{1}{4v}\frac{\partial z}{\partial v}\right)$$

$$= \frac{1}{4u}\left\{-\frac{1}{4u^2}\frac{\partial z}{\partial u} + \frac{1}{4u}\frac{\partial^2 z}{\partial u^2} - \frac{1}{4v}\frac{\partial^2 z}{\partial u \partial v}\right\}$$

$$+ \frac{1}{4v}\left\{\frac{1}{4u}\frac{\partial^2 z}{\partial v \partial u} + \frac{1}{4v^2}\frac{\partial z}{\partial v} - \frac{1}{4v}\frac{\partial^2 z}{\partial v^2}\right\}$$

$$= \frac{1}{16u^2}\frac{\partial^2 z}{\partial u^2} - \frac{1}{16v^2}\frac{\partial^2 z}{\partial v^2} - \frac{1}{16u^3}\frac{\partial z}{\partial u} + \frac{1}{16v^3}\frac{\partial z}{\partial v}.$$

The given equation therefore becomes

$$\frac{1}{u^2}\frac{\partial^2 z}{\partial u^2} - \frac{1}{u^3}\frac{\partial z}{\partial u} = \frac{1}{v^2}\frac{\partial^2 z}{\partial v^2} - \frac{1}{v^3}\frac{\partial z}{\partial v}.$$

Example 6.22.

Introduce polar co-ordinates into the equation

$$\frac{\partial^2 z}{\partial x^2} + \frac{\partial^2 z}{\partial y^2} = 0.$$

We have that

$$\frac{\partial}{\partial x} = \frac{\partial r}{\partial x}\frac{\partial}{\partial r} + \frac{\partial \theta}{\partial x}\frac{\partial}{\partial \theta}$$

$$\frac{\partial}{\partial y} = \frac{\partial r}{\partial y}\frac{\partial}{\partial r} + \frac{\partial \theta}{\partial y}\frac{\partial}{\partial \theta}.$$

Recall that

$$\left.\begin{array}{l} x = r\cos\theta \\ y = r\sin\theta \end{array}\right\}$$

and hence

$$\begin{pmatrix} \dfrac{\partial r}{\partial x} & \dfrac{\partial r}{\partial y} \\ \dfrac{\partial \theta}{\partial x} & \dfrac{\partial \theta}{\partial y} \end{pmatrix} = \begin{pmatrix} \dfrac{\partial x}{\partial r} & \dfrac{\partial x}{\partial \theta} \\ \dfrac{\partial y}{\partial r} & \dfrac{\partial y}{\partial \theta} \end{pmatrix}^{-1} = \begin{pmatrix} \cos\theta & -r\sin\theta \\ \sin\theta & r\cos\theta \end{pmatrix}^{-1}.$$

It follows that

$$\begin{pmatrix} \dfrac{\partial r}{\partial x} & \dfrac{\partial r}{\partial y} \\ \dfrac{\partial \theta}{\partial x} & \dfrac{\partial \theta}{\partial y} \end{pmatrix} = \frac{1}{r(\cos^2\theta + \sin^2\theta)}\begin{pmatrix} r\cos\theta & r\sin\theta \\ -\sin\theta & \cos\theta \end{pmatrix}$$

$$= \begin{pmatrix} \cos\theta & \sin\theta \\ -\dfrac{\sin\theta}{r} & \dfrac{\cos\theta}{r} \end{pmatrix}.$$

Thus

$$\frac{\partial}{\partial x} = \cos\theta\frac{\partial}{\partial r} - \frac{\sin\theta}{r}\frac{\partial}{\partial \theta}$$

$$\frac{\partial}{\partial y} = \sin\theta\frac{\partial}{\partial r} + \frac{\cos\theta}{r}\frac{\partial}{\partial \theta}.$$

We conclude that

$$\frac{\partial^2 z}{\partial x^2} = \frac{\partial}{\partial x}\left(\frac{\partial z}{\partial x}\right) = \left(\cos\theta\frac{\partial}{\partial r} - \frac{\sin\theta}{r}\frac{\partial}{\partial\theta}\right)\left(\cos\theta\frac{\partial z}{\partial r} - \frac{\sin\theta}{r}\frac{\partial z}{\partial\theta}\right)$$

$$= \cos^2\theta\frac{\partial^2 z}{\partial r^2} + \frac{\cos\theta\sin\theta}{r^2}\frac{\partial z}{\partial\theta} - \frac{\cos\theta\sin\theta}{r}\frac{\partial^2 z}{\partial r\partial\theta} + \frac{\sin^2\theta}{r}\frac{\partial z}{\partial r}$$

$$- \frac{\sin\theta\cos\theta}{r}\frac{\partial^2 r}{\partial\theta\partial r} + \frac{\sin\theta\cos\theta}{r^2}\frac{\partial z}{\partial\theta} + \frac{\sin^2\theta}{r^2}\frac{\partial^2 z}{\partial\theta^2}$$

$$\frac{\partial^2 z}{\partial y^2} = \frac{\partial}{\partial y}\left(\frac{\partial z}{\partial y}\right) = \left(\sin\theta\frac{\partial}{\partial r} + \frac{\cos\theta}{r}\frac{\partial}{\partial\theta}\right)\left(\sin\theta\frac{\partial z}{\partial r} + \frac{\cos\theta}{r}\frac{\partial z}{\partial\theta}\right)$$

$$= \sin^2\theta\frac{\partial^2 z}{\partial r^2} - \frac{\sin\theta\cos\theta}{r^2}\frac{\partial z}{\partial\theta} + \frac{\sin\theta\cos\theta}{r}\frac{\partial^2 z}{\partial r\partial\theta} + \frac{\cos^2\theta}{r}\frac{\partial z}{\partial r}$$

$$+ \frac{\cos\theta\sin\theta}{r}\frac{\partial^2 z}{\partial\theta\partial r} - \frac{\cos\theta\sin\theta}{r^2}\frac{\partial z}{\partial\theta} + \frac{\cos^2\theta}{r^2}\frac{\partial^2 z}{\partial\theta^2}$$

Hence the given partial differential equation reduces to

$$\frac{\partial^2 f}{\partial r^2} + \frac{1}{r}\frac{\partial f}{\partial r} + \frac{1}{r^2}\frac{\partial^2 f}{\partial\theta^2} = 0.$$

Exercises 6.23

1. Find formulae for local inverses to the function $f:\mathbb{R}\to\mathbb{R}$ defined by $y = f(x) = \sin x$ at the points

(i) $x = \pi/6$ (ii) $x = 5\pi/6 = \pi - (\pi/6)$ (iii) $x = -7\pi/6 = -\pi - (\pi/6)$.

Compute the derivatives of these local inverses directly from your formulae. Evaluate these derivatives at the point $y = \frac{1}{2}$. Verify in each case that

$$\frac{dx}{dy} = \left(\frac{dy}{dx}\right)^{-1}.$$

Explain why no local inverse exists at the point $x = \pi/2$.

2.* Find formulae for local inverses to the function $f:\mathbb{R}^2\to\mathbb{R}^2$ defined by

$$\begin{pmatrix} u \\ v \end{pmatrix} = f(x,y) = \begin{pmatrix} x^2 + y^2 \\ x^2 - y^2 \end{pmatrix}$$

at the points

(i) $\begin{pmatrix} x \\ y \end{pmatrix} = \begin{pmatrix} 2 \\ 1 \end{pmatrix}$ (ii) $\begin{pmatrix} x \\ y \end{pmatrix} = \begin{pmatrix} -2 \\ 1 \end{pmatrix}$ (iii) $\begin{pmatrix} x \\ y \end{pmatrix} = \begin{pmatrix} 2 \\ -1 \end{pmatrix}$ (iv) $\begin{pmatrix} x \\ y \end{pmatrix} = \begin{pmatrix} -2 \\ -1 \end{pmatrix}$.

Compute the derivatives of these local inverses directly from your formulae. Evaluate these derivatives at the point $(u, v)^T = (5, 3)^T$. Verify in each case that

$$\begin{pmatrix} \dfrac{\partial x}{\partial u} & \dfrac{\partial x}{\partial v} \\[2mm] \dfrac{\partial y}{\partial u} & \dfrac{\partial y}{\partial v} \end{pmatrix} = \begin{pmatrix} \dfrac{\partial u}{\partial x} & \dfrac{\partial u}{\partial y} \\[2mm] \dfrac{\partial v}{\partial x} & \dfrac{\partial v}{\partial y} \end{pmatrix}^{-1}.$$

3. A function $f: \mathbb{R}^2 \to \mathbb{R}^2$ is defined by

$$\begin{pmatrix} u \\ v \end{pmatrix} = f(x, y) = \begin{pmatrix} xe^y \\ xe^{-y} \end{pmatrix}.$$

Find the derivative of this function. Determine the points at which f has a local inverse and find the derivative of this local inverse at these points. Evaluate

$$\left(\frac{\partial u}{\partial x} \right)_y \quad \text{and} \quad \left(\frac{\partial x}{\partial u} \right)_v.$$

4.* A function $f: \mathbb{R}^2 \to \mathbb{R}^2$ is defined by

$$\begin{pmatrix} u \\ v \end{pmatrix} = f(x, y) = \begin{pmatrix} x^3 + y^2 \\ x^2 - y^3 \end{pmatrix}.$$

Find the derivative of this function. Determine the points at which f has a local inverse and find the derivative of this local inverse at these points. Evaluate

$$\left(\frac{\partial v}{\partial y} \right)_x \quad \text{and} \quad \left(\frac{\partial y}{\partial v} \right)_u.$$

5. The equations

$$\left. \begin{array}{l} u = x + y \\ v = x - y \end{array} \right\}$$

are used to change the co-ordinate system in the $(x, y)^T$ plane. Sketch the new co-ordinate grid. Express

$$\frac{\partial^2 z}{\partial x^2}$$

in terms of the new co-ordinates.

6.* The equations

$$u = \tfrac{1}{2}(x^2 - y^2) \quad \Big\}$$
$$v = xy$$

are used to change the co-ordinate system in the first quadrant of the $(x, y)^T$ plane. Sketch the new co-ordinate grid. Express

$$\frac{\partial^2 z}{\partial x^2}$$

in terms of the new co-ordinates.

7. The equations

$$x = u + v \quad \Big\}$$
$$y = u - v$$

are used to change the co-ordinate system in the $(x, y)^T$ plane. Prove that

$$\frac{\partial^2 z}{\partial x^2} - \frac{\partial^2 z}{\partial y^2} = \frac{\partial^2 z}{\partial u \partial v}.$$

8.* The equations

$$x = \tfrac{1}{2}(u^2 - v^2) \quad \Big\}$$
$$y = uv$$

are used to change the co-ordinate system. Express

(i) $\quad x\dfrac{\partial z}{\partial x} + y\dfrac{\partial z}{\partial y}$

(ii) $\quad \dfrac{\partial^2 z}{\partial x^2} + \dfrac{\partial^2 z}{\partial y^2}$

in terms of the new co-ordinates.

9. Let $(r, \theta)^T$ be the polar co-ordinates of a point $(x, y)^T$. Prove that, for any function $f : \mathbb{R}^2 \to \mathbb{R}$,

$$\frac{\partial f}{\partial r} = \cos \theta \frac{\partial f}{\partial x} + \sin \theta \frac{\partial f}{\partial y} \quad \Bigg\}$$
$$\frac{1}{r}\frac{\partial f}{\partial \theta} = -\sin \theta \frac{\partial f}{\partial x} + \cos \theta \frac{\partial f}{\partial y}.$$

What is the geometric significance of these quantities? [Hint: Put $\mathbf{e}_1 = (\cos\theta, \sin\theta)^T$, $\mathbf{e}_2 = (-\sin\theta, \cos\theta)^T$ and consider $\langle\nabla f, \mathbf{e}_1\rangle$, $\langle\nabla f, \mathbf{e}_2\rangle$.]

10.* A function $f:\mathbb{R}^3 \to \mathbb{R}^3$ is defined by the formulae

(i) $\quad x = r\sin\theta\cos\phi$

$\quad\quad y = r\sin\theta\sin\phi$

$\quad\quad z = r\cos\theta$

where $(x, y, z) = f(r, \theta, \phi)$ Find the Jacobian of this function and determine its critical points.

Proceed in a similar way with the formulae

(ii) $\quad x = r\cos\theta \qquad$ (iii) $x = u^2 v^{-1}$

$\quad\quad y = r\sin\theta \qquad\qquad\quad y = v^2 w^{-1}$

$\quad\quad z = Z \qquad\qquad\qquad\quad z = w^2 u^{-1}.$

SOME APPLICATIONS (OPTIONAL)

6.24 Contract curve

We discussed the Edgeworth box and the contract curve in §2.25.

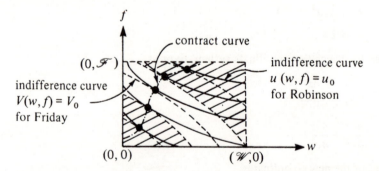

Observe that each point on the contract curve is a critical point of the function $\phi:\mathbb{R}^2 \to \mathbb{R}^2$ defined by

$$\begin{pmatrix} x \\ y \end{pmatrix} = \phi(w, f) = \begin{pmatrix} u(w, f) \\ V(w, f) \end{pmatrix}.$$

It follows that the Jacobian is zero on the contract curve — i.e.

$$\begin{vmatrix} \dfrac{\partial x}{\partial w} & \dfrac{\partial x}{\partial f} \\[2mm] \dfrac{\partial y}{\partial w} & \dfrac{\partial y}{\partial f} \end{vmatrix} = 0.$$

We can use this fact to obtain an equation for the contract curve. Consider, for example, the case

$$x = u(w,f) = wf^2$$
$$\left. y = V(w,f) = (2-w)^2(1-f) \right\}$$

which we studied in §2.25. On the contract curve,

$$0 = \begin{vmatrix} \dfrac{\partial x}{\partial w} & \dfrac{\partial x}{\partial f} \\[2mm] \dfrac{\partial y}{\partial w} & \dfrac{\partial y}{\partial f} \end{vmatrix} = \begin{vmatrix} f^2 & 2wf \\[2mm] -2(2-w)(1-f) & -(2-w)^2 \end{vmatrix}$$
$$= -(2-w)f\{f(2-w)-4w(1-f)\}.$$

Hence the contract curve has equation $3wf - 4w + 2f = 0$. (Only the part outside the shaded region is relevant.)

7

Implicit functions

The implicit function theorem is very easy to apply in practice provided that the material on vector derivatives given in chapter 4 has been properly assimilated. We are therefore able to deal with it quite quickly in this chapter. Some understanding of §7.2 is necessary for the conclusion of the implicit function theorem to be meaningful but too much time should not be spent on this section; especially in respect of the theoretical discussion at the end of the section which links the proof of the implicit function theorem with the discussion of local inverses given in the previous chapter.

7.1 Implicit differentiation

Consider the equation

$$y^2 - x^3 - x^2 = 0.$$

Suppose that $y = g(x)$ is a solution of this equation for y in terms of x — i.e. $(g(x))^2 - x^3 - x^2 = 0$. We can then calculate $g'(x)$ using the technique of implicit differentiation. We have that

$$0 = \frac{d}{dx}(y^2 - x^3 - x^2)$$

$$= 2y\frac{dy}{dx} - 3x^2 - 2x$$

and hence

$$\frac{dy}{dx} = \frac{3x^2 + 2x}{2y}$$

i.e.

$$g'(x) = \frac{3x^2 + 2x}{2y} = \frac{3x^2 + 2x}{2g(x)}.$$

The same technique can be applied in the general case using the chain rule. Suppose that $\phi:\mathbb{R}^2 \to \mathbb{R}$ and that

$$\phi(x,y) = 0$$

has the solution $y = g(x)$ for y in terms of x. Then

$$0 = \frac{d\phi}{dx} = \frac{\partial\phi}{\partial x}\frac{dx}{dx} + \frac{\partial\phi}{\partial y}\frac{dy}{dx}$$

and so

$$\frac{dy}{dx} = -\left(\frac{\partial\phi}{\partial y}\right)^{-1}\frac{\partial\phi}{\partial x}$$

i.e.

$$g'(x) = -\frac{\phi_x(x,y)}{\phi_y(x,y)} = -\frac{\phi_x(x,g(x))}{\phi_y(x,g(x))}.$$

(In the case when $\phi(x,y) = y^2 - x^3 - x^2$ we have that $\phi_x(x,y) = -3x^2 - 2x$ and $\phi_y(x,y) = 2y$.)

7.2 Implicit functions

The account of implicit differentiation given above begs various questions. In particular, the account takes for granted that the equation $\phi(x,y) = 0$ *implicitly* defines a function $y = g(x)$ and that this function may be differentiated. But it is clear that, in general, it will *not* always be possible to solve the equation $\phi(x,y) = 0$ for y in terms of x. Moreover, where it is possible to solve the equation for y in terms of x, there may well be *several* solutions. Finally, there is no guarantee that a given solution will be differentiable.

Consider, for example, the case

$$\phi(x,y) = y^2 - x^3 - x^2.$$

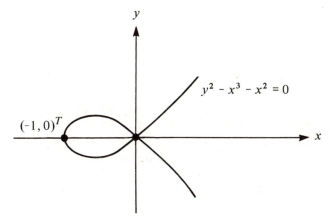

It is evident from the diagram that for values of x satisfying $x < -1$, the equation $y^2 - x^3 - x^2 = 0$ has no (real) solutions for y in terms of x. Furthermore, for all other values of x (except $x = -1$ and $x = 0$), the equation $y^2 - x^3 - x^2 = 0$ has *two* solutions for y in terms of x.

The diagrams below illustrate three different functions $g : [-1, \infty) \to \mathbb{R}$ which are given implicitly by $y^2 - x^3 - x^2 = 0$ – i.e. which satisfy $(g(x))^2 - x^3 - x^2 = 0$.

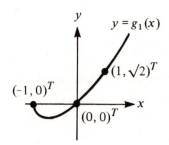

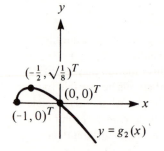

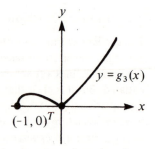

This leads us to the notion of a local solution. Suppose that $\phi : \mathbb{R}^2 \to \mathbb{R}$. Then we say that $g : I \to \mathbb{R}$ is a *local solution* of

$$\phi(x, y) = 0$$

provided that

$$\phi(x, g(x)) = 0$$

for each x in the interval I.

If ξ is an interior point of the interval I and $\eta = g(\xi)$, then we say that $g : I \to \mathbb{R}$ is a local solution of $\phi(x, y) = 0$ *at the point* $(\xi, \eta)^T$.

Consider, for example, the functions g_1, g_2 and g_3 sketched above. The function g_1 is a local solution of $y^2 - x^3 - x^2 = 0$ at the point $(1, \sqrt{2})^T$. The function g_2 is a local solution of $y^2 - x^3 - x^2 = 0$ at the point $(-\frac{1}{2}, 1/\sqrt{8})^T$. All three functions are local solutions of $y^2 - x^3 - x^2 = 0$ at the point $(0, 0)^T$. Note that *no* local solution of $y^2 - x^3 - x^2 = 0$ exists at the point $(-1, 0)^T$. This is because $(-1, 0)^T$ is not an interior point of $[-1, \infty)$ which is the largest interval on which a local inverse can be defined.

Observe that the 'bad points' on $y^2 - x^3 - x^2 = 0$ are $(-1, 0)^T$ and $(0, 0)^T$. At these points there are either no local solutions or too many local solutions. At all other points $(\xi, \eta)^T$ on $y^2 - x^3 - x^2 = 0$ there is an essentially unique differentiable local solution ('unique' in the sense that any two local solutions at $(\xi, \eta)^T$ are the same for values of x sufficiently close to ξ although the intervals I on which they are defined may not be the same).

Notice that, at the 'bad points' $(-1, 0)^T$ and $(0, 0)^T$, we have that

$$\frac{\partial \phi}{\partial y} = 2y = 0$$

and so the formula

$$\frac{dy}{dx} = -\left(\frac{\partial \phi}{\partial y}\right)^{-1}\left(\frac{\partial \phi}{\partial x}\right)$$

which we obtained in §7.1 ceases to make sense. It is therefore not surprising that we should find something peculiar going on at these points.

The above discussion would seem to indicate that the appropriate condition for the existence of a local solution to $\phi(x, y) = 0$ at a point $(\xi, \eta)^T$ satisfying $\phi(\xi, \eta) = 0$ is that

$$\frac{\partial \phi}{\partial y} \neq 0$$

when $(x, y)^T = (\xi, \eta)^T$. It is quite easy to check that this is correct using the results on the existence of local inverses given in the previous chapter.

Let $F: \mathbb{R}^2 \to \mathbb{R}^2$ be the function given by

$$F(\mathbf{z}) = F(x, y) = (x, \phi(x, y))$$

where $\mathbf{z} = (x, y)^T$. Then

$$\frac{dF}{d\mathbf{z}} = \begin{pmatrix} \dfrac{\partial x}{\partial x} & \dfrac{\partial x}{\partial y} \\ \dfrac{\partial \phi}{\partial x} & \dfrac{\partial \phi}{\partial y} \end{pmatrix} = \begin{pmatrix} 1 & 0 \\ \dfrac{\partial \phi}{\partial x} & \dfrac{\partial \phi}{\partial y} \end{pmatrix}$$

Thus the Jacobian is

$$\det\left(\frac{dF}{d\mathbf{z}}\right) = \frac{\partial \phi}{\partial y}.$$

It follows that the critical points of F occur where $\phi_y = 0$. Thus, if $\phi_y(\xi, \eta) \neq 0$, then F admits a local inverse G at the point $(\xi, \eta)^T$. This means that

$$(u, v)^T = F(x, y) \quad \text{if and only if} \quad (x, y)^T = G(u, v)$$

provided $(x, y)^T$ and $(u, v)^T$ are appropriately restricted.

Now consider values of x and y which satisfy $\phi(x, y) = 0$. Then

$$F(x, y) = (x, \phi(x, y))^T = (x, 0)^T$$

is equivalent to

$$(x, y)^T = G(x, 0) = (G_1(x, 0), G_2(x, 0))^T.$$

In particular, $y = G_2(x, 0) = g(x)$ is a unique local solution of $\phi(x, y) = 0$ at $(\xi, \eta)^T$.

7.3   Implicit function theorem

The implicit function theorem generalizes the above discussion. Suppose that $\phi : \mathbb{R}^{n+m} \to \mathbb{R}^m$ has a continuous derivative and consider the equation

$$\phi(\mathbf{x}, \mathbf{y}) = \mathbf{0}$$

in which $\mathbf{x}$ is an $n \times 1$ column vector and $\mathbf{y}$ is an $m \times 1$ column vector. Under what circumstances does there exist a unique local differentiable solution $\mathbf{y} = g(\mathbf{x})$ at the point $(\boldsymbol{\xi}, \boldsymbol{\eta})^T$?

The implicit function theorem asserts that sufficient conditions are that $\phi(\boldsymbol{\xi}, \boldsymbol{\eta}) = 0$ and

$$\det \left(\frac{\partial \phi}{\partial \mathbf{y}} \right) \neq 0$$

at the point $(\boldsymbol{\xi}, \boldsymbol{\eta})^T$. Under these conditions there is a unique local differentiable solution $\mathbf{y} = g(\mathbf{x})$ of the equation $\phi(\mathbf{x}, \mathbf{y}) = 0$ and its derivative is given by

$$\frac{d\mathbf{y}}{d\mathbf{x}} = -\left(\frac{\partial \phi}{\partial \mathbf{y}} \right)^{-1} \frac{\partial \phi}{\partial \mathbf{x}}.$$

Example 7.4.
Consider the equation

$$x^2 + y^2 = 4.$$

To apply the implicit function theorem, we write $\phi(x, y) = x^2 + y^2 - 4$. Then

$$\frac{\partial \phi}{\partial y} = 2y.$$

It follows that a unique local differentiable solution $y = g(x)$ of $x^2 + y^2 = 4$ exists at each point $(\xi, \eta)^T$ satisfying $\xi^2 + \eta^2 = 4$ *except* $(-2, 0)^T$ and $(2, 0)^T$. The derivative of this function is given by

$$\frac{dy}{dx} = -\left(\frac{\partial \phi}{\partial y}\right)^{-1}\left(\frac{\partial \phi}{\partial x}\right) = \frac{-2y}{2x}.$$

Thus the derivative of the local solution $y = g_1(\dot{x})$ at $(0, 2)^T$ is

$$g_1'(0) = \frac{-2 \times 0}{2 \times 2} = 0$$

while the derivative of the local solution $y = g_2(x)$ at $(\sqrt{2}, -\sqrt{2})^T$ is

$$g_2'(\sqrt{2}) = \frac{-2 \times (-\sqrt{2})}{2 \times \sqrt{2}} = 1.$$

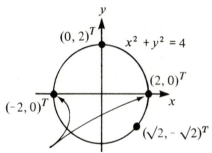

No local solutions

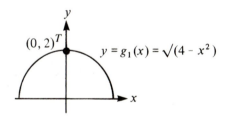

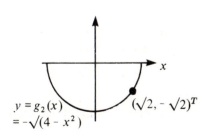

Example 7.5.

Consider the equation

$$xz^3 + y^2z - 2xy = 0.$$

This is satisfied by $(x, y, z)^T = (1, 1, 1)^T$. Can we find a unique local differentiable solution $z = g(x, y)$ at this point?

To apply the implicit function theorem, we put $\mathbf{u} = (x, y)^T$ and write

$$\phi(\mathbf{u}, z) = \phi(x, y, z) = xz^3 + y^2z - 2xy.$$

Then

$$\frac{\partial \phi}{\partial z} = 3xz^2 + y^2.$$

At the point $(x, y, z)^T = (1, 1, 1)^T$, this is equal to $3 + 1 = 4$. Since this is non-zero, it follows that a unique local differentiable solution $z = g(x, y)$ exists at the point $(1, 1, 1)^T$.

The derivative of $z = g(\mathbf{u}) = g(x, y)$ is given by

$$\frac{dz}{d\mathbf{u}} = -\left(\frac{\partial \phi}{\partial z}\right)^{-1}\left(\frac{\partial \phi}{\partial \mathbf{u}}\right)$$

i.e.

$$\left(\frac{\partial z}{\partial x}, \frac{\partial z}{\partial y}\right) = -\left(\frac{\partial \phi}{\partial z}\right)^{-1}\left(\frac{\partial \phi}{\partial x}, \frac{\partial \phi}{\partial y}\right)$$

$$= -\frac{1}{3xz^2 + y^2}(z^3 - 2y, 2yz - 2x).$$

In particular,

$$g'(1, 1) = (\tfrac{1}{4}, 0).$$

Example 7.6.

The equations

$$\begin{aligned} xuy + uyv + 2yvz &= 0 \\ 2uyv + yvz + vzx &= 0 \end{aligned} \Bigg\}$$

are satisfied by $(x, y, z, u, v)^T = (1, 1, -1, 1, 1)^T$. Can we find a unique local differentiable solution

$$(u, v)^T = g(x, y, z)$$

at this point?

To apply the implicit function theorem, we write $\mathbf{t} = (u, v)^T$, $\mathbf{s} = (x, y, z)^T$ and put

$$\phi(\mathbf{s}, \mathbf{t}) = \phi(x, y, z, u, v) = \begin{pmatrix} xuy + uyv + 2yvz \\ 2uyv + yvz + vzx \end{pmatrix}.$$

Then

$$\frac{\partial \phi}{\partial t} = \begin{pmatrix} \dfrac{\partial \phi_1}{\partial u} & \dfrac{\partial \phi_1}{\partial v} \\[2ex] \dfrac{\partial \phi_2}{\partial u} & \dfrac{\partial \phi_2}{\partial v} \end{pmatrix}$$

$$= \begin{pmatrix} xy + yv & uy + 2yz \\[1ex] 2yv & 2uy + yz + zx \end{pmatrix}$$

and so

$$\det \left(\frac{\partial \phi}{\partial u} \right) = \begin{vmatrix} 2 & -1 \\ 2 & 0 \end{vmatrix} = 2 \neq 0$$

at the point $(x, y, z, u, v)^T = (1, 1, -1, 1, 1)^T$. Thus a unique local differentiable solution exists at this point. Its derivative is given by

$$\frac{dt}{ds} = \begin{pmatrix} \dfrac{\partial u}{\partial x} & \dfrac{\partial u}{\partial y} & \dfrac{\partial u}{\partial z} \\[2ex] \dfrac{\partial v}{\partial x} & \dfrac{\partial v}{\partial y} & \dfrac{\partial v}{\partial z} \end{pmatrix}$$

$$= -\left(\frac{\partial \phi}{\partial t} \right)^{-1} \left(\frac{\partial \phi}{\partial s} \right).$$

At the point $(x, y, z, u, v)^T = (1, 1, -1, 1, 1)^T$,

$$\left(\frac{\partial \phi}{\partial t} \right)^{-1} = \begin{pmatrix} \dfrac{\partial \phi_1}{\partial u} & \dfrac{\partial \phi_1}{\partial v} \\[2ex] \dfrac{\partial \phi_2}{\partial u} & \dfrac{\partial \phi_2}{\partial v} \end{pmatrix}^{-1} = \begin{pmatrix} 2 & -1 \\ 2 & 0 \end{pmatrix}^{-1} = \frac{1}{2}\begin{pmatrix} 0 & 1 \\ -2 & 2 \end{pmatrix}$$

and

$$\frac{\partial \phi}{\partial s} = \begin{pmatrix} \dfrac{\partial \phi_1}{\partial x} & \dfrac{\partial \phi_1}{\partial y} & \dfrac{\partial \phi_1}{\partial z} \\[2ex] \dfrac{\partial \phi_2}{\partial x} & \dfrac{\partial \phi_2}{\partial y} & \dfrac{\partial \phi_2}{\partial z} \end{pmatrix} = \begin{pmatrix} uy & xu + uv + 2vz & 2yv \\[1ex] vz & 2uv + vz & yv + vx \end{pmatrix}$$

$$= \begin{pmatrix} 1 & 0 & 2 \\ -1 & 1 & 2 \end{pmatrix}.$$

Thus

$$g'(1, 1, -1) = -\frac{1}{2}\begin{pmatrix} 0 & 1 \\ -2 & 2 \end{pmatrix}\begin{pmatrix} 1 & 0 & 2 \\ -1 & 1 & 2 \end{pmatrix}$$

$$= \begin{pmatrix} \frac{1}{2} & -\frac{1}{2} & -1 \\ 2 & -1 & 0 \end{pmatrix}.$$

In this particular case it is quite simple to find a formula for the function g and hence to check this result. If $y \neq 0$ and $v \neq 0$, the original equations reduce to

$$\left. \begin{array}{l} xu + uv + 2vz = 0 \\ 2uy + yz + zx = 0. \end{array} \right\}$$

From the latter equation

$$u = -\frac{z(x+y)}{2y}.$$

Substitute this in the other equation. Then

$$\frac{-z(x+y)}{2y}x - \frac{z(x+y)}{2y}v + 2vz = 0.$$

If $z \neq 0$ and $x \neq 3y$,

$$v\left(2 - \frac{(x+y)}{2y}\right) = \frac{(x+y)x}{2y}$$

$$v = \frac{(x+y)x}{3y-x}.$$

It follows that

$$\begin{pmatrix} \dfrac{\partial u}{\partial x} & \dfrac{\partial u}{\partial y} & \dfrac{\partial u}{\partial z} \\[2ex] \dfrac{\partial v}{\partial x} & \dfrac{\partial v}{\partial y} & \dfrac{\partial v}{\partial z} \end{pmatrix} = \begin{pmatrix} \dfrac{-z}{2y} & \dfrac{zx}{2y^2} & \dfrac{-(x+y)}{2y} \\[2ex] \dfrac{(2x+y)(3y-x)+(x+y)x}{(3y-x)^2} & \dfrac{-4x^2}{(3y-x)^2} & 0 \end{pmatrix}$$

$$= \begin{pmatrix} \frac{1}{2} & -\frac{1}{2} & -1 \\ 2 & -1 & 0 \end{pmatrix}$$

at the point $(x, y, z)^T = (1, 1, -1)^T$.

Exercises 7.7

1. Sketch the curve defined by the equation

$$y^2 = x^2(1-x^2).$$

Find the points $(\xi, \eta)^T$ on this curve at which $y^2 = x^2(1-x^2)$ has a unique local differentiable solution $y = g(x)$ and evaluate $g'(\xi)$.

Find formulae for local solutions at $(1/\sqrt{2}, \frac{1}{2})^T$ and $(-1/\sqrt{2}, -\frac{1}{2})^T$.

2.* Find the points $(\xi, \eta)^T$ on the curve

$$y^2(1-y^2) = x^2(1-x^2)$$

at which the equation has a unique local differentiable solution $y = g(x)$ and evaluate $g'(\xi)$.

Find formulae for local solutions at $(\frac{1}{2}, \frac{1}{2})^T$, $(\frac{1}{2}, -\frac{1}{2})^T$, $(\frac{1}{2}, \sqrt{3}/2)^T$ and $(\frac{1}{2}, -\sqrt{3}/2)^T$.

3. Show that the point $(x, y)^T = (-1, 1)^T$ lies on the curve

$$\sin(\pi x^2 y) = 1 + xy^2.$$

Prove that a unique local differentiable solution $y = g(x)$ exists at $(-1, 1)^T$ and evaluate $g'(-1)$.

4.* Show that the point $(x, y)^T = (\sqrt{e}, 1/\sqrt{e})^T$ lies on the curve

$$xy - \log x + \log y = 0 \quad (x > 0, y > 0).$$

Prove that a unique local differentiable solution $y = g(x)$ exists at $(\sqrt{e}, 1/\sqrt{e})^T$. Evaluate $g'(x)$ and $g''(x)$ and hence show that g has a local maximum at $x = \sqrt{e}$.

5. Find a point $(\xi, \eta, \zeta)^T$ at which the equation

$$\sin(yz) + \sin(zx) + \sin(xy) = 0$$

has a unique local differentiable solution $z = g(x, y)$. Find $g'(\xi, \eta)$.

6.* Show that the point $(x, y, z)^T = (1, 1, -1)^T$ lies on the curve

$$xy^2 + yz^2 + zx^2 = 1.$$

Prove that a unique local differentiable solution $z = g(x, y)$ exists at the point $(1, 1, -1)^T$ and find $g'(1, 1)$.

Also obtain a formula for $g(x, y)$.

7. Show that the equations

$$\left. \begin{array}{r} xy^2 + y^2z^3 + z^4x^5 = 1 \\ zx^2 + x^2y^3 + y^4z^5 = -1 \end{array} \right\}$$

admit a unique local differentiable solution

$$\binom{x}{y} = g(z)$$

at $(x, y, z)^T = (1, 1, -1)^T$. Evaluate $g'(-1)$.

8.* Show that the equations

$$x^4 + (x + z)y^3 - 3 = 0$$
$$x^4 + (2x + 3z)y^3 - 6 = 0$$

admit a unique local differentiable solution

$$\binom{y}{z} = g(x)$$

at any point $(\xi, \eta, \zeta)^T$ which satisfies the equations. Find a formula for $g'(x)$.

9. Show that there exists a unique local differentiable solution

$$\begin{pmatrix} u \\ v \\ w \end{pmatrix} = g(x, y, z)$$

to the equations

$$x^2 + u + e^v = 0$$
$$y^2 + v + e^w = 0$$
$$z^2 + w + e^u = 0$$

at any point which satisfies the equations. Find a formula for $g'(x, y, z)$.

10.* Show that there exists a unique local differentiable solution

$$\binom{u}{v} = g(x, y)$$

to the equations

$$x = e^u \cos v$$
$$v = e^y \sin x$$

at any point which satisfies the equations. Find a formula for $g'(x, y)$.

SOME APPLICATIONS (OPTIONAL)

7.8   Shadow prices

We have seen that in constrained optimization problems of the type: find

$$\max f(\mathbf{x})$$

subject to the constraint

$$g(\mathbf{x}) = \mathbf{0},$$

it is often a good idea to form the Lagrangian

$$L = f(\mathbf{x}) + \boldsymbol{\lambda}^T g(\mathbf{x})$$

(see §4.16). The stationary points of L are then found by solving the simultaneous equations

$$\begin{cases} \dfrac{\partial L}{\partial \mathbf{x}} = 0 \\[2mm] \dfrac{\partial L}{\partial \boldsymbol{\lambda}} = 0 - \text{i.e.} \, g(\mathbf{x}) = \mathbf{0}. \end{cases}$$

If $\mathbf{x} = \tilde{\mathbf{x}}$ and $\boldsymbol{\lambda} = \tilde{\boldsymbol{\lambda}}$ satisfy the equations, then $\tilde{\mathbf{x}}$ is a constrained stationary point for the original problem. An added advantage to the use of the Lagrangian is that the vector $\boldsymbol{\lambda} = \tilde{\boldsymbol{\lambda}}$ is also an interesting quantity. The reasons for this are discussed below.

Suppose that $\mathbf{x}$ represents a bundle of outputs from a production process and that $f(\mathbf{x})$ represents the profit derived from the sale of $\mathbf{x}$. We shall suppose that the input to the production process is a given commodity bundle $\mathbf{b}$ of raw materials. The various outputs $\mathbf{x}$ which can be produced using the given input $\mathbf{b}$ are specified by the equation $G(\mathbf{x}) = \mathbf{b}$. The production manager therefore seeks to maximize $f(\mathbf{x})$ subject to the constraint

$$g(\mathbf{x}) = \mathbf{b} - G(\mathbf{x}) = \mathbf{0}.$$

This scenario, of course, is very similar to that described in §1.36 and continued in §1.37, §4.28 and §5.15.

The optimum output $\tilde{\mathbf{x}}$ and the corresponding $\tilde{\boldsymbol{\lambda}}$ satisfy the Lagrangian equations

$$(1) \quad \begin{aligned} \phi_1(\mathbf{x}, \boldsymbol{\lambda}, \mathbf{b}) &= \frac{df}{d\mathbf{x}} - \boldsymbol{\lambda}^T \frac{dG}{d\mathbf{x}} = 0 \\[2mm] \phi_2(\mathbf{x}, \boldsymbol{\lambda}, \mathbf{b}) &= \mathbf{b} - G(\mathbf{x}) = \mathbf{0}. \end{aligned} \Bigg\}$$

If we wish to know how $\tilde{\mathbf{x}}$ and $\tilde{\boldsymbol{\lambda}}$ depend on the given input $\mathbf{b}$, it is necessary to solve these equations to obtain $\tilde{\mathbf{x}}$ and $\tilde{\boldsymbol{\lambda}}$ as functions of $\mathbf{b}$. The implicit function theorem requires that

$$\det \begin{pmatrix} \dfrac{\partial\phi_1}{\partial x} & \dfrac{\partial\phi_1}{\partial \lambda} \\[2ex] \dfrac{\partial\phi_2}{\partial x} & \dfrac{\partial\phi_2}{\partial \lambda} \end{pmatrix} = \det \begin{pmatrix} \dfrac{\partial^2 L}{\partial x^2} & \left(\dfrac{\partial g}{\partial x}\right)^{T} \\[2ex] \dfrac{\partial g}{\partial x} & 0 \end{pmatrix} \neq 0.$$

If the second order conditions for $\tilde{x}$ to be a local maximum for $f(x)$ subject to the constraint $g(x) = 0$ are satisfied (see §5.16), then this determinant is indeed non-zero. A unique local differentiable solution

$$\left. \begin{aligned} \tilde{x} &= h(b) \\ \tilde{\lambda} &= k(b) \end{aligned} \right\}$$

of the equations (1) then exists and

$$\begin{pmatrix} \dfrac{d\tilde{x}}{db} \\[2ex] \dfrac{d\tilde{\lambda}}{db} \end{pmatrix} = - \begin{pmatrix} \dfrac{\partial\phi_1}{\partial x} & \dfrac{\partial\phi_1}{\partial \lambda} \\[2ex] \dfrac{\partial\phi_2}{\partial x} & \dfrac{\partial\phi_2}{\partial \lambda} \end{pmatrix}^{-1} \begin{pmatrix} \dfrac{\partial\phi_1}{\partial b} \\[2ex] \dfrac{\partial\phi_2}{\partial b} \end{pmatrix}$$

which we rewrite as

$$\begin{pmatrix} \dfrac{\partial\phi_1}{\partial x} & \dfrac{\partial\phi_1}{\partial \lambda} \\[2ex] \dfrac{\partial\phi_2}{\partial x} & \dfrac{\partial\phi_2}{\partial \lambda} \end{pmatrix} \begin{pmatrix} \dfrac{d\tilde{x}}{db} \\[2ex] \dfrac{d\tilde{\lambda}}{db} \end{pmatrix} = - \begin{pmatrix} \dfrac{\partial\phi_1}{\partial b} \\[2ex] \dfrac{\partial\phi_2}{\partial b} \end{pmatrix}$$

This yields that

$$\frac{\partial\phi_2}{\partial x}\frac{d\tilde{x}}{db} + \frac{\partial\phi_2}{\partial \lambda}\frac{d\tilde{\lambda}}{db} = -\frac{\partial\phi_2}{\partial b}$$

$$\frac{dG}{dx}\frac{d\tilde{x}}{db} = I.$$

This last result is more easily obtained by differentiating the equation $G(\tilde{x}(b)) = b$ with respect to b using the chain rule.

If we are interested in how a change in the input vector b will affect our profit given that we produce the optimal output $\tilde{x}$, then we need to calculate

$$\frac{d}{db}\{f(\tilde{x}(b))\} = \frac{df}{dx}\frac{d\tilde{x}}{db}$$

$$= \tilde{\lambda}^{T}\frac{dG}{dx}\frac{d\tilde{x}}{db} \quad \text{(using equations (1))}$$

$$= \tilde{\lambda}^{T}I = \tilde{\lambda}^{T}.$$

Using the first two terms of the Taylor series expression, we have that

$$f(\tilde{\mathbf{x}}(\mathbf{b} + \delta\mathbf{b})) = f(\tilde{\mathbf{x}}(\mathbf{b})) + \tilde{\boldsymbol{\lambda}}^T\delta\mathbf{b} + \ldots.$$

It follows that if one could purchase the commodity bundle $\delta\mathbf{b}$ of inputs, given the price vector $\tilde{\boldsymbol{\lambda}}$, then the cost of so doing would be cancelled out by the extra profit which would be obtained. We call $\tilde{\boldsymbol{\lambda}}$ the vector of *shadow prices* for the input commodities. One would be just as ready to start selling off one's stock $\mathbf{b}$ at the shadow prices $\tilde{\boldsymbol{\lambda}}$ as one would be to begin a program of optimal production.

8

Differentials

This is a comparatively late stage to introduce the subject of differentials but there are good reasons for the delay. Readers will probably already have met the notation dx and dy in applied subjects and have been told that these represent 'very small changes in x and y'. There are even those who tell their students that dx and dy stand for 'infinitesimally small changes in x and y'. Such statements about differentials do not help very much in explaining why manipulations with differentials give correct answers and, in some contexts, are a positive hindrance. In this chapter we have tried to provide a more accurate account of the nature of differentials without attempting anything in the way of a systematic theoretical discussion. In studying this account, readers may find it necessary to put aside some of the pre-conceptions they perhaps have about differentials. Those who find this hard to do may take comfort in the fact that everything which is done in this book using differentials may also be done by means of techniques described in other chapters. For instance, three alternative methods of solving a problem are given in example 8.3 before the method of differentials is used. However, this is not a sound reason for neglecting differentials. Their use in both applied and theoretical work is too widespread for this to be sensible.

8.1 Matrix algebra and linear systems[□]

Matrix algebra may be regarded as a neat way of summarizing the properties of linear functions. These in turn characterize the properties of systems of linear equations.

For example, to say that an $n \times n$ matrix A is non-singular with inverse A^{-1} means that the linear function $L : \mathbb{R}^n \to \mathbb{R}^n$ defined by

$$\mathbf{y} = L(\mathbf{x}) = A\mathbf{x}$$

has an inverse function $M : \mathbb{R}^n \to \mathbb{R}^n$ and that this is given by

$$\mathbf{x} = M(\mathbf{y}) = A^{-1}\mathbf{y}.$$

This in turn is equivalent to the assertion that the system of linear equations

$$\left.\begin{aligned}
y_1 &= a_{11}x_1 + a_{12}x_2 + \ldots + a_{1n}x_n \\
y_2 &= a_{21}x_1 + a_{22}x_2 + \ldots + a_{2n}x_n \\
&\ \vdots \\
y_n &= a_{n1}x_1 + a_{n2}x_2 + \ldots + a_{nn}x_n.
\end{aligned}\right\}$$

has a unique solution **x** for every **y** and this solution is given by

$$\left.\begin{aligned}
x_1 &= b_{11}y_1 + b_{12}y_2 + \ldots + b_{1n}y_n \\
x_2 &= b_{21}y_1 + b_{22}y_2 + \ldots + b_{2n}y_n \\
&\ \vdots \\
x_n &= b_{n1}y_1 + b_{n2}y_2 + \ldots + b_{nn}y_n.
\end{aligned}\right\}$$

where $B = A^{-1}$.

Although it is often easiest to translate problems involving systems of linear equations into matrix language, this is by no means invariably true. Consider, for example, the system

$$\left.\begin{aligned}
y_1 &= x_1 + x_2 \\
y_2 &= x_1
\end{aligned}\right\}$$

This can be shown to have the unique solution

$$\left.\begin{aligned}
x_1 &= y_2 \\
x_2 &= y_1 - y_2.
\end{aligned}\right\}$$

simply by substituting $x_1 = y_2$ in the equation $y_1 = x_1 + x_2$. There is therefore no need to consider whether the matrix

$$A = \begin{pmatrix} 1 & 1 \\ 1 & 0 \end{pmatrix}$$

is non-singular nor to compute its inverse. The fact that the equations have a unique solution guarantees that A is non-singular and the form of the solution shows that

$$A^{-1} = \begin{pmatrix} 0 & 1 \\ 1 & -1 \end{pmatrix}.$$

Similar situations occur when dealing with the derivative of a function $f : \mathbb{R}^n \to \mathbb{R}^m$. It is often easier not to work directly with the $m \times n$ matrix $f'(\mathbf{x})$ but to use instead the system of linear equations which this matrix defines. The subject of this chapter is the special notation which is used for this purpose.

8.2   Differentials

Suppose that $f: \mathbb{R}^n \to \mathbb{R}^m$ is differentiable at $x \in \mathbb{R}^n$. Its *derivative* $f'(x)$ is then an $m \times n$ matrix and hence defines a linear function $L: \mathbb{R}^n \to \mathbb{R}^m$. This linear function is called the *differential* of f at x. We use the notation

(1) $dy = L(dx) = f'(x)dx$

In this expression dx denotes an arbitrary vector in $\mathbb{R}^n$ and dy represents the corresponding vector in $\mathbb{R}^m$. Thus dx and dy are *variables*. We choose to denote these variables by dx and dy because (1) can then be rewritten in the form

(2) $dy = \left(\dfrac{dy}{dx}\right)dx$

and this makes it impossible to forget what dx and dy represent. (It is, of course, *not* true that (2) holds *because* the dx terms 'cancel out'. On the contrary, in so far as the dx terms can be said to 'cancel out', it is because the notation has been rigged to give this appearance.)

Written as a system of linear equations, (2) becomes

$$(3) \quad \left. \begin{aligned}
dy_1 &= \frac{\partial y_1}{\partial x_1} dx_1 + \frac{\partial y_1}{\partial x_2} dx_2 + \ldots + \frac{\partial y_1}{\partial x_n} dx_n \\
dy_2 &= \frac{\partial y_2}{\partial x_1} dx_1 + \frac{\partial y_2}{\partial x_2} dx_2 + \ldots + \frac{\partial y_2}{\partial x_n} dx_n \\
&\quad \vdots \\
dy_m &= \frac{\partial y_m}{\partial x_1} dx_1 + \frac{\partial y_m}{\partial x_2} dx_2 + \ldots + \frac{\partial y_m}{\partial x_n} dx_n.
\end{aligned} \right\}$$

The notion of a differential is actually a more fundamental mathematical concept than that of a derivative. In §4.4, a function $f: \mathbb{R}^n \to \mathbb{R}^m$ was said to be differentiable at the point x if and only if the graph of the function has a tangent flat at the point x. The diagram below illustrates a function $f: \mathbb{R} \to \mathbb{R}$ which is differentiable at x. If dx and dy axes are introduced as illustrated, then the equation of the tangent line takes the form

$dy = mdx.$

The definition of a derivative given in §4.4 then reduces to the requirement that $f'(x) = m$.

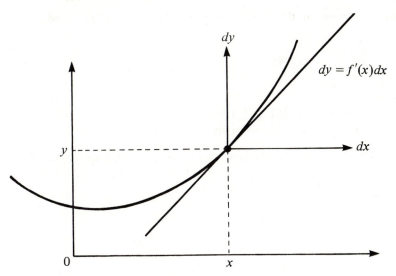

$$dy$$

$$dy = f'(x)dx$$

$$dx$$

$$y$$

$$0 \qquad x$$

Example 8.3.

Express

$$\left(\frac{\partial z}{\partial x}\right)_y$$

in terms of partial derivatives of z with respect to u and v when

(4) $$\left.\begin{array}{l} x = u^2 + v^2 \\ y = u^2 - v^2. \end{array}\right\}$$

This problem was considered in example 6.20. We have that

$$\frac{\partial}{\partial x} = \frac{\partial u}{\partial x}\frac{\partial}{\partial u} + \frac{\partial v}{\partial x}\frac{\partial}{\partial v}$$

and

$$\begin{pmatrix} \dfrac{\partial u}{\partial x} & \dfrac{\partial u}{\partial y} \\[2mm] \dfrac{\partial v}{\partial x} & \dfrac{\partial v}{\partial y} \end{pmatrix} = \begin{pmatrix} \dfrac{\partial x}{\partial u} & \dfrac{\partial x}{\partial v} \\[2mm] \dfrac{\partial y}{\partial u} & \dfrac{\partial y}{\partial v} \end{pmatrix}^{-1}$$

(5) $$= \begin{pmatrix} 2u & 2v \\ 2u & -2v \end{pmatrix}^{-1} = \begin{pmatrix} \dfrac{1}{4u} & \dfrac{1}{4u} \\[3mm] \dfrac{1}{4v} & -\dfrac{1}{4v} \end{pmatrix}.$$

Thus

$$\frac{\partial z}{\partial x} = \frac{1}{4u}\frac{\partial z}{\partial u} + \frac{1}{4v}\frac{\partial z}{\partial v}$$

or, more precisely,

$$\left(\frac{\partial z}{\partial x}\right)_y = \frac{1}{4u}\left(\frac{\partial z}{\partial u}\right)_v + \frac{1}{4v}\left(\frac{\partial z}{\partial v}\right)_u$$

But this is not the only way in which one can proceed. Another possible approach is to begin by solving equations (4). If we are interested only in positive u and v, the required solutions are

$$u = \sqrt{\left(\frac{x+y}{2}\right)}$$

$$v = \sqrt{\left(\frac{x-y}{2}\right)}.$$

Hence

$$\frac{\partial u}{\partial x} = \frac{1}{2}\left(\frac{x+y}{2}\right)^{-1/2}\frac{1}{2} = \frac{1}{4u}$$

$$\frac{\partial v}{\partial x} = \frac{1}{2}\left(\frac{x-y}{2}\right)^{-1/2}\frac{1}{2} = \frac{1}{4v}$$

and thus

$$\frac{\partial z}{\partial x} = \frac{\partial u}{\partial x}\frac{\partial z}{\partial u} + \frac{\partial v}{\partial x}\frac{\partial z}{\partial v}$$

$$= \frac{1}{4u}\frac{\partial z}{\partial u} + \frac{1}{4v}\frac{\partial z}{\partial v}.$$

This method suffers from the defect that not all sets of equations are as easy to solve as (4). Moreover, it is by no means always clear which local solution one wishes to consider.

It is also possible to proceed using the method of implicit differentiation. Differentiating equations (4) as they stand partially with respect to x and y, we obtain that

$$1 = 2u\frac{\partial u}{\partial x} + 2v\frac{\partial v}{\partial x}$$

$$0 = 2u\frac{\partial u}{\partial x} - 2v\frac{\partial v}{\partial x}.$$

Thus, first adding these equations and then subtracting them,

$$1 = 4u\frac{\partial u}{\partial x} \quad ; \quad 1 = 4v\frac{\partial v}{\partial x}$$

and so

$$\frac{\partial z}{\partial x} = \frac{\partial u}{\partial x}\frac{\partial z}{\partial u} + \frac{\partial v}{\partial x}\frac{\partial z}{\partial v}$$

$$= \frac{1}{4u}\frac{\partial z}{\partial u} + \frac{1}{4v}\frac{\partial z}{\partial v}.$$

Finally, we may use differentials. In this problem, equations (3) of §8.2 take the form

$$dx = \frac{\partial x}{\partial u}du + \frac{\partial x}{\partial v}dv$$

$$dy = \frac{\partial y}{\partial u}du + \frac{\partial y}{\partial v}dv$$

i.e.

$$(6) \qquad \left.\begin{array}{l} dx = 2udu + 2vdv \\ dy = 2udu - 2vdv. \end{array}\right\}$$

These equations have the solution

$$\left.\begin{array}{l} du = \dfrac{1}{4u}dx + \dfrac{1}{4u}dy \\[2mm] dv = \dfrac{1}{4v}dx - \dfrac{1}{4v}dy. \end{array}\right\}$$

But we also have that

$$\left.\begin{array}{l} du = \dfrac{\partial u}{\partial x}dx + \dfrac{\partial u}{\partial y}dy \\[2mm] dv = \dfrac{\partial v}{\partial x}dx + \dfrac{\partial v}{\partial y}dy. \end{array}\right\}$$

Since the final two systems of equations must hold for all dx and dy, it follows that

$$\frac{\partial u}{\partial x} = \frac{1}{4u} \quad \text{and} \quad \frac{\partial v}{\partial x} = \frac{1}{4v}$$

and thus

$$\frac{\partial z}{\partial x} = \frac{1}{4u}\frac{\partial z}{\partial u} + \frac{1}{4v}\frac{\partial z}{\partial v}.$$

Which of these various methods is the most appropriate to use depends on the problem in hand. In particular, one might as well use the first method rather than the fourth (or the third) if one proposes to solve the system (6) of linear equations by inverting the matrix as in (5).

Example 8.4.

The energy u, pressure p, volume v and temperature T of a gas are so related that each may be expressed as a function of any two of the others. Prove that

$$\left.\begin{array}{l}\left(\dfrac{\partial u}{\partial v}\right)_T\left(\dfrac{\partial v}{\partial p}\right)_T = \left(\dfrac{\partial u}{\partial p}\right)_T \\[3ex] \left(\dfrac{\partial u}{\partial v}\right)_T\left(\dfrac{\partial v}{\partial T}\right)_p = \left(\dfrac{\partial u}{\partial T}\right)_p - \left(\dfrac{\partial u}{\partial T}\right)_v. \end{array}\right\}$$

We are given that functions $f:\mathbb{R}^2 \to \mathbb{R}$, $g:\mathbb{R}^2 \to \mathbb{R}$ and $h:\mathbb{R}^2 \to \mathbb{R}$ exist such that

$$u = f(v, T) = g(p, T)$$
$$v = h(p, T).$$

Introducing differentials, we obtain that

$$du = \left(\frac{\partial u}{\partial v}\right)_T dv + \left(\frac{\partial u}{\partial T}\right)_v dT$$

$$du = \left(\frac{\partial u}{\partial p}\right)_T dp + \left(\frac{\partial u}{\partial T}\right)_p dT$$

$$(7) \quad dv = \left(\frac{\partial v}{\partial p}\right)_T dp + \left(\frac{\partial v}{\partial T}\right)_p dT.$$

Eliminate du from the first two equations. Then

$$(8) \quad \left(\frac{\partial u}{\partial v}\right)_T dv = \left(\frac{\partial u}{\partial p}\right)_T dp + \left\{\left(\frac{\partial u}{\partial T}\right)_p - \left(\frac{\partial u}{\partial T}\right)_v\right\} dT.$$

Equations (7) and (8) hold for all dp and dT and thus

$$\left(\frac{\partial v}{\partial p}\right)_T = \left(\frac{\partial u}{\partial p}\right)_T \bigg/ \left(\frac{\partial u}{\partial v}\right)_T$$

$$\left(\frac{\partial v}{\partial T}\right)_p = \left\{\left(\frac{\partial u}{\partial T}\right)_p - \left(\frac{\partial u}{\partial T}\right)_v\right\} \bigg/ \left(\frac{\partial u}{\partial v}\right)_T.$$

Note that the first of these two results may be obtained directly from the implicit function theorem. (Why?)

8.5 Stationary points

If $f: \mathbb{R}^n \to \mathbb{R}$ is differentiable at $\mathbf{x} = (x_1, x_2, \ldots, x_n)^T$, then its tangent hyperplane has equation

$$dy = f'(\mathbf{x})d\mathbf{x}$$

$$= \frac{\partial f}{\partial x_1}dx_1 + \frac{\partial f}{\partial x_2}dx_2 + \ldots + \frac{\partial f}{\partial x_n}dx_n$$

provided that the $dx_1, dx_2, \ldots, dx_n$ and dy axes are introduced as indicated for the case $n = 2$ in the diagram below:

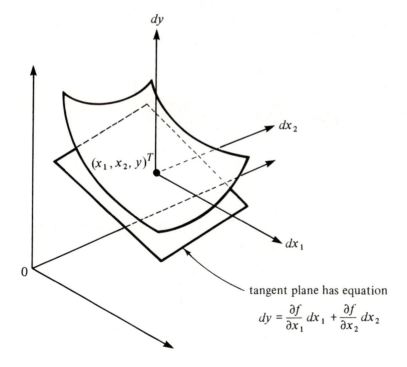

tangent plane has equation

$$dy = \frac{\partial f}{\partial x_1}dx_1 + \frac{\partial f}{\partial x_2}dx_2$$

At a stationary point, the tangent hyperplane is horizontal — i.e. it reduces to the form

$$dy = 0.$$

Thus, at a stationary point

$$\frac{\partial f}{\partial x_1} dx_1 + \frac{\partial f}{\partial x_2} dx_2 + \ldots + \frac{\partial f}{\partial x_n} dx_n = 0.$$

This must hold for all $dx_1, dx_2, \ldots, dx_n$ and so we obtain the familiar condition

$$\frac{\partial f}{\partial x_1} = \frac{\partial f}{\partial x_2} = \ldots = \frac{\partial f}{\partial x_n} = 0$$

i.e.

$$f'(\mathbf{x}) = \mathbf{0}.$$

Consider next the case of constrained optimization. Let $f: \mathbb{R}^n \to \mathbb{R}$ and let $g: \mathbb{R}^n \to \mathbb{R}^m$ and suppose that we are interested in the constrained stationary points of f when the constraint is $g(\mathbf{x}) = \mathbf{0}$. As before we require that

$$(1) \quad 0 = dy = \frac{\partial f}{\partial x_1} dx_1 + \frac{\partial f}{\partial x_2} dx_2 + \ldots + \frac{\partial f}{\partial x_n} dx_n$$

but no longer is it true that this must hold for *all* $dx_1, dx_2, \ldots, dx_n$. In the presence of the constraint, we require only that $dy = 0$ for values of dx_1, $dx_2, \ldots, dx_n$ satisfying

$$(2) \quad \left.\begin{array}{l} 0 = \dfrac{\partial g_1}{\partial x_1} dx_1 + \dfrac{\partial g_1}{\partial x_2} dx_2 + \ldots + \dfrac{\partial g_1}{\partial x_n} dx_n \\[2ex] 0 = \dfrac{\partial g_2}{\partial x_1} dx_1 + \dfrac{\partial g_2}{\partial x_2} dx_2 + \ldots + \dfrac{\partial g_2}{\partial x_n} dx_n \\[1ex] \vdots \\[1ex] 0 = \dfrac{\partial g_m}{\partial x_1} dx_1 + \dfrac{\partial g_m}{\partial x_2} dx_2 + \ldots + \dfrac{\partial g_m}{\partial x_n} dx_n \end{array}\right\}$$

Multiplying these equations by $\lambda_1, \lambda_2, \ldots, \lambda_m$ and adding, we obtain that

$$0 = \left(\frac{\partial f}{\partial x_1} + \lambda_1 \frac{\partial g_1}{\partial x_1} + \ldots + \lambda_m \frac{\partial g_m}{\partial x_n} \right) dx_1$$

$$+ \ldots + \left(\frac{\partial f}{\partial x_n} + \lambda_1 \frac{\partial g_1}{\partial x_n} + \ldots + \lambda_m \frac{\partial g_m}{\partial x_n} \right) dx_n.$$

Choose $\lambda_1, \lambda_2, \ldots, \lambda_m$ to make the last m coefficients in this expression zero. The first $n - m$ coefficients must then also be zero because we can choose $dx_1, dx_2, \ldots, dx_{n-m}$ freely without violating equations (2). We therefore obtain the Lagrangian conditions

$$\frac{\partial f}{\partial x_1} + \lambda_1 \frac{\partial g_1}{\partial x_1} + \ldots + \lambda_m \frac{\partial g_m}{\partial x_1} = 0$$

$$\frac{\partial f}{\partial x_2} + \lambda_1 \frac{\partial g_1}{\partial x_2} + \ldots + \lambda_m \frac{\partial g_m}{\partial x_2} = 0$$

$$\vdots$$

$$\frac{\partial f}{\partial x_n} + \lambda_1 \frac{\partial g_1}{\partial x_n} + \ldots + \lambda_m \frac{\partial g_m}{\partial x_n} = 0$$

i.e.

$$f'(\mathbf{x}) + \boldsymbol{\lambda}^T g'(\mathbf{x}) = \mathbf{0}.$$

8.6 Small changes

It is *not* true that

$$dy = f'(\mathbf{x})d\mathbf{x}$$

represents the 'small change' in **y** caused by a 'small change' $d\mathbf{x}$ in **x**. What is meant when an assertion of this type is made is that, if $\delta\mathbf{x}$ is small (i.e. each of $\delta x_1, \delta x_2, \ldots, \delta x_n$ is small) and $\delta\mathbf{y} = f(\mathbf{x} + \delta\mathbf{x}) - f(\mathbf{x})$ is the 'small change' in **y** induced by the 'small change' $\delta\mathbf{x}$ in **x**, then

$$\delta\mathbf{y} \doteq f'(\mathbf{x})\delta\mathbf{x}$$

i.e. $\delta\mathbf{y}$ is *approximately* equal to $f'(\mathbf{x})\delta\mathbf{x}$. This follows from Taylor's theorem:

$$f(\mathbf{x} + \delta\mathbf{x}) = f(\mathbf{x}) + \frac{1}{1!}f'(\mathbf{x})\delta\mathbf{x} + \ldots.$$

A better approximation, of course, may be obtained by taking the second order terms in Taylor's series as well as the first order terms.

Example 8.7.
The cosine rule for a triangle asserts that

$$a^2 = b^2 + c^2 - 2bc \cos A.$$

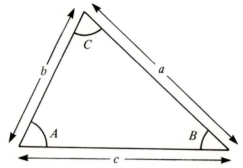

Small changes δA, δb and δc are made in A, b and c respectively. Find an approximation for the consequent change δa in a. We have that

$$2a\,da = 2b\,db + 2c\,dc - 2b\cos A\,dc - 2c\cos A\,db + 2bc\sin A\,dA.$$

Hence

$$a\,da = (b - c\cos A)\,db + (c - b\cos A)\,dc + bc\sin A\,dA.$$

We conclude that

$$\delta a \doteqdot \left(\frac{b - c\cos A}{a}\right)\delta b + \left(\frac{c - b\cos A}{a}\right)\delta c + \frac{bc\sin A}{a}\delta A.$$

Exercises 8.8

1. The equations

$$\left.\begin{array}{l} f(x, y, z) = 0 \\ g(x, y, z) = 0 \end{array}\right\}$$

can be solved simultaneously to give y as a function of x – i.e. $y = h(x)$. Obtain a formula for $h'(x)$ in terms of the partial derivatives of f and g.

2.* The equations

$$\left.\begin{array}{l} f(x, y, z, t) = 0 \\ g(x, y, z, t) = 0 \end{array}\right\}$$

can be solved simultaneously to give z and t as functions of x and y. Find

$$\left(\frac{\partial z}{\partial y}\right)_x$$

first by using the implicit function theorem and then by using differentials.

3. Given the equations

$$x = \tfrac{1}{2}(u^2 - v^2)$$
$$y = uv$$

find

$$\left(\frac{\partial u}{\partial x}\right)_y \quad \text{and} \quad \left(\frac{\partial u}{\partial y}\right)_x$$

using differentials.

4.* Given the equations

$$x = u^2 - v$$
$$y = uv^2,$$

find

$$\left(\frac{\partial u}{\partial x}\right)_y \quad \text{and} \quad \left(\frac{\partial v}{\partial x}\right)_y$$

by at least four different methods including the use of differentials.

5. With the assumptions of example 8.4, prove that

$$
\begin{vmatrix}
\left(\dfrac{\partial T}{\partial p}\right)_u & -1 & \left(\dfrac{\partial T}{\partial u}\right)_p & 0 \\[2ex]
\left(\dfrac{\partial T}{\partial p}\right)_v & -1 & 0 & \left(\dfrac{\partial T}{\partial v}\right)_p \\[2ex]
-1 & \left(\dfrac{\partial p}{\partial T}\right)_v & 0 & \left(\dfrac{\partial p}{\partial v}\right)_T \\[2ex]
-1 & \left(\dfrac{\partial p}{\partial T}\right)_u & \left(\dfrac{\partial p}{\partial u}\right)_T & 0
\end{vmatrix} = 0
$$

6.* If $f(x, y, z) = 0$, prove that

$$\left(\frac{\partial x}{\partial y}\right)_z \left(\frac{\partial y}{\partial z}\right)_x \left(\frac{\partial z}{\partial x}\right)_y = -1$$

7. Given the equations

$$x = \tfrac{1}{2}(u^2 - v^2)$$
$$y = uv,$$

find an approximation for the small change δv in v caused by a small change δx in x and a small change δu in u.

8.* The sine rule for a triangle (see example 8.7) asserts that

$$\frac{\sin a}{A} = \frac{\sin b}{B} = \frac{\sin c}{C}$$

Find an approximation for the small change δb in b caused by small changes $\delta a, \delta A$ and δB in a, A and B respectively.

SOME APPLICATIONS (OPTIONAL)

8.9   Slutsky equations

An individual has a quantity I of money to spend (his *income*) and demands a commodity bundle $\mathbf{x}$ which maximizes his utility $u = \phi(\mathbf{x})$ subject to the budget constraint $\mathbf{p}^T\mathbf{x} = I$ where $\mathbf{p}$ is the *price* vector.

Suppose that the bundle which maximises utility subject to the budget constraint is $\mathbf{x}^*$. This will depend on $\mathbf{p}$ and I and so we may write

$$(1) \quad \mathbf{x}^* = f(\mathbf{p}, I).$$

The resulting utility is

$$(2) \quad u^* = \phi(\mathbf{x}^*) = \phi(f(\mathbf{p}, I)).$$

Eliminating I between these equations yields

$$(3) \quad \mathbf{x}^* = g(\mathbf{p}, u^*).$$

From equation (1) we obtain that

$$(4) \quad d\mathbf{x}^* = \left(\frac{\partial \mathbf{x}^*}{\partial \mathbf{p}}\right) d\mathbf{p} + \left(\frac{\partial \mathbf{x}^*}{\partial I}\right)_{\mathbf{p}}^T dI$$

and from equation (3) we obtain that

$$(5) \quad d\mathbf{x}^* = \left(\frac{\partial \mathbf{x}^*}{\partial \mathbf{p}}\right)_{u^*} d\mathbf{p} + \left(\frac{\partial \mathbf{x}^*}{\partial u^*}\right)_{\mathbf{p}}^T du^*.$$

Next we recall that $\mathbf{x}^*$ (and the associated shadow price λ^*) must satisfy the Lagrangian conditions

$$\left.\begin{array}{c} \phi'(\mathbf{x}^*) - \lambda^*\mathbf{p}^T = \mathbf{0} \\ \mathbf{p}^T\mathbf{x}^* = I. \end{array}\right\}$$

From the final equation, we obtain that $\mathbf{p}^T d\mathbf{x}^* + \mathbf{x}^{*T} d\mathbf{p} = dI$ and therefore, by equation (2),

$$du^* = \phi'(\mathbf{x}^*)d\mathbf{x}^*$$
$$= \lambda^*\mathbf{p}^T d\mathbf{x}^*$$
$$= \lambda^*(dI - \mathbf{x}^{*T}d\mathbf{p}).$$

We substitute this result in (5) and obtain

$$d\mathbf{x}^* = \left\{\left(\frac{\partial \mathbf{x}^*}{\partial \mathbf{p}}\right)_{u^*} - \lambda^*\left(\frac{\partial \mathbf{x}^*}{\partial u^*}\right)_{\mathbf{p}}^T \mathbf{x}^{*T}\right\}d\mathbf{p} + \lambda^*\left(\frac{\partial \mathbf{x}^*}{\partial u^*}\right)_{\mathbf{p}}^T dI.$$

Compare this equation with equation (4). Since both equations hold for all $d\mathbf{p}$ and dI, we have that

$$\left(\frac{\partial \mathbf{x}^*}{\partial \mathbf{p}}\right)_I = \left(\frac{\partial \mathbf{x}^*}{\partial \mathbf{p}}\right)_{u^*} - \lambda^*\left(\frac{\partial \mathbf{x}^*}{\partial u^*}\right)_{\mathbf{p}}^T \mathbf{x}^{*T}$$

$$\left(\frac{\partial \mathbf{x}^*}{\partial I}\right)_{\mathbf{p}}^T = \lambda^*\left(\frac{\partial \mathbf{x}^*}{\partial u^*}\right)_{\mathbf{p}}^T.$$

Substituting the final equation in the previous equation, we obtain

$$\left(\frac{\partial \mathbf{x}^*}{\partial \mathbf{p}}\right)_I = \left(\frac{\partial \mathbf{x}^*}{\partial \mathbf{p}}\right)_{u^*} - \left(\frac{\partial \mathbf{x}^*}{\partial I}\right)_{\mathbf{p}}^T \mathbf{x}^{*T}.$$

This is a matrix equation and so summarizes a block of equations of the form

$$\left(\frac{\partial x_i^*}{\partial p_j}\right)_I = \left(\frac{\partial x_i^*}{\partial p_j}\right)_{u^*} - \left(\frac{\partial x_i^*}{\partial I}\right)_{\mathbf{p}} x_j^* \quad (i, j = 1, 2, \ldots, n)$$

total effect　　　　　　　　substitution effect　　　income effect

These are the *Slutsky equations*. If the price vector $\mathbf{p}$ is changed, this will lead to a change in the demand $\mathbf{x}^*$. This change will depend on the tastes of the individual and on his income. The Slutsky equation separates these two effects and gives the rate at which demand for the ith commodity x_i increases with respect to the price p_j of the jth commodity in terms of a 'substitution effect' and an 'income effect'. Note that the former describes how demand changes with price, *given* that income is adjusted to keep utility constant.

9

Sums and integrals

This chapter is chiefly about integrating real-valued functions of one real variable. Because an integral is a generalization of the idea of a sum, it is also convenient to include some discussion of summation in this chapter. It is assumed that readers have some previous knowledge of these subjects. In particular, §9.2–§9.20 inclusive consists of an accelerated account of elementary integration theory which a newcomer to the topic would find difficult to assimilate adequately unless they had an unusual aptitude for mathematics. However, experience shows that students are often very rusty on the techniques of integration and it is strongly advised that exercise 9.24 be used as a check on how well the reader recalls the relevant material before moving on to more advanced work. It is also suggested that all readers study §9.5 with some care. The notation for indefinite integrals can be very confusing if imperfectly understood.

The material given in §9.25–§9.30 about infinite series and integrals over infinite ranges will probably be new to most readers of this book. It is not essential for most of what follows and is best omitted if found at all difficult to understand. The same applies to §9.35 on power series. This is not to say that these are unimportant subjects: only that it may be advisable to leave their study until a later date. There remains §9.33 on differentiating integrals. Although this may well be new, it is not difficult and the technique can be very useful in some contexts.

9.1 Sums[□]

We use the notation

$$\sum_{k=1}^{n} a_k = a_1 + a_2 + a_3 + \ldots + a_{n-1} + a_n.$$

Thus, for example,

$$\sum_{k=1}^{5} k^2 = 1^2 + 2^2 + 3^2 + 4^2 + 5^2$$
$$= 1 + 4 + 9 + 16 + 25 = 55.$$

Some familiar but very useful summation formulae are given below:

$$\tfrac{1}{2} n (n + 1) = \sum_{k=1}^{n} k = 1 + 2 + 3 + \ldots + n$$

$$\frac{1-x^{n+1}}{1-x} = \sum_{k=0}^{n} x^k = 1 + x + x^2 + \ldots + x^n \quad (x \neq 1)$$

$$(1+x)^n = \sum_{k=0}^{n} \binom{n}{k} x^k = 1 + nx + \frac{n(n-1)}{2!} x^2 + \ldots + x^n$$

The first of these formulae is the formula for the sum of an 'arithmetic progression'. The formula can be verified in various ways of which the diagrams below indicate one:

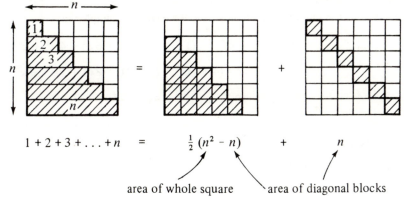

$$1 + 2 + 3 + \ldots + n \quad = \quad \tfrac{1}{2} (n^2 - n) \quad + \quad n$$

area of whole square   area of diagonal blocks

The second formula is that for the sum of a 'geometric progression'. This is most easily verified algebraically as below.

$$(1 - x)(1 + x + x^2 + \ldots + x^n)$$
$$= 1(1 + x + x^2 + \ldots + x^n) - x(1 + x + x^2 + \ldots + x^n)$$
$$= 1 + x + x^2 + \ldots + x^n - x - x^2 - x^3 - \ldots - x^{n+1}$$
$$= 1 - x^{n+1}$$

The third formula is the *binomial theorem*. The number

$$\binom{n}{k} = \frac{n!}{k!(n-k)!}$$

is called a *binomial coefficient*. These are often most easily calculated using 'Pascal's triangle'.

					1							$\binom{0}{0}$

<table>
<tr><td colspan="6" align="center">1</td><td colspan="7" align="center">$\binom{0}{0}$</td></tr>
<tr><td colspan="6" align="center">1 1</td><td colspan="7" align="center">$\binom{1}{0}$ $\binom{1}{1}$</td></tr>
<tr><td colspan="6" align="center">1 2 1</td><td colspan="7" align="center">$\binom{2}{0}$ $\binom{2}{1}$ $\binom{2}{2}$</td></tr>
<tr><td colspan="6" align="center">1 3 3 1</td><td colspan="7" align="center">$\binom{3}{0}$ $\binom{3}{1}$ $\binom{3}{2}$ $\binom{3}{3}$</td></tr>
<tr><td colspan="6" align="center">1 4 6 4 1</td><td colspan="7" align="center">$\binom{4}{0}$ $\binom{4}{1}$ $\binom{4}{2}$ $\binom{4}{3}$ $\binom{4}{4}$</td></tr>
<tr><td colspan="6" align="center">1 5 10 10 5 1</td><td colspan="7" align="center">$\binom{5}{0}$ $\binom{5}{1}$ $\binom{5}{2}$ $\binom{5}{3}$ $\binom{5}{4}$ $\binom{5}{5}$</td></tr>
<tr><td colspan="6" align="center">1 6 15 20 15 6 1</td><td colspan="7" align="center">$\binom{6}{0}$ $\binom{6}{1}$ $\binom{6}{2}$ $\binom{6}{3}$ $\binom{6}{4}$ $\binom{6}{5}$ $\binom{6}{6}$</td></tr>
</table>

The numbers in the left-hand triangle are the values of the binomial co-efficients listed in the right-hand triangle. Note that each number in the Pascal triangle (except for the ones) is the sum of the two numbers above it. This explains why the binomial theorem is true. We have for example that

$$(1+x)^5 = (1+x)(1+x)^4 = (1+x)(1+4x+6x^2+4x^3+x^4)$$
$$= 1+(1+4)x+(4+6)x^2+(6+4)x^3+(4+1)x^4+x^5$$
$$= 1+5x+10x^2+10x^3+5x^4+x^5.$$

9.2 Integrals[□]

If $b \geqslant a$, the notation

$$\int_a^b f(x)dx$$

may be interpreted as meaning the area under the graph of $y = f(x)$ between $x = a$ and $x = b$. It must be understood, however, that area beneath the x axis counts as *negative*. Thus, in the diagram below,

$$\int_a^b f(x)dx = A - B + C.$$

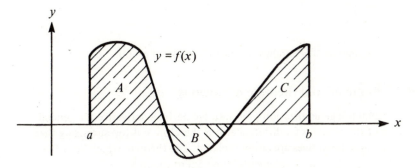

If $b < a$, we define

$$\int_a^b f(x)dx = -\int_b^a f(x)dx.$$

The integral sign is an elongated 'S' from the word 'sum'. It is not hard to see what the connexion is between integrals and sums. In the diagram below, the interval $[a, b]$ has been split into a large number of small sub-intervals. The kth interval is of length δx_k and contains the point x_k. If the function is continuous, it follows that the shaded region below will then be approximately a rectangle of area $f(x_k)\delta x_k$ and hence

$$\sum_k f(x_k)\delta x_k$$

will be approximately equal to the area under the graph between $x = a$ and $x = b$.

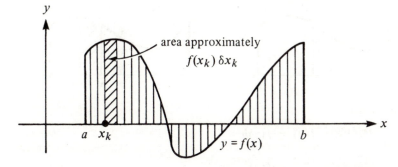

As more and more sub-intervals, each of smaller and smaller length, are taken, the corresponding approximating sums get closer and closer to the value of the integral

$$\int_a^b f(x)dx.$$

Thus an integral is a generalization of the notion of a sum.

9.3   Fundamental theorem of calculus[口]

An integral is a generalization of a sum. Similarly, a derivative is a general-isation of the notion of a difference. It is therefore not too surprising that integrals and derivatives are closely related. The theorem which links these two ideas is called the fundamental theorem of calculus.

A *primitive* (or an *anti-derivative*) of f is a function F such that $F'(x) = f(x)$. The first half of the fundamental theorem of calculus asserts that the integral

$$I(x) = \int_a^x f(t)dt$$

is a primitive for f – i.e.

$$I'(x) = \frac{d}{dx} \int_a^x f(t)dt = f(x)$$

To justify this result, we observe that the shaded area $I(x + h) - I(x)$ in the right-hand diagram below is nearly a rectangle of area $f(x)h$ (provided that f is continuous).

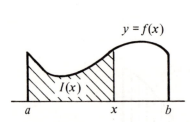
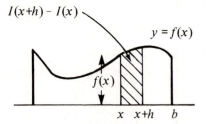

It follows that

$$I(x + h) - I(x) \doteqdot f(x)h$$

$$\frac{I(x + h) - I(x)}{h} \doteqdot f(x).$$

It can be shown that this approximation improves as h gets smaller (provided that f is continuous) and, in the limit,

$$I'(x) = \lim_{h \to 0} \left(\frac{I(x+h) - I(x)}{h} \right) = f(x).$$

It is important to note that I is not the only primitive for f. For example, given any constant ξ, the function J defined by

$$J(x) = \int_{\xi}^{x} f(t)dt$$

is also a primitive for f. Observe that

$$J(x) = \int_{a}^{x} f(t)dt + \int_{\xi}^{a} f(t)dt$$

and thus $J(x) = I(x) + c$ where c is a *constant*. This result holds in general. Any primitive F for the function f has the property that

$$F(x) = \int_{a}^{x} f(t)dt + c$$

for some constant c. In particular

$$F(b) - F(a) = \left\{ \int_{a}^{b} f(t)dt + c \right\} - \left\{ \int_{a}^{a} f(t)dt + c \right\}.$$

This leads to the second half of the fundamental theorem of calculus – i.e.

If F is any primitive for f, then

$$\int_{a}^{b} f(t)dt = \left[F(x) \right]_{a}^{b} = F(b) - F(a)$$

This last result provides a very convenient way of calculating an integral providing that one can guess a primitive for the function f.

Example 9.4.[□]
Calculate

$$\int_{1}^{2} x^2\,dx.$$

We begin by guessing a primitive. The function $F(x) = \frac{1}{3}x^3$ comes to mind. To verify that this is a primitive, we differentiate and obtain

$$F'(x) = \frac{1}{3} \cdot 3x^2 = x^2 = f(x).$$

It follows from the fundamental theorem of calculus that

$$\int_1^2 x^2\,dx = \left[\frac{1}{3}\cdot x^3\right]_1^2 = \frac{1}{3}\cdot 2^3 - \frac{1}{3}1^3 = \frac{7}{3}.$$

Although the method given above is very useful it is not adequate for calculating *all* integrals. Consider, for example, the integral

$$\int_1^2 e^{-x^2/2}\,dx$$

It is true that there *is* a function F such that

$$F'(x) = e^{-x^2/2}$$

but this function F is *not* expressible in terms of simple functions (like the exponential, logarithm or trigonometric functions). However hard one may try to guess F, there is therefore no prospect whatever of any success if attention is confined to these simple functions.

9.5   Notation[□]

The notation

$$\int f(x)dx$$

is widely used to denote a primitive of the function $f(x)$. For this reason a primitive is often called an *indefinite integral.*

Although it has advantages, this notation can sometimes lead to confusion. One cause of confusion is that x is placed as though it were a 'dummy' variable of integration whereas it actually corresponds to an upper limit of an integral – i.e. for some constant ξ,

$$\int f(x)dx = \int_\xi^x f(t)dt.$$

The symbol t on the right-hand side is the 'dummy' variable of integration. Any other symbol (*except*, paradoxically, x) would do equally well for this 'dummy' variable.

The first half of the fundamental theorem of calculus can be written in the form

$$\frac{d}{dx}\left\{\int f(x)dx\right\} = f(x)$$

but one should not suppose that this involves 'differentiating with respect to a dummy variable of integration'. It is simply a shorthand for

$$\frac{d}{dx}\left\{\int_{\xi}^{x}f(t)dt\right\} = f(x).$$

To write

$$\frac{d}{dx}\left\{\int_{a}^{b}f(x)dx\right\} = f(x)$$

would of course be nonsensical. The integral

$$\int_{a}^{b}f(x)dx$$

is a *constant* which could equally well be written as

$$\int_{a}^{b}f(y)dy.$$

Examples 9.6[¤]

(i) $\quad \dfrac{d}{dx}\displaystyle\int_{0}^{x}\dfrac{dt}{1+t^{100}} = \dfrac{1}{1+x^{100}}$

(ii) $\quad \dfrac{d}{dx}\displaystyle\int_{99}^{x}\dfrac{dt}{1+t^{100}} = \dfrac{1}{1+x^{100}}.$

These results are instances of the first half of the fundamental theorem of calculus which may be expressed in the form

$$\frac{d}{dx}\int\frac{dx}{1+x^{100}} = \frac{1}{1+x^{100}}$$

but it would be *nonsense* to write

$$\frac{d}{dt}\int_{0}^{x}\frac{dt}{1+t^{100}} = \frac{1}{1+t^{100}}.$$

Note also that

(iii) $\quad \dfrac{d}{dx}\displaystyle\int_{x}^{1}\dfrac{dt}{1+t^{100}} = \dfrac{d}{dx}\left(-\int_{1}^{x}\dfrac{dt}{1+t^{100}}\right) = -\dfrac{1}{1+x^{100}}.$

A further problem with the 'indefinite integral' notation is that it tempts one to forget the *constant of integration*. The notation

$$\int f(x)dx$$

means a primitive for $f(x)$ – i.e. a function whose derivative is $f(x)$. But $f(x)$ has *many* primitives. The general form for a primitive is

$$\int f(x)dx + c$$

where c is a constant (the constant of integration).

Example 9.7.[□]
From §6.8 we know that

$$\frac{d}{dx}\{\arcsin x\} = \frac{1}{\sqrt{(1-x^2)}} \qquad (-1 < x < 1)$$

and so it follows that we may write

$$\int \frac{dx}{\sqrt{(1-x^2)}} = \arcsin x$$

But it is equally true that

$$\frac{d}{dx}\{-\arccos x\} = \frac{1}{\sqrt{(1-x^2)}} \qquad (-1 < x < 1)$$

and so we may also write

$$\int \frac{dx}{\sqrt{(1-x^2)}} = -\arccos x$$

But we should *not* fall into the trap of forgetting the constant of integration and deducing that $\arcsin x = -\arccos x$. In fact

$$\arcsin x = -\arccos x + \frac{\pi}{2}.$$

9.8   Standard integrals[□]

From the formulae for derivatives given in §2.3 and §6.8 we obtain the following list of 'standard integrals' – i.e. commonly occurring primitives. These *must* be known by heart. More extensive lists will be found in the backs of books of tables but the standard integrals given here, together with the integration techniques given later, will be found adequate for most purposes:

$$\int x^{\alpha}\, dx = \frac{1}{\alpha+1} x^{\alpha+1} \qquad (\alpha \neq -1)$$

$$\int \frac{dx}{ax+b} = \frac{1}{a} \log\,(ax+b) \qquad (a \neq 0)$$

$$\int e^{ax}\, dx = \frac{1}{a} e^{ax} \qquad (a \neq 0)$$

$$\int \cos(ax)\, dx = \frac{1}{a} \sin(ax) \qquad (a \neq 0)$$

$$\int \sin(ax)\, dx = -\frac{1}{a} \cos(ax) \qquad (a \neq 0)$$

$$\int \frac{dx}{a^2 + x^2} = \frac{1}{a} \arctan\!\left(\frac{x}{a}\right) \qquad (a \neq 0)$$

$$\int \frac{dx}{\sqrt{(a^2 - x^2)}} = \arcsin\!\left(\frac{x}{a}\right) \qquad (a \neq 0)$$

Each of these results should be checked by differentiating the right-hand side to confirm that the integrand is obtained. For example,

$$\frac{d}{dx}\left\{\frac{1}{a} \arctan\!\left(\frac{x}{a}\right)\right\} = \frac{1}{a} \cdot \frac{1}{1 + (x/a)^2} \cdot \frac{1}{a} = \frac{1}{a^2 + x^2}.$$

Remember that the functions on the right-hand side are only *one* of the primitives for the integrands on the left-hand side. The general form of the primitive is obtained by adding a constant of integration.

9.9 Partial fractions[□]

A *polynomial* is an expression of the form

$$P(x) = a_n x^n + a_{n-1} x^{n-1} + \ldots + a_1 x + a_0.$$

If $a_n \neq 0$, the polynomial is said to be of *degree* n. A polynomial of degree n has n *roots* $\xi_1, \xi_2, \ldots, \xi_n$. This means that $P(x)$ may be rewritten in the form

$$P(x) = a_n(x - \xi_1)(x - \xi_2) \ldots (x - \xi_n).$$

In general, however, the roots may be complex numbers (see chapter 12).

A *rational function R* is the quotient of two polynomials – i.e.

$$R(x) = \frac{P(x)}{Q(x)}$$

where P and Q are polynomials. If the degree of P is less than the degree of Q and the roots $\xi_1, \xi_2, \ldots, \xi_n$ of Q are all distinct, then R admits a *partial fraction* expansion of the form

$$\frac{P(x)}{Q(x)} = \frac{A_1}{x - \xi_1} + \frac{A_2}{x - \xi_2} + \ldots + \frac{A_n}{x - \xi_n}.$$

The number A_k may be calculated from the formula

$$A_k = \lim_{x \to \xi_k} (x - \xi_k)\frac{P(x)}{Q(x)}.$$

The partial fraction expansion may be used to evaluate $\int R(x)dx$. Thus,

$$\int \frac{P(x)}{Q(x)}dx = A_1 \int \frac{dx}{x - \xi_1} + A_2 \int \frac{dx}{x - \xi_2} + \ldots + A_n \int \frac{dx}{x - \xi_n}$$

$$= A_1 \log(x - \xi_1) + A_2 \log(x - \xi_2) + \ldots + A_n \log(x - \xi_n).$$

Example 9.10.$^{\square}$

We have that

$$\frac{x - 1}{x^2 - 5x + 6} = \frac{(x - 1)}{(x - 2)(x - 3)} = \frac{A}{x - 2} + \frac{B}{x - 3}$$

where

$$A = \lim_{x \to 2} (x - 2)\frac{(x - 1)}{(x - 2)(x - 3)} = \lim_{x \to 2} \frac{(x - 1)}{(x - 3)} = -1$$

$$B = \lim_{x \to 3} (x - 3)\frac{(x - 1)}{(x - 2)(x - 3)} = \lim_{x \to 3} \frac{(x - 1)}{(x - 2)} = 2.$$

Thus

$$\int \frac{x - 1}{x^2 - 5x + 6}dx = \int \left(\frac{-1}{x - 2} + \frac{2}{x - 3}\right) dx$$

$$= -\log(x - 2) + 2\log(x - 3)$$

$$= \log(x - 2)^{-1} + \log(x - 3)^2$$

$$= \log\frac{(x - 3)^2}{(x - 2)}.$$

In particular,

$$\int_4^5 \frac{x - 1}{x^2 - 5x + 6}dx = \left[\log\frac{(x - 3)^2}{(x - 2)}\right]_4^5$$

$$= \log\frac{4}{3} - \log\frac{1}{2}$$

$$= \log\frac{4}{3} \cdot 2 = \log\frac{8}{3}.$$

If the degree of P is less than the degree of Q but the roots of Q are *not* all distinct, then the partial fraction expansion given above does *not* apply. If ξ_k is a root of Q with multiplicity m (i.e. ξ_k is repeated m times), then the term

$$\frac{A_k}{x - \xi_k}$$

in the partial fraction expansion must be replaced by a term of the form

$$\frac{B_1}{x - \xi_k} + \frac{B_2}{(x - \xi_k)^2} + \ldots + \frac{B_m}{(x - \xi_k)^m}.$$

Fortunately, in practice m is seldom greater than 2.

Example 9.11.

We have that

$$\frac{x + 1}{(x - 2)(x - 3)^2} = \frac{A}{x - 2} + \frac{B}{(x - 3)} + \frac{C}{(x - 3)^2}.$$

The coefficient A may be calculated from the formula

$$A = \lim_{x \to 2} (x - 2)\frac{(x + 1)}{(x - 2)(x - 3)^2} = \lim_{x \to 2} \frac{(x + 1)}{(x - 3)^2} = 3.$$

Similarly

$$C = \lim_{x \to 3} (x - 3)^2 \frac{(x + 1)}{(x - 2)(x - 3)^2} = \lim_{x \to 3} \frac{(x + 1)}{(x - 2)} = 4.$$

Then

$$\frac{B}{x - 3} = \frac{x + 1}{(x - 2)(x - 3)^2} - \frac{3}{x - 2} - \frac{4}{(x - 3)^2}.$$

Taking $x = 4$ we obtain that $B = -3$ (or, more easily, multiply through by $(x - 3)$ and let $x \to +\infty$).

It follows that

$$\int \frac{x + 1}{(x - 2)(x - 3)^2} dx = \int \left(\frac{3}{x - 2} - \frac{3}{(x - 3)} + \frac{4}{(x - 3)^2}\right) dx$$

$$= 3 \log (x - 2) - 3 \log (x - 3) - \frac{4}{(x - 3)}$$

$$= \log \left(\frac{x - 2}{x - 3}\right)^3 - \frac{4}{x - 3}.$$

In particular,

$$\int_4^5 \frac{x+1}{(x-2)(x-3)^2}\,dx = \left[\log\left(\frac{x-2}{x-3}\right)^3 - \frac{4}{x-3}\right]_4^5$$

$$= \left(\log\left(\frac{3}{2}\right)^3 - \frac{4}{2}\right) - \left(\log\left(\frac{2}{1}\right)^3 - \frac{4}{1}\right)$$

$$= \log\left(\frac{27}{64}\right) + 2.$$

If the degree of P is *not* less than the degree of Q, then one begins by writing

$$\frac{P(x)}{Q(x)} = S(x) + \frac{P_1(x)}{Q_1(x)}$$

where $S(x)$ is a polynomial and the degree of P_1 *is* less than the degree of Q_1.

Example 9.12.
We have that

$$\frac{x+1}{x+2} = \frac{x+2-1}{x+2} = 1 - \frac{1}{x+2}.$$

Hence

$$\int \frac{x+1}{x+2}\,dx = \int\left(1 - \frac{1}{x+2}\right)dx$$

$$= x - \log(x+2).$$

In particular,

$$\int_4^5 \frac{x+1}{x+2}\,dx = \frac{5}{4}[x - \log(x+2)]$$

$$= (5 - \log 7) - (4 - \log 6)$$

$$= 1 + \log\left(\frac{6}{7}\right).$$

9.13 Completing the square

If $a \neq 0$, we have that

$$ax^2 + bx + c = \frac{1}{4a}(4a^2x^2 + 4abx + 4ac)$$

$$= \frac{1}{4a}\{(2ax + b)^2 - b^2 + 4ac\}$$

This algebraic manipulation is called 'completing the square'. Its most familiar application is in finding the roots of quadratic equations. If $ax^2 + bx + c = 0$, then

$$(2ax + b)^2 - b^2 + 4ac = 0.$$

Hence

$$2ax + b = \pm\sqrt{(b^2 - 4ac)}$$

$$x = \frac{-b \pm \sqrt{(b^2 - 4ac)}}{2a}.$$

The example below gives another application.

Example 9.14.[□]
One can use partial fractions to calculate

$$\int \frac{dx}{x^2 + x + 1}$$

but then complex numbers intrude. An alternative is to complete the square in the denominator. Then

$$\int \frac{dx}{x^2 + x + 1} = \int \frac{dx}{(x + \tfrac{1}{2})^2 + \tfrac{3}{4}}$$

$$= (\tfrac{3}{4})^{-1/2} \arctan\left\{(\tfrac{3}{4})^{-1/2}\left(x + \frac{1}{2}\right)\right\}.$$

In particular,

$$-\int_0^1 \frac{dx}{x^2 + x + 1} = \left[(\tfrac{3}{4})^{-1/2} \arctan\left\{(\tfrac{3}{4})^{-1/2}\left(x + \frac{1}{2}\right)\right\}\right]_0^1$$

$$= \frac{2}{\sqrt{3}} \arctan\left(\frac{2}{\sqrt{3}} \cdot \frac{3}{2}\right) - \frac{2}{\sqrt{3}} \arctan\left(\frac{2}{\sqrt{3}} \cdot \frac{1}{2}\right)$$

$$= \frac{2}{\sqrt{3}}\left(\arctan\sqrt{3} - \arctan\frac{1}{\sqrt{3}}\right)$$

$$= \frac{2}{\sqrt{3}}\left(\frac{\pi}{3} - \frac{\pi}{6}\right) = \frac{\pi}{3\sqrt{3}}$$

because $\tan(\pi/6) = 1/\sqrt{3}$ and $\tan(\pi/3) = \sqrt{3}$.

9.15 Change of variable[□]

Suppose that $x = \phi(y)$. Then

$$dx = \phi'(y)dy$$

and so the formula

$$\int_a^b f(x)\,dx = \int_A^B f(\phi(y))\,\phi'(y)\,dy$$

where

$$b = \phi(B) \text{ and } a = \phi(A)$$

for changing the variable in an integral is very natural.

To justify the formula we observe that, if F is a primitive for f, then the left-hand side is equal to

$$\int_a^b f(x)\,dx = {}_a^b[F(x)] = F(b) - F(a) = {}_A^B[F(\phi(x))]$$

But the formula for differentiating a 'function of a function' (§2.3) implies that

$$\frac{d}{dy}\{F(\phi(y))\} = F'(\phi(y))\phi'(y) = f(\phi(y))\phi'(y)$$

and so we also have that

$$\int_a^b f(\phi(y))\phi'(y)\,dy = {}_A^B[F(\phi(y))].$$

Example 9.16.[H]
Consider the integral

$$\int_1^4 \frac{\cos\sqrt{t}}{\sqrt{t}}\,dt.$$

A natural change of variable is $u = \sqrt{t}$. Then

$$du = \tfrac{1}{2}t^{-1/2}\,dt.$$

Also $u = 2$ when $t = 4$ and $u = 1$ when $t = 1$. Thus

$$\int_1^4 \frac{\cos\sqrt{t}}{\sqrt{t}}\,dt = \int_1^2 \frac{\cos\sqrt{t}}{\sqrt{t}}\,2\sqrt{t}\,du$$

$$= \int_1^2 2\cos u\,du$$

$$= {}_1^2[2\sin u] = 2(\sin 2 - \sin 1).$$

Example 9.17.

Integrals involving $\sqrt{(1-t^2)}$ can often be reduced to something manageable by the change of variable $t = \sin\theta$. For example,

$$\int_0^1 \sqrt{(1-t^2)}\,dt = \int_0^{\pi/2} \sqrt{(1-\sin^2\theta)} \cos\theta\,d\theta$$

$$= \int_0^{\pi/2} \cos^2\theta\,d\theta$$

$$= \int_0^{\pi/2} \frac{\cos 2\theta + 1}{2}\,d\theta$$

$$= {}^{\pi/2}\left[\frac{\sin 2\theta}{4} + \frac{1}{2}\theta\right]_0$$

$$= \frac{\pi}{4}.$$

(Similarly the change of variable $t = \tan\theta$ is sometimes useful with integrals involving $\sqrt{(1+t^2)}$ because $1 + \tan^2\theta = \sec^2\theta$.)

Example 9.18.

Often integrals occur in which the integrand looks like the right-hand side of the formula for changing the variable of an integral. Such integrals can be evaluated immediately provided one knows a primitive F for the function f because

$$\int_a^b f(\phi(y))\phi'(y)\,dy = {}^b_a[F(\phi(y))].$$

A particularly common example is the case with $f(x) = 1/x$. We then have that

$$\int_a^b \frac{\phi'(y)}{\phi(y)}\,dy = {}^b_a[\log\phi(y)] = \log\left(\frac{\phi(b)}{\phi(a)}\right).$$

Example 9.19.

The integral

$$\int_0^1 \frac{2x+3}{x^2+3x+2}\,dx$$

can be evaluated using partial fractions. A better method is to observe that, if $\phi(x) = x^2 + 3x + 2$, then $\phi'(x) = 2x + 3$. Hence

$$\int_0^1 \frac{2x+3}{x^2+3x+2} dx = \tfrac{1}{0}[\log(x^2+3x+2)]$$

$$= \log\frac{6}{2} = \log 3.$$

The method is essentially the same as introducing the change of variable $y = x^2 + 3x + 2$. Then $dy = (2x+3)dx$ and so

$$\int_0^1 \frac{2x+3}{x^2+3x+2} dx = \int_2^6 \frac{dy}{y} = \tfrac{6}{2}[\log y] = \log\frac{6}{2}.$$

9.20   Integration by parts[¤]

If U is any primitive for u, the formula for differentiating a product (§2.3) gives

$$\frac{d}{dx}\{U(x)v(x)\} = U'(x)v(x) + U(x)v'(x)$$

$$= u(x)v(x) + U(x)v'(x)$$

and hence

$$\int_a^b u(x)v(x)\,dx = \tfrac{b}{a}[U(x)v(x)] - \int_a^b U(x)\,v'(x)\,dx$$

which is the formula for '*integrating by parts*'.

Example 9.21.[¤]
We have that

$$\int_1^2 x\cos x\,dx = [x\sin x] - \int_1^2 1\sin x\,dx$$

$$= \tfrac{2}{1}[x\sin x] - \tfrac{2}{1}[-\cos x]$$

$$= (2\sin 2 + \cos 2) - (1\sin 1 + \cos 1).$$

Example 9.22.[¤]
We have that

$$\int_a^b \log x\,dx = \int_a^b 1\cdot\log x\,dx$$

$$= \,_a^b[x \log x] - \int_a^b x \cdot \frac{1}{x} dx$$

$$= \,_a^b[x \log x - x]$$

Note that $x \log x - x$ is therefore a primitive for $\log x$.

Example 9.23.[$\boxdot$]
Let

$$I_{mn} = \int_0^1 x^m(1-x)^n dx.$$

Then

$$I_{mn} = \left[\frac{x^{m+1}}{(m+1)}(1-x)^n\right]_0^1 - \int_0^1 \frac{x^{m+1}}{m+1}\{-n(1-x)^{n-1}\} dx.$$

Thus, provided $n \geqslant 1$,

$$I_{mn} = \frac{n}{m+1}I_{m+1, n-1}$$

$$= \frac{n(n-1)}{(m+1)(m+2)}I_{m+2, n-2}$$

$$\ldots$$

$$= \frac{n(n-1)\ldots 1}{(m+1)(m+2)\ldots(m+n)}I_{m+n, 0}$$

$$= \frac{m!n!}{(m+n)!}\int_0^1 x^{m+n} dx = \frac{m!n!}{(m+n+1)!}$$

and so

$$I_{mn} = \binom{m+n+1}{m}^{-1}.$$

Exercises 9.24.[$\boxdot$]

1. Use Pascal's triangle to obtain an expansion for $(1+x)^8$.
2.* Prove that, for $k = 1, 2, 3, \ldots, n+1$,

$$\frac{(n+1)!}{k!(n+1-k)!} = \frac{n!}{k!(n-k)!} + \frac{n!}{(k-1)!(n-k+1)!}$$

and explain what this has to do with Pascal's triangle.

3.† Explain why

$$\int_0^1 f(x)dx = \lim_{n \to \infty} \sum_{k=0}^{n-1} f\left(\frac{k}{n}\right)\frac{1}{n}$$

provided that f is continuous. Hence evaluate

$$\int_0^1 e^x dx$$

without using the fundamental theorem of calculus.

4.*† Justify the formula

$$\int_0^1 (ax + b)dx = \lim_{n \to \infty} \sum_{k=0}^{n-1} \left(\frac{ak}{n} + b\right)\frac{1}{n}$$

and hence evaluate the integral without using the fundamental theorem of calculus.

5. Evaluate

(i) $\displaystyle\int_{-1/2}^{1/2} \frac{dy}{\sqrt{(1 + y^2)}}$ (ii) $\displaystyle\int_{-1/2}^{1/2} \frac{dy}{1 - y^2}$

using exercises 6.9 (3) and (4).

6.* Show that

(i) $\displaystyle\frac{d}{dx}\left\{\log\left(\tan\frac{1}{2}x\right)\right\} = \text{cosec } x$ (ii) $\displaystyle\frac{d}{dx}\{e^{-x^2/2}\} = -xe^{-x^2/2}$

and hence evaluate

(i) $\displaystyle\int_{\pi/4}^{3\pi/4} \text{cosec } x \, dx$ (ii) $\displaystyle\int_0^x xe^{-x^2/2} \, dx.$

7. Since the integrand is non-negative, the integral

$$\int_{\pi/4}^{3\pi/4} \sec^2 x \, dx$$

should also be non-negative. But

$$\int_{\pi/4}^{3\pi/4} \sec^2 x \, dx = {}^{3\pi/4}_{\pi/4}[\tan x] = -1 - 1 = -2.$$

Draw graphs of $y = \sec^2 x$ and $y = \tan x$ in the relevant range and hence find the error in this reasoning.

8.* Evaluate

(i) $\dfrac{d}{dx}\displaystyle\int_0^x e^{-t^2}\,dt$ (ii) $\dfrac{d}{dx}\displaystyle\int_1^x e^{-t^2}\,dt$ (iii) $\dfrac{d}{dx}\displaystyle\int_x^2 e^{-t^2}\,dt.$

9. Use partial fractions to evaluate the integral

$$\int_2^3 \frac{x^3}{x^2+x-2}\,dx.$$

10.* Use partial fractions to evaluate the integrals

(i) $\displaystyle\int_2^3 \frac{dx}{x^2+x-2}$ (ii) $\displaystyle\int_2^3 \frac{dx}{(1-x^2)^2}$

(iii) $\displaystyle\int_2^3 \frac{x\,dx}{x^2+x-2}$ (iv) $\displaystyle\int_2^3 \frac{x^2\,dx}{x^2+x-2}.$

11. Complete the square in the denominator and hence evaluate

$$\int_0^1 \frac{dx}{\sqrt{(7-2x-x^2)}}.$$

12.* Complete the square in the denominator and hence evaluate

$$\int_0^1 \frac{dx}{x^2+2x+7}.$$

13. Evaluate the integral

$$\int_0^{\pi/3} \cos^3\theta\,d\theta$$

using the formula $\cos 3\theta = 4\cos^3\theta - 3\cos\theta$.

14.* Evaluate the integrals

(i) $\displaystyle\int_0^{\pi/2} \cos^2\theta\,d\theta$ (ii) $\displaystyle\int_0^{\pi/2} \sin^2\theta\,d\theta$

using the formulae $\cos 2\theta = 2\cos^2\theta - 1 = 1 - 2\sin^2\theta$.

15. Make an appropriate change of variable in the integral

$$\int_1^2 \frac{\sqrt{(1+\sqrt{y})}}{\sqrt{y}}\,dy$$

and hence evaluate the integral.

16.* Make an appropriate change of variable in the integrals

(i) $\displaystyle\int_{-1}^{\sqrt{2}} \frac{x\,dx}{(x^2+1)^3}$ (ii) $\displaystyle\int_{1}^{2} \frac{dt}{\sqrt{(3+5t)}}$ (iii) $\displaystyle\int_{0}^{1} x\sqrt{(1-x^2)}\,dx$

and hence evaluate them. [Hint: For (iii), see example 9.17.]

17. Express $\sin\theta$ and $\cos\theta$ in terms of $t = \tan\frac{1}{2}\theta$. Evaluate the integrals below by means of this change of variable.

(i) $\displaystyle\int_{\pi/4}^{\pi/2} \frac{d\theta}{\sin\theta}$ (ii) $\displaystyle\int_{0}^{\pi/4} \frac{d\theta}{\cos\theta}$ (iii) $\displaystyle\int_{\pi/6}^{\pi/3} \frac{d\theta}{\sin\theta + \cos\theta}$.

18.* Evaluate the integral

$$\int_{-1/2}^{1/2} \frac{dt}{\sqrt{(1+t^2)}}$$

using the change of variable $t = \tan\theta$. Also evaluate

$$\int_{-1/2}^{1/2} \frac{t\,dt}{\sqrt{(1+t^2)}}.$$

19. Evaluate the integral

$$\int_{0}^{1} \frac{3x^2 + 2x + 1}{x^3 + x^2 + x + 1}\,dx.$$

20.* Evaluate the integral

$$\int_{2}^{3} \frac{x\,dx}{(1-x^2)^2}$$

(i) by using partial fractions (ii) by writing $\phi(x) = 1 - x^2$ and observing that $x = -\frac{1}{2}\phi'(x)$.

21. Integrate by parts and hence evaluate

(i) $\displaystyle\int_{0}^{\pi} x^3 \cos x\,dx$ (ii) $\displaystyle\int_{0}^{X} x^2 e^{-x}\,dx$.

22.* Integrate

$$I_m = \int_{0}^{\pi} \sin^m x\,dx = \int_{0}^{\pi} (\sin x)(\sin x)^{m-1}\,dx$$

by parts and hence show that

$$I_m = \left(\frac{m-1}{m}\right) I_{m-2} \qquad (m > 2).$$

Evaluate I_4 and I_5.

9.25   Infinite sums and integrals†

In this section we discuss very briefly the idea of an 'infinite sum' or *series*

$$\sum_{n=1}^{\infty} a_n.$$

Similar remarks apply to the idea of an 'infinite integral'

$$\int_{\xi}^{\infty} f(x)dx.$$

The series is said to *converge* if and only if there is a number s such that

$$\sum_{n=1}^{N} a_n \to s \quad \text{as} \quad N \to \infty$$

in which case we write

$$\sum_{n=1}^{\infty} a_n = s.$$

If no such number s exists, then the series is said to *diverge*.

Similarly, the 'infinite integral' is said to *converge* if and only if there is a number I such that

$$\int_{\xi}^{X} f(x)dx \to I \quad \text{as} \quad X \to \infty$$

in which case we write

$$\int_{\xi}^{\infty} f(x)dx = I.$$

If no such number I exists, the 'infinite integral' is said to *diverge*.

Examples 9.26†
(i) Suppose that $|x| < 1$. Then $x^N \to 0$ as $N \to \infty$ and so

$$\sum_{n=0}^{N} x^n = \frac{1 - x^{N+1}}{1 - x} \to \frac{1}{1 - x} \quad \text{as} \quad N \to \infty.$$

Hence

$$\sum_{n=0}^{\infty} x^n = \frac{1}{1 - x} \qquad (|x| < 1).$$

(ii) suppose that $|x| > 1$. Then $x^N \to \infty$ as $N \to \infty$ if $x > 1$ and x^N 'oscillates' if $x < -1$. In neither case does there exist a number s such that

$$\frac{1 - x^{N+1}}{1 - x} \to s \quad \text{as} \quad N \to \infty.$$

(The symbol ∞ does not, of course, stand for a *number*.) It follows that, when $|x| > 1$, the series

$$\sum_{n=0}^{\infty} x^n$$

diverges. The series also diverges when $x = 1$ or $x = -1$. (Why?)

Examples 9.27[†]

(i) Suppose that $\alpha > 1$. Then $X^{1-\alpha} \to 0$ as $X \to \infty$ and so

$$\int_1^X \frac{dx}{x^\alpha} = \left[\frac{1}{1 - \alpha} x^{1-\alpha} \right]_1^X$$

$$= \frac{1}{\alpha - 1} \{1 - X^{1-\alpha}\}$$

$$\to \frac{1}{\alpha - 1} \quad \text{as} \quad X \to \infty.$$

Hence

$$\int_1^\infty \frac{dx}{x^\alpha} = \frac{1}{\alpha - 1} \qquad (\alpha > 1).$$

(ii) Suppose $\alpha < 1$. Then $X^{1-\alpha} \to \infty$ as $X \to \infty$ and so the 'infinite integral'

$$\int_1^\infty \frac{dx}{x^\alpha}$$

diverges when $\alpha < 1$.

(iii) Suppose $\alpha = 1$. We then have to consider

$$\int_1^X \frac{dx}{x} = \Big[\log x \Big]_1^X = \log X \to \infty \quad \text{as} \quad X \to \infty.$$

Again the 'infinite integral' *diverges*.

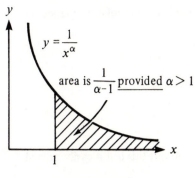 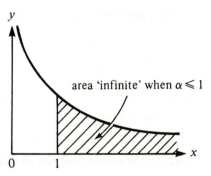

Example 9.28.[†]
The series

$$\sum_{n=1}^{\infty} \frac{1}{n^{\alpha}}$$

is represented by the shaded area in the diagram.

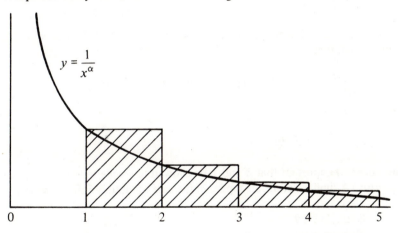

It is therefore not hard to see why the series is convergent if and only if the 'infinite integral'

$$\int_{1}^{\infty} \frac{dx}{x^{\alpha}}$$

is convergent. Referring to the previous example, it follows that

$$\sum_{n=1}^{\infty} \frac{1}{n^{\alpha}}$$

converges when $\alpha > 1$ and diverges when $\alpha \geqslant 1$. In particular, the series

$$\sum_{n=1}^{\infty} \frac{1}{n}$$

diverges while the series

$$\sum_{n=1}^{\infty} \frac{1}{n^2}$$

converges. (The sum of the latter series is $\pi^2/6$ but this is not easily proved.)

One sometimes has to deal with 'doubly-infinite series' of the type

$$\sum_{n=-\infty}^{\infty} a_n$$

or with 'doubly-infinite integrals' of the form

$$\int_{-\infty}^{\infty} f(x)dx.$$

The 'doubly-infinite integral' is said to converge if and only if there exists a number I such that

$$\int_{-X}^{Y} f(x)dx \to I \quad \text{as} \quad X \to \infty \quad \text{and} \quad Y \to \infty$$

in which case we write

$$I = \int_{-\infty}^{\infty} f(x)dx.$$

(Note that it is important that X and Y be allowed to recede to ∞ *independently*.) A similar definition applies in the case of 'doubly-infinite series'.

Example 9.29.[†]
We have that

$$\int_{-\infty}^{\infty} \frac{dx}{1+x^2} = \pi$$

because

$$\int_{-X}^{Y} \frac{dx}{1+x^2} = [\arctan x]_{-X}^{Y}$$

$$= \arctan Y - \arctan(-X)$$

$$\to \frac{\pi}{2} - \left(-\frac{\pi}{2}\right) \quad \text{as} \quad X \to \infty \quad \text{and} \quad Y \to \infty$$

(see the graph of $x = \arctan y$ given in §6.8).

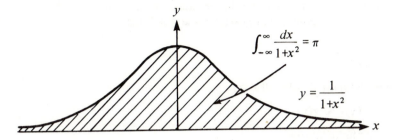

$$\int_{-\infty}^{\infty} \frac{dx}{1+x^2} = \pi$$

$$y = \frac{1}{1+x^2}$$

Note that

$$\int_{-\infty}^{\infty} \frac{x}{1+x^2}\, dx$$

diverges. We have that

$$\int_{-X}^{Y} \frac{x}{1+x^2}\, dx = \left[\frac{1}{2}\log(1+x^2)\right]_{-X}^{Y}$$

$$= \frac{1}{2}\{\log(1+Y^2) - \log(1+X^2)\}.$$

But this tends to ∞ as $Y \to \infty$ and tends to $-\infty$ as $X \to \infty$.

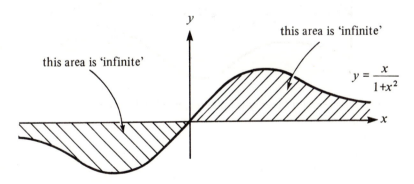

this area is 'infinite'

this area is 'infinite'

$$y = \frac{x}{1+x^2}$$

It is true that

$$\lim_{Z \to \infty} \int_{-Z}^{Z} \frac{x}{1+x^2}\, dx = 0$$

But this is not a significant fact since the upper and lower limits of integration need to recede to ∞ and $-\infty$ *independently* for the 'doubly-infinite integral' to be defined.

9.30   Dominated convergence†

We begin with the *comparison test.* In the case of integrals this takes the following form: suppose that $g(x) \geqslant 0$ for $x \geqslant \xi$ and that

$$\int_{\xi}^{\infty} g(x)dx$$

converges. Suppose that for all $x \geqslant \xi$

$$|f(x)| \leqslant g(x).$$

Then the 'infinite integral'

$$\int_{\xi}^{\infty} f(x)dx$$

converges also.

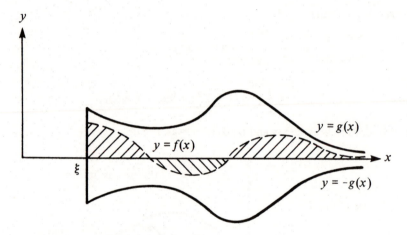

Similar results hold for 'doubly-infinite integrals' and for series.

Example 9.31.†
We have that

$$\int_{1}^{X} e^{-x}dx = \left[-e^{-x}\right]_{1}^{X} = e^{-1} - e^{-X}$$

$$\to e^{-1} \quad \text{as} \quad X \to \infty.$$

It follows that the integral

$$\int_{1}^{\infty} e^{-x}dx$$

converges. But $x^2 \geqslant x$ for $x \geqslant 1$. Hence $-x^2 \leqslant -x$ for $x \geqslant 1$ and therefore

$$e^{-x^2} \leqslant e^{-x} \qquad (x \geqslant 1).$$

From the comparison test, it follows that the integral

$$\int_1^\infty e^{-x^2} dx$$

converges.

The convergence of

$$\int_{-\infty}^\infty e^{-x^2} dx$$

can be established in a similar way by splitting the range of integration into the three parts $[-\infty, -1]$, $[-1, 1]$ and $[1, \infty]$ and examining the integral over each of these intervals separately.

Suppose again that $g(x) \geqslant 0$ for $x \geqslant \xi$ and that

$$\int_\xi^\infty g(x) dx$$

converges. Suppose also that

(1) $|f(x, y)| \leqslant g(x)$

for all $x \geqslant \xi$ and all y in some interval I for which η is an interior point. It follows from the comparison test that the integral

(2) $\displaystyle\int_\xi^\infty f(x, y) dx$

converges for each y in the interval I. In view of the inequality (1), we say that the convergence is *dominated*.

It is by no means guaranteed that

$$\lim_{y \to \eta} \int_\xi^\infty f(x, y) dx = \int_\xi^\infty \lim_{y \to \eta} f(x, y) dx$$

even for those cases for which

$$\lim_{y \to \eta} f(x, y)$$

exists for each $x \geqslant \xi$ (see exercise 9.37 (5)). However, if the convergence of (2) is *dominated*, then the reversal of the limit and integration signs *is* legitimate.

A similar problem which is more directly relevant to our concerns in this book is the question of the reversal of differentiation and integration signs. Under what circumstances is it true that

$$\frac{d}{dy} \int_{\xi}^{\infty} f(x,y)dx = \int_{\xi}^{\infty} \frac{\partial f}{\partial y}(x,y)dx?$$

Here the relevant criterion is that the convergence of

(3) $$\int_{\xi}^{\infty} \frac{\partial f}{\partial y}(x,y)dx$$

be dominated.

Note that similar results hold for integrals over *any* range of integration (although when the limits of integration are finite, proving domination in (3) is usually too trivial to bother with). Similar results also hold for series.

Example 9.32.[†]
Consider

$$I(y) = \int_{0}^{\infty} \cos{(xy)}e^{-x^2/2} \, dx.$$

We have that

$$I'(y) = \frac{d}{dy} \int_{0}^{\infty} \cos{(xy)}e^{-x^2/2} \, dx$$

(4) $$= \int_{0}^{\infty} \frac{\partial}{\partial y} \cos{(xy)}e^{-x^2/2} \, dx$$

$$= \int_{0}^{\infty} -x \sin{(xy)}e^{-x^2/2} \, dx.$$

Integrating by parts, we obtain that

$$I'(y) = [e^{-x^2/2} \sin{(xy)}]_{0}^{\infty} - \int_{0}^{\infty} e^{-x^2/2} \, (y \cos{xy})dx$$

$$= -yI(y).$$

Thus

$$\frac{I'(y)}{I(y)} = -y$$

and so

$$\int_{0}^{Y} \frac{I'(y)}{I(y)}dy = -\frac{1}{2}Y^2$$

$$\log I(Y) - \log I(0) = -\tfrac{1}{2}Y^2$$

$$I(Y) = I(0)e^{-Y^2/2}.$$

In the next chapter we shall see that $I(0) = \sqrt{(\pi/2)}$ and so

$$I(Y) = \sqrt{\left(\frac{\pi}{2}\right)} e^{-Y^2/2}.$$

Strictly speaking, we should justify the step leading to (4) although few applied mathematicians would give much attention to this point. One needs to show that the convergence of

$$\int_0^\infty -x \sin (xy) e^{-x^2/2} \, dx$$

is dominated for all y. This follows from the fact that

$$|-x \sin (xy) e^{-x^2/2}| \leqslant x e^{-x^2/2} \qquad (x \geqslant 0).$$

9.33 Differentiating integrals

Suppose that

$$I(y) = \int_a^y f(x) dx.$$

From the fundamental theorem of calculus,

$$\frac{dI}{dy} = \frac{d}{dy} \int_a^y f(x) dx = f(y).$$

If $y = \phi(z)$, we have that

$$\frac{dI}{dz} = \frac{dI}{dy} \frac{dy}{dz} = f(y)\phi'(z) = f(\phi(z))\phi'(z).$$

Hence

$$\frac{d}{dz} \int_a^{\phi(z)} f(x) dx = f(\phi(z))\phi'(z).$$

On a somewhat different tack, we have from §9.30 that, under fairly general conditions,

$$\frac{d}{dy} \int_a^b f(x, y) dx = \int_a^b \frac{\partial f}{\partial y}(x, y) dx.$$

These results may be combined to obtain the general formula

$$\boxed{\frac{d}{dy} \int_a^{\phi(y)} f(x, y) \, dx = f(\phi(y), y) \, \phi'(y) + \int_a^{\phi(y)} \frac{\partial f}{\partial y}(x, y) \, dx}$$

To prove this we write

$$I(u, v) = \int_a^u f(x, v)dx.$$

If $u = \phi(y)$ and $v = y$, it follows from the *chain rule* that

$$\frac{dI}{dy} = \frac{\partial I}{\partial u}\frac{du}{dy} + \frac{\partial I}{\partial v}\frac{dv}{dy}$$

$$= f(u, v)\phi'(y) + \int_a^u \frac{\partial f}{\partial v}(x, v)dx$$

$$= f(\phi(y), y)\phi'(y) + \int_a^{\phi(y)} \frac{\partial f}{\partial y}(x, y)dx.$$

Example 9.34

$$\frac{d}{dy}\int_0^{y^2}(x + y)^2\,dx = (y^2 + y)^2\,2y + \int_0^{y^2}\frac{\partial}{\partial y}(x + y)^2\,dx$$

$$= (y^2 + y)^2\,2y + \int_0^{y^2} 2(x + y)dx.$$

9.35   Power series†

A *power series* about the point ξ is an expression of the form

$$\sum_{n=0}^{\infty} a_n(x - \xi)^n.$$

We met examples of power series with $\xi = 0$ in §4.22 when considering the Taylor series expansions of some elementary functions.

The set of values of x for which a power series converges is always an interval with midpoint ξ. This interval I is called the *interval of convergence* of the power series. This interval may be the set $\mathbb{R}$ of *all* real numbers or it may consist of *just* the single point ξ. Between these two extremes is the case when the endpoints are of the form $\xi - R$ and $\xi + R$ where $R > 0$. Note that the power series may *or may not* converge at these endpoints.

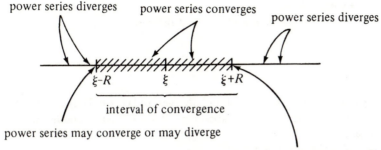

power series diverges   power series converges   power series diverges

interval of convergence

power series may converge or may diverge

power series may converge or may diverge

The number R is called the *radius of convergence* of the power series and can sometimes be found using the formulae,

$$\frac{1}{R} = \lim_{n \to \infty} \left| \frac{a_{n+1}}{a_n} \right| \quad ; \quad \frac{1}{R} = \lim_{n \to \infty} |a_n|^{1/n}.$$

If I consists just of ξ, then $R = 0$. If I consists of all real numbers, then we say that R is 'infinite'. (The formulae given above for R remain valid provided one is willing to work with the conventions $1/0 = \infty$ and $1/\infty = 0$ but *do not expect these conventions to work elsewhere*.)

Differentiating and integrating power series is particularly easy. If a and b are any points in the interval of convergence (i.e. the power series converges for $x = a$ and $x = b$), then

$$\int_a^b \sum_{n=0}^{\infty} a_n(x-\xi)^n dx = \sum_{n=0}^{\infty} \int_a^b a_n (x-\xi)^n dx$$

and, if x is any point in the interior of the interval of convergence (*not* an endpoint)

$$\frac{d}{dx} \sum_{n=0}^{\infty} a_n (x-\xi)^n = \sum_{n=0}^{\infty} \frac{d}{dx} a_n (x-\xi)^n$$

(These results may be proved using the idea of dominated convergence introduced in §9.30.)

Example 9.36.†
From example 9.26 we know that the power series

$$\sum_{n=0}^{\infty} x^n$$

converges to $(1-x)^{-1}$ for $|x|<1$ and diverges for $|x| \geqslant 1$. Thus its interval of convergence is the set of all x satisfying $-1<x<1$ and its radius of convergence is $R=1$.

If $-1<y<1$, it follows that

$$\int_0^y \frac{dx}{1-x} = \int_0^y \left(\sum_{n=0}^{\infty} x^n \right) dx = \sum_{n=0}^{\infty} \int_0^y x^n dx.$$

Thus

$$\left[-\log(1-x)\right]_0^y = \sum_{n=0}^{\infty} \frac{y^{n+1}}{(n+1)}$$

i.e.

$$\log(1-y) = -y - \frac{y^2}{2} - \frac{y^3}{3} - \dots$$

provided $-1<y<1$.

Also, if $-1<x<1$,

$$\frac{d}{dx}\left(\frac{1}{1-x}\right) = \frac{d}{dx}\left(\sum_{n=0}^{\infty} x^n\right) = \sum_{n=0}^{\infty} \frac{d}{dx} x^n = \sum_{n=0}^{\infty} nx^{n-1}$$

i.e.

$$\frac{1}{(1-x)^2} = \sum_{n=0}^{\infty} nx^{n-1}$$

and so

$$\frac{x}{(1-x)^2} = x + 2x^2 + 3x^3 + 4x^4 + \dots$$

provided $-1<x<1$.

Exercises 9.37

1.† The *Gamma function* is defined for $y>0$ by

$$\Gamma(y) = \int_0^{\infty} x^{y-1} e^{-x} dx.$$

Prove that the integral for $\Gamma(n)$ converges for each natural number $n=1,2,3\dots$. [Hint: exercise 6.9 (7).] Integrate by parts to show that

$$\Gamma(n + 1) = n\Gamma(n) \qquad (n = 1, 2, 3, \ldots)$$

and hence prove that $\Gamma(n + 1) = n!$

2.[†] Explain why the integrals

$$\int_0^\infty (\cos bx)e^{-ax}\,dx \; ; \qquad \int_0^\infty (\sin bx)e^{-ax}\,dx$$

converge for all $a > 0$ and all b. Integrate twice by parts to prove that

(i) $\quad \displaystyle\int_0^\infty (\cos bx)e^{-ax}\,dx = \frac{a}{a^2 + b^2}$

(ii) $\quad \displaystyle\int_0^\infty (\sin bx)e^{-ax}\,dx = \frac{b}{a^2 + b^2}.$

3.[†] By considering

$$\frac{d^n}{dy^n} \int_0^1 x^y\,dx,$$

prove that

$$\int_0^1 (\log x)^n\,dx = (-1)^n n!$$

4.*[†] By considering

$$\frac{d^n}{dy^n} \int_0^\infty e^{-yx}\,dx$$

give an alternative proof that $\Gamma(n + 1) = n!$ (See question 1).

5.[†] Show that

$$\lim_{n \to \infty} \int_0^1 (n + 1)x^n\,dx \neq \int_0^1 \lim_{n \to \infty} (n + 1)x^n \; dx.$$

6.* Calculate

(i) $\quad \displaystyle\frac{d}{dy} \int_0^{\log y} \sin (xy^2)\,dx$
(ii) $\quad \displaystyle\frac{d}{dy} \int_0^{y^3} \log (x + y^2)\,dx$

using the formula of §9.33.

7. Find a formula for

$$\frac{d}{dy} \int_{a(y)}^{b(y)} f(x, y)\,dx.$$

8.* A function f satisfies the integral equation

$$f(y) = \int_{-\infty}^{y} e^{x-y} f(x) dx.$$

Prove that $f'(y) = 0$ and deduce that f is constant.

9.† By integrating the equation

$$\frac{1}{1+x^2} = 1 - x^2 + x^4 - x^6 + \dots,$$

show that

$$\arctan y = y - \frac{y^3}{3} + \frac{y^5}{5} - \frac{y^7}{7} + \dots$$

provided $-1 < y < 1$.

10.*† By differentiating appropriate power series, prove that

$$\frac{x(1+x^2)}{(1-x^2)^2} = 1^2 x + 2^2 x^2 + 3^2 x^3 + 4^2 x^4 + \dots$$

provided $-1 < x < 1$.

11.† Find the radii of convergence of the following power series:

(i) $\sum_{n=0}^{\infty} n^2 x^n$ (ii) $\sum_{n=0}^{\infty} 2^n x^n$ (iii) $\sum_{n=0}^{\infty} n^n x^n$.

12.*† Find the radii of convergence of the following power series

(i) $\sum_{n=0}^{\infty} n^{-2} x^n$ (ii) $\sum_{n=0}^{\infty} 2^{-n} x^n$ (iii) $\sum_{n=0}^{\infty} n^{-n} x^n$.

SOME APPLICATIONS (OPTIONAL)

9.38   Probability

Suppose that it is not known for certain which of a number of possible events will occur. In certain circumstances it is sensible to quantify the uncertainty involved by attaching a *probability* to each possible event. Such a probability is a real number p satisfying $0 \leqslant p \leqslant 1$. To say that an event has probability .013, for example, means that, if the circumstances generating the event were to be repeated many times, then the event in question would be observed thirteen times in every thousand, on average.

Note that an impossible event has probability zero. But if there are an infinite number of events to be considered, it may well be that an event has zero probability without being impossible. For example, it is not impossible that a fair coin will *always* come down heads when tossed: but this event has *zero* probability.

There are various rules for manipulating probabilities. The most important of these are the two which follow.

(1) If E and F are two events which cannot *both* happen, then

$$\text{Prob } (E \text{ or } F) = \text{Prob } (E) + \text{Prob } (F).$$

(2) If E and F are independent events, then

$$\text{Prob } (E \text{ and } F) = \text{Prob } (E) \times \text{Prob } (F).$$

Consider, for example, two horses 'Punter's Folly' and 'Gambler's Ruin'. The probability that 'Punter's Folly' will win its race is $\frac{1}{2}$. The probability that 'Gambler's Ruin' will win its race is $\frac{1}{4}$. If the horses are running in the same race (and the possibility of a tie is ignored), then the probability that at least one of the two horses will win is

$$\tfrac{1}{2} + \tfrac{1}{4} = \tfrac{3}{4}.$$

If the horses are running in different (and independent) races, then the probability that *both* will win is

$$\tfrac{1}{2} \times \tfrac{1}{4} = \tfrac{1}{8}.$$

9.39 Probability density functions

A real number X whose value depends on which event out of a set of possible events actually happens is called a *random variable*. A good example of a random variable is the amount of money a gambler brings home from a casino after having used a prearranged betting system.

We shall first consider random variables X which can only take integer values (i.e. $0, \pm 1, \pm 2, \dots$). Such random variables are called *discrete*. The function f defined by

$$f(x) = \text{Prob } (X = x)$$

is called the probability density function for X. The notation Prob $(X = x)$ means the probability that the random variable X takes the value x. In order for f to qualify as a probability density function for X, it is necessary that

(i) $f(x) \geqslant 0 \qquad (x = 0, \pm 1, \pm 2, \dots)$

(ii) $\displaystyle\sum_{x=-\infty}^{\infty} f(x) = 1.$

The second condition expresses the requirement that it is certain that X will take one of the available values of x. Observe also that

$$\text{Prob } (a \leqslant X \leqslant b) = \sum_{x=a}^{b} f(x).$$

This latter observation leads us to the consideration of random variables which may take any real value. Such random variables are called *continuous*. If it is true that, for each $a \leqslant b$,

$$\text{Prob } (a \leqslant X \leqslant b) = \int_{a}^{b} f(x)dx,$$

then the function f is called the *probability density function* for the continuous random variable X. In order that f qualifies as a probability density function for a continuous random variable X, it is necessary that

(i) $f(x) \geqslant 0$ (all x)

(ii) $\int_{-\infty}^{\infty} f(x)dx = 1.$

The second condition expresses the requirement that it is certain that X will take some real value.

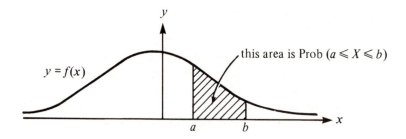

In both the discrete and the continuous case, the *probability distribution function* F for X is defined by $F(y) = \text{Prob } (X \leqslant y)$. In the discrete case

$$F(y) = \sum_{x=-\infty}^{y} f(x)$$

and in the continuous case,

$$F(y) = \int_{-\infty}^{y} f(x)dx.$$

9.40 Binomial distribution

A weighted coin comes down heads with probability p and tails with probability $q = 1 - p$. (The toss of such a coin is called a Bernoulli trial.) A discrete random variable X is obtained by tossing the coin n times and observing the total number of heads which results. What is the probability density function for X?

We need to calculate $f(x) = \text{Prob}\,(X = x)$ for $x = 0, 1, 2, \ldots, n$. (If x is not one of these values, then $f(x) = 0$.) The tables below enumerate the possibilities in the case $n = 4$:

1. $\underline{X = 4}$

Result of 4 trials	*HHHH*
Probability	*pppp*

2. $\underline{X = 3}$

Result of 4 trials	*HHHT*	*HHTH*	*HTHH*	*THHH*
Probability	*pppq*	*ppqp*	*pqpp*	*qppp*

3. $\underline{X = 2}$

Result of 4 trials	*HHTT*	*HTHT*	*HTTH*	*THHT*	*THTH*	*TTHH*
Probability	*ppqq*	*pqpq*	*pqqp*	*qppq*	*qpqp*	*qqpp*

4. $\underline{X = 1}$

Result of 4 trials	*HTTT*	*THTT*	*TTHT*	*TTTH*
Probability	*pqqq*	*qpqq*	*qqpq*	*qqqp*

5. $\underline{X = 0}$

Result of 4 trials	*TTTT*
Probability	*qqqq*

The probabilities attached to each possible event are computed on the assumption that each trial is *independent* of the others. Since the events listed are mutually exclusive, we obtain for the case $n = 4$ that

$$\text{Prob}\,(X = 4) = pppp$$

$$\text{Prob}\,(X = 3) = pppq + ppqp + pqpp + qppp$$

$$\text{Prob}\,(X = 2) = ppqq + pqpq + pqqp + qppq + qpqp + qqpp$$

$$\text{Prob}\,(X = 1) = pqqq + qpqq + qqpq + qqqp$$

$$\text{Prob}\,(X = 0) = qqqq.$$

Now consider the following proof of the Binomial theorem in the case $n = 4$. We have that

$$
\begin{aligned}
(p + q)^4 &= (p + q)^2 (p + q)(p + q) \\
&= (p + q)^2 \{p(p + q) + q(p + q)\} \\
&= (p + q)(p + q)(pp + pq + qp + qq) \\
&= (p + q)\{p(pp + pq + qp + qq) + q(pp + pq + qp + qq)\} \\
&= (p + q)(ppp + ppq + pqp + pqq + qpp + qpq + qqp + qqq) \\
&= (pppp) + (pppq + ppqp + pqpp + qppp) \\
&\quad + (ppqq + pqpq + pqqp + qppq + qpqp + qqpp) \\
&\quad + (pqqq + qpqq + qqpq + qqqp) + (qqqq) \\
&= \binom{4}{0}p^4 + \binom{4}{1}p^3 q + \binom{4}{2}p^2 q^2 + \binom{4}{3}pq^3 + \binom{4}{4}q^4 .
\end{aligned}
$$

Comparing the final two lines, we see that, in the case $n = 4$,

$$f(x) = \text{Prob}\,(X = x) = \binom{n}{x}p^x q^{n-x}.$$

The same proof works for *all* values of n. (Another way of obtaining the same result is by observing that

$$\binom{n}{x}$$

is the total number of ways that x *H*s and $n - x$ *T*s can be arranged in a row.)

For obvious reasons, the random variable X (equal to the number of heads which appear in n trials) is said to have the *binomial distribution*. Note that $f(x) \geqslant 0$ $(x = 0, 1, 2, \ldots, n)$ and, since $p + q = 1$,

$$1 = (p + q)^n = \sum_{x=0}^{n} \binom{n}{x}p^x q^{n-x} = \sum_{x=0}^{n} f(x).$$

9.41 Poisson distribution

It often happens that it is necessary to consider a binomially distributed random variable in the case when n is very large. It is then often easier to work with a simple approximation to the Binomial distribution rather than with the Binomial distribution itself. Consider, for example, the following problem.

During the Second World War, the fall of rockets on London was to a large extent random. Suppose for simplicity that any two sites of equal area in London were equally likely to be hit. If n rockets fell on London and London is divided into n regions of equal area (so that, on average, each region will be hit by one rocket), what is the probability that a given region was actually hit exactly ten times?

The probability p of the given region being hit by any particular rocket is $p = 1/n$. The required probability is therefore

$$\binom{n}{10}\left(\frac{1}{n}\right)^{10}\left(\frac{n-1}{n}\right)^{n-10}.$$

For large values of n this is clearly an awkward quantity. We therefore seek a simple approximation.

Stirling's approximation for $m!$ is

$$m! \sim \sqrt{(2\pi)}m^{m-1/2}e^{-m}.$$

(Here $\sim$ means that the ratio of the two sides is nearly one when m is large.) We shall use this to obtain an approximation for

$$\binom{n}{x}p^{x}q^{n-x}$$

in the case when $p = \lambda n$ (where λ is a constant). We have that

$$\binom{n}{x}p^{x}q^{n-x} = \frac{n!}{x!(n-x)!}\left(\frac{\lambda}{n}\right)^{x}\left(1-\frac{\lambda}{n}\right)^{n-x}$$

$$\sim \frac{\sqrt{(2\pi)}n^{n-1/2}e^{-n}}{x!\sqrt{(2\pi)}(n-x)^{n-x-1/2}e^{-(n-x)}}\left(\frac{\lambda}{n}\right)^{x}\left(1-\frac{\lambda}{n}\right)^{n-x}$$

$$= \frac{\lambda^{x}}{x!}\frac{\left(1-\dfrac{x}{n}\right)^{x+1/2}\left(1-\dfrac{\lambda}{n}\right)^{n}}{\left(1-\dfrac{x}{n}\right)^{n}e^{x}\left(1-\dfrac{\lambda}{n}\right)^{x}}$$

$$\to \frac{\lambda^{x}}{x!}e^{-\lambda} \quad \text{as} \quad n \to \infty$$

For the last step, one needs to know that

$$\left(1 + \frac{y}{n}\right)^n \to e^y \quad \text{as} \quad n \to \infty.$$

If λ is constant and n is large, then this argument shows that, if X is the number of 'successes' in n independent trials with the probability of 'success' in each individual trial equal to $p = \lambda/n$, then

$$\text{Prob } (X = x) \doteqdot \frac{\lambda^x}{x!} e^{-\lambda}.$$

In our particular problem, $\lambda = 1$ and $x = 10$ and so the required probability is approximately $1/e\,10!$

Observe that, if $\lambda > 0$, then the function

$$f(x) = \frac{\lambda^x e^{-\lambda}}{x!} \qquad (x = 0, 1, 2, \dots)$$

qualifies as the probability density function for a discrete random variable because $f(x) \geqslant 0 \; (x = 0, 1, \dots)$ and

$$\sum_{x=0}^{\infty} f(x) = \sum_{x=0}^{\infty} \frac{\lambda^x e^{-\lambda}}{x!}$$

$$= e^{-\lambda}\left(1 + \frac{\lambda}{1!} + \frac{\lambda^2}{2!} + \frac{\lambda^3}{3!} + \dots\right)$$

$$= e^{-\lambda} e^{\lambda} = 1.$$

A random variable Y with this probability density function is said to have a *Poisson distribution* (with parameter λ). The random variable X of our example is therefore *approximately* Poisson.

9.42   Mean

The mean μ of a discrete random variable X with probability density function f is given by

$$\mu = \sum_{x=-\infty}^{\infty} xf(x).$$

Suppose that $X_1, X_2, X_3, \dots$ are independent random variables, all of which have probability density function f. Such a sequence may be generated by repeating the trial which leads to X over and over again. We shall think of

X_k as the amount a gambler wins at the kth trial. His *average* winnings over n trials will then be

$$A_n = \frac{X_1 + X_2 + \ldots + X_n}{n}.$$

The law of large numbers asserts that

$$A_n \to \mu \quad \text{as} \quad n \to \infty$$

with probability one. (This does not mean that $A_n \to \mu$ as $n \to \infty$ *always*: only that $A_n \nrightarrow \mu$ as $n \to \infty$ happens too seldom to deserve a positive probability.) Thus, in the long run, the gambler's average winnings will almost certainly be approximately μ. This leads us to call μ the *expectation* (or expected value) of the random variable X and we write

$$\mathscr{E}(X) = \mu = \sum_{x=-\infty}^{\infty} xf(x).$$

For a binomially distributed random variable, for example, we have that

$$\mu = \sum_{x=0}^{n} x \binom{n}{x} p^x q^{n-x}.$$

The value of μ can be calculated as follows. We have that

$$(p+q)^n = \sum_{x=0}^{n} \binom{n}{x} p^x q^{n-x}.$$

Hence

$$n(p+q)^{n-1} = \frac{\partial}{\partial p}(p+q)^n = \frac{\partial}{\partial p} \sum_{x=0}^{n} \binom{n}{x} p^x q^{n-x}$$

$$= \sum_{x=0}^{n} x \binom{n}{x} p^{x-1} q^{n-x}$$

and therefore

$$\mu = \sum_{x=0}^{n} x \binom{n}{x} p^x q^{n-x} = np(p+q)^{n-1} = np.$$

The mean for the Poisson distribution is more easily calculated. We have that

$$\mu = \sum_{x=0}^{\infty} x\frac{\lambda^x}{x!} e^{-\lambda} = e^{-\lambda} \sum_{x=1}^{\infty} \frac{\lambda^x}{(x-1)!}$$

$$= \lambda e^{-\lambda}\left(1 + \frac{\lambda}{1!} + \frac{\lambda^2}{2!} + \ldots\right) = \lambda e^{-\lambda}e^{\lambda} = \lambda.$$

In the case of a *continuous* random variable X with probability density function f, we define the mean μ (or expectation $\mathscr{E}(X)$) by

$$\mathscr{E}(X) = \mu = \int_{-\infty}^{\infty} xf(x)dx.$$

The interpretation is identical with that for the discrete case.

9.43 Variance

The *variance* Var (X) of a random variable X is defined by

$$\begin{aligned} \text{Var}(X) &= \mathscr{E}(X - \mu)^2 \\ &= \mathscr{E}(X^2 - 2\mu X + \mu^2) \\ &= \mathscr{E}(X^2) - 2\mu\mathscr{E}(X) + \mu^2\mathscr{E}(1) \quad \text{(see §10.21)} \\ &= \mathscr{E}(X^2) - 2\mu^2 + \mu^2 \\ &= \mathscr{E}(X^2) - \mu^2 \end{aligned}$$

The variance is a measure of the degree of dispersion of the random variable X. It indicates the extent to which the probability density function clusters around the mean. For this reason, it is usual to write

$$\text{Var}(X) = \sigma^2$$

where the number σ is called the *standard deviation* of X. One would be surprised to find a random variable with small standard deviation taking values far from its mean and not at all surprised in the case of a random variable with large standard deviation.

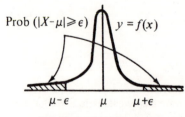

Small standard deviation

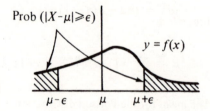

Large standard deviation

For a discrete random variable, the variance is given by

$$\sigma^2 = \sum_{x=-\infty}^{\infty} (x-\mu)^2 f(x)$$

$$= \left(\sum_{x=-\infty}^{\infty} x^2 f(x) \right) - \mu^2 .$$

In the case of a binomially distributed random variable we therefore have to calculate

$$\sum_{x=0}^{n} x^2 \binom{n}{x} p^x q^{n-x}$$

From §9.42,

$$np(p+q)^{n-1} = \sum_{x=0}^{n} x \binom{n}{x} p^x q^{n-x}$$

and so

$$n(n-1)p(p+q)^{n-2} = \frac{\partial}{\partial q} \sum_{x=0}^{n} x \binom{n}{x} p^x q^{n-x}$$

$$= \sum_{x=0}^{n} x(n-x) \binom{n}{x} p^x q^{n-x-1} .$$

Thus

$$\sum_{x=0}^{n} x^2 \binom{n}{x} p^x q^{n-x} = n \sum_{x=0}^{n} x \binom{n}{x} p^x q^{n-x} - n(n-1)pq(p+q)^{n-2}$$

$$= n\mu - n(n-1)pq.$$

It follows that

$$\sigma^2 = (n\mu - n(n-1)pq) - \mu^2$$

$$= n^2 p - n^2 pq + npq - n^2 p^2$$

$$= n^2 p(1-p) - n^2 pq + npq$$

$$= npq.$$

The Poisson distribution is again more easily dealt with. We have that

$$\sum_{x=0}^{\infty} x^2 \frac{\lambda^x}{x!} e^{-\lambda} = \sum_{x=0}^{\infty} x(x-1)\frac{\lambda^x}{x!} e^{-\lambda} + \sum_{x=0}^{\infty} x\frac{\lambda^x}{x!} e^{-\lambda}$$

$$= \lambda^2 e^{-\lambda} \sum_{x=2}^{\infty} \frac{\lambda^{x-2}}{(x-2)!} + \mu$$

$$= \lambda^2 + \lambda.$$

Hence

$$\sigma^2 = (\lambda^2 + \lambda) - \mu^2 = \lambda.$$

9.44 Standardized random variables

If a random variable X has mean μ and variance σ^2 and

$$Y = \frac{X - \mu}{\sigma},$$

then

$$\mathscr{E}(Y) = \mathscr{E}\left(\frac{X - \mu}{\sigma}\right) = \frac{1}{\sigma}\{\mathscr{E}(X) - \mu\} = 0;$$

$$\text{Var}(Y) = \mathscr{E}((Y - o)^2) = \mathscr{E}\left(\left(\frac{X - \mu}{\sigma}\right)^2\right)$$

$$= \frac{1}{\sigma^2}\mathscr{E}((X - \mu)^2) = \frac{\sigma^2}{\sigma^2} = 1.$$

It follows that the random variable Y has mean 0 and variance 1. Such a random variable is said to be 'standardized'.

9.45 Normal distribution

In §9.41 we saw that the Poisson distribution provides a simple approximation to the binomial distribution in the case when n is large and $p \doteq \lambda/n$ where λ is constant. A more pressing problem is to find a simple approximation for the case when n is large and p is constant.

Suppose that the probability density function for the discrete random variable X_n is binomial with parameters n and p. Consider the associated standardized random variable

$$Y_n = \frac{X_n - \mu_n}{\sigma_n} = \frac{X_n - pn}{\sqrt{(npq)}}.$$

We shall illustrate the probability density function for Y_n by drawing a histogram. The histogram below is for Y_4 in the case $p = \frac{1}{4}$. The *area* of each rectangle represents the probability that Y_n is equal to the midpoint of its base. The mean and standard deviation for Y_4 in the case $p = \frac{1}{4}$ are 1 and $\sqrt{(3/4)} = .87$ respectively. Thus, for example, since $(4 - 1)/.87 = 3.45$,

$$\text{Prob } (Y_4 = 3.45) = \text{Prob } (X_4 = 4) = (\tfrac{1}{4})^4 = .0039$$

and thus the area of the rectangle whose base has midpoint 3.45 is equal to .0039.

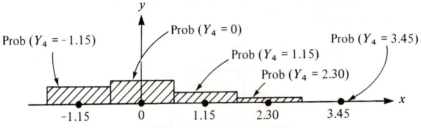

The histograms below (drawn to a different scale) show the probability density functions for Y_{10}, Y_{30} and Y_{90} in the cases $p = \frac{1}{5}$ and $p = \frac{1}{2}$.

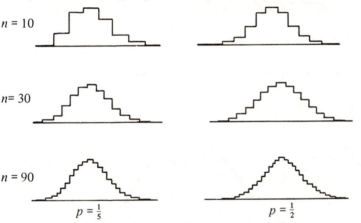

In both cases the histograms approach a smooth bell-shaped curve as n becomes large

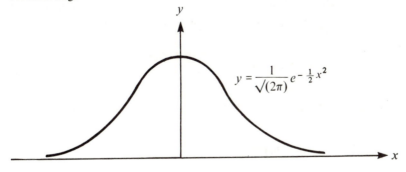

$$y = \frac{1}{\sqrt{(2\pi)}} e^{-\frac{1}{2}x^2}$$

In fact, for *all* values of p with $0 < p < 1$, the appropriate histograms approach the *same* curve and this curve has equation

$$y = \frac{1}{\sqrt{(2\pi)}} e^{-x^2/2}$$

This can be proved with an argument similar to that used in §9.41.

The area shaded in the histogram for Y_n drawn below is equal to Prob $(Y_n \leqslant z)$.

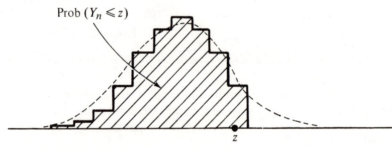

From §9.2, it follows that Prob $(Y_n \leqslant z)$ is

$$\text{Prob } (Y_n \leqslant z) \doteqdot \frac{1}{\sqrt{(2\pi)}} \int_{-\infty}^{z} e^{-x^2/2} \, dx$$

The latter quantity, of course, is that illustrated below:

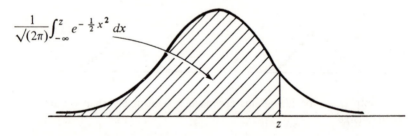

$$\frac{1}{\sqrt{(2\pi)}} \int_{-\infty}^{z} e^{-\frac{1}{2}x^2} \, dx$$

If $\sigma > 0$, a continuous random variable X with probability density function

$$f(x) = \frac{1}{\sigma\sqrt{(2\pi)}} e^{-\frac{1}{2}\left(\frac{x-\mu}{\sigma}\right)^2}$$

is said to have the *normal distribution*. The mean of this random variable is

$$\mathcal{E}(X) = \frac{1}{\sigma\sqrt{(2\pi)}} \int_{-\infty}^{\infty} x e^{-\frac{1}{2}\left(\frac{x-\mu}{\sigma}\right)^2} \, dx$$

$$= \frac{1}{\sqrt{(2\pi)}} \int_{-\infty}^{\infty} (\sigma t + \mu) e^{-t^2/2} \, dt$$

$$= \frac{\sigma}{\sqrt{(2\pi)}} \int_{-\infty}^{\infty} t e^{-t^2/2} \, dt + \frac{\mu}{\sqrt{(2\pi)}} \int_{-\infty}^{\infty} e^{-t^2/2} \, dt$$

where we have introduced the change of variable $x = \sigma t + \mu$ which makes $dx = \sigma dt$. But

$$\int_{-T_1}^{T_2} t e^{-t^2/2} \, dt = {}_{-T_1}^{T_2} [-e^{-t^2/2}] = e^{-T_1^2/2} - e^{-T_2^2/2}$$

$$\to 0 \quad \text{as} \quad T_1 \to \infty \quad \text{and} \quad T_2 \to \infty$$

and it is shown in the next chapter that

$$(1) \qquad \int_{-\infty}^{\infty} e^{-t^2/2} \, dt = \sqrt{(2\pi)}.$$

It follows that $\mathscr{E}(X) = \mu$ (as one might reasonably expect from the use of the symbol μ). Similarly Var $(X) = \sigma^2$. To prove this, we begin by making the change of variable $t = yu$ in (1). Then

$$\int_{-\infty}^{\infty} e^{-y^2 u^2/2} \, du = \frac{\sqrt{(2\pi)}}{y}.$$

Hence

$$\frac{-\sqrt{(2\pi)}}{y^2} = \frac{d}{dy} \int_{-\infty}^{\infty} e^{-y^2 u^2/2} \, du = \int_{-\infty}^{\infty} \frac{\partial}{\partial y} e^{-y^2 u^2/2} \, du$$

$$= \int_{-\infty}^{\infty} -yu^2 e^{-y^2 u^2/2} \, du.$$

Taking $y = 1$, we obtain that

$$\int_{-\infty}^{\infty} u^2 e^{-u^2/2} \, du = \sqrt{(2\pi)}.$$

Thus

$$\text{Var } (X) = \mathscr{E}((X - \mu)^2)$$

$$= \frac{1}{\sigma \sqrt{(2\pi)}} \int_{-\infty}^{\infty} (x - \mu)^2 \, e^{-\frac{1}{2}\left(\frac{x-\mu}{\sigma}\right)^2} \, dx$$

$$= \frac{1}{\sqrt{(2\pi)}} \int_{-\infty}^{\infty} \sigma^2 t^2 e^{-t^2/2} \, dt = \sigma^2.$$

The function n defined by

$$n(x) = \frac{1}{\sqrt{(2\pi)}} e^{-x^2/2}$$

is therefore the probability density function of a 'standardized' normal distribution. Returning to the problem of approximating binomial random variables, we see that the 'standardized' binomial random variable Y_n is approximately the 'standardized' normal random variable Y provided n is large – i.e.

$$\text{Prob } (Y_n \leqslant z) \doteqdot \text{Prob } (Y \leqslant z).$$

Suppose that one has to calculate $\text{Prob } (W \leqslant w)$ where W is a binomial random variable with parameters p and n. One should then note that

$$\text{Prob } (W \leqslant w) = \text{Prob}\left(\frac{W - np}{\sqrt{(npq)}} \leqslant \frac{w - np}{\sqrt{(npq)}}\right)$$

$$= \text{Prob}\left(Y \leqslant \frac{w - np}{\sqrt{(npq)}}\right) = \frac{1}{\sqrt{(2\pi)}} \int_{-\infty}^{\frac{w - np}{\sqrt{(npq)}}} e^{-x^2/2} \, dx$$

Since tables for

$$N(z) = \frac{1}{\sqrt{(2\pi)}} \int_{-\infty}^{z} e^{-x^2/2} \, dx$$

are readily available, this means that it is easy to obtain an approximation to $\text{Prob } (W \leqslant w)$. How good is this approximation? Our explanation only indicates that it is good for large values of n but the further p is from $\frac{1}{2}$, the larger n has to be. A useful rule of thumb is that the approximation will be adequate for practical purposes when $\sigma_n^2 = np(1 - p) > 10$. (If $p = \frac{1}{2}$, this means that $n > 40$.)

9.46   Sums of random variables

If X and Y are two random variables it is *always* the case that $\mathscr{E}(X + Y) = \mathscr{E}(X) + \mathscr{E}(Y)$. If X and Y are *independent* random variables, then it is also true that $\mathscr{E}(XY) = \mathscr{E}(X)\mathscr{E}(Y)$. (See §10.21 and §10.22.)

Now suppose that $X_1, X_2, X_3, \ldots$ is a sequence of independent random variables each of which has the same probability distribution with mean μ and variance σ^2. We can then easily calculate the mean and variance of the random variable

$$S_n = X_1 + X_2 + \ldots + X_n.$$

We have that

$$\mathscr{E}(S_n) = \mathscr{E}(X_1 + X_2 + \ldots + X_n)$$

$$= \mathscr{E}(X_1) + \mathscr{E}(X_2) + \ldots + \mathscr{E}(X_n)$$

$$= n\mu.$$

We illustrate the proof that

$$\text{Var}(S_n) = \text{Var}(X_1) + \text{Var}(X_2) + \ldots + \text{Var}(X_n) = n\sigma^2$$

for the case $n = 3$. Note that

$$
\begin{aligned}
\mathscr{E}((X_1 + X_2 + X_3)^2) &= \mathscr{E}(X_1^2 + X_2^2 + X_3^2 + 2X_1X_2 + 2X_1X_3 + 2X_2X_3) \\
&= \mathscr{E}(X_1^2) + \mathscr{E}(X_2^2) + \mathscr{E}(X_3^2) + 2\mathscr{E}(X_1)\mathscr{E}(X_2) \\
&\quad + 2\mathscr{E}(X_1)\mathscr{E}(X_3) + 2\mathscr{E}(X_2)\mathscr{E}(X_3) \\
&= \text{Var}(X_1) + \text{Var}(X_2) + \text{Var}(X_3) + \mathscr{E}(X_1)^2 \\
&\quad + \mathscr{E}(X_2)^2 + \mathscr{E}(X_3)^2 + 2\mathscr{E}(X_1)\mathscr{E}(X_2) \\
&\quad + 2\mathscr{E}(X_1)\mathscr{E}(X_3) + 2\mathscr{E}(X_2)\mathscr{E}(X_3) \\
&= \text{Var}(X_1) + \text{Var}(X_2) + \text{Var}(X_3) \\
&\quad + \{\mathscr{E}(X_1 + X_2 + X_3)\}^2.
\end{aligned}
$$

Thus
$$\text{Var}(S_3) = \text{Var}(X_1) + \text{Var}(X_2) + \text{Var}(X_3).$$

Suppose, for example, that a weighted coin is tossed n times with probability p on each occasion that it comes down heads and $1 - p = q$ that it comes down tails. Let X_k be 1 if the kth trial produces a head and 0 if it produces a tail. Then

$$\mu = \mathscr{E}(X_k) = 1p + 0(1-p) = p$$
$$
\begin{aligned}
\sigma^2 &= \mathscr{E}(X_k^2) - \mu^2 \\
&= \{1^2 p + 0^2(1-p)\} - \mu^2 = p - p^2 = pq.
\end{aligned}
$$

From the previous discussion it follows that the random variable $S_n = X_1 + X_2 + \ldots + X_n$ has mean and variance

$$
\left.
\begin{aligned}
\mathscr{E}(S_n) &= n\mu = np \\
\text{Var}(S_n) &= n\sigma^2 = npq.
\end{aligned}
\right\}
$$

But S is of course a binomial random variable with parameters p and n. These calculations therefore provide an alternative method of computing the mean and variance of a binomial random variable.

Returning to our arbitrary sequence $X_1, X_2, \ldots$ of independent random variables, all of which have the same probability distribution, we introduce the 'standardized' random variable

$$T_n = \frac{S_n - n\mu}{\sigma\sqrt{n}}.$$

It is a remarkable fact that under quite mild conditions it is *always* true that

$$\text{Prob}\,(T_n \leqslant t) = \frac{1}{\sqrt{(2\pi)}} \int_{-\infty}^{t} e^{-x^2/2}\, dx$$

– i.e. the random variable T is approximately equal to the 'standardized' normal random variable Y when n is large. This result is called the 'central limit theorem'. It remains valid even when $X_1, X_2, \ldots$ do not have quite the same probability distribution. Since random variables observed in practice often arise from the superposition of large numbers of random errors, the central limit theorem makes it clear why many random variables turn out to be normally distributed. The use of the word 'normal' indicates just how pervasive these random variables are.

Going back to our example in which X_k is the random variable which is equal to 1 if the kth toss of a weighted coin yields heads and equal to 0 if tails, we note that the discussion of §9.45 is concerned with a simple special case of the central limit theorem. As an application, observe that

$$\text{Prob}\left(\frac{S_n}{n} - p \leqslant -\epsilon\right) = \text{Prob}\,(S_n - np \leqslant -n\epsilon)$$

$$= \text{Prob}\left(\frac{S_n - np}{\sqrt{(npq)}} \leqslant -\epsilon\sqrt{\left(\frac{n}{pq}\right)}\right)$$

$$\doteq \frac{1}{\sqrt{(2\pi)}} \int_{-\infty}^{-\epsilon\sqrt{(n/pq)}} e^{-x^2/2}\, dx$$

$$\leqslant \frac{1}{\sqrt{(2\pi)}} \int_{-\infty}^{-\epsilon\sqrt{(n/pq)}} \frac{2}{x^2}\, dx = \frac{2}{\epsilon}\sqrt{\left(\frac{pq}{n}\right)}$$

because $e^y = 1 + y + \frac{1}{2}y^2 + \ldots \geqslant y \ (y \geqslant 0)$. A similar inequality may be obtained for

$$\text{Prob}\left(\frac{S_n}{n} - p \geqslant \epsilon\right).$$

and it follows that

$$\text{Prob}\left(-\epsilon < \frac{S_n}{n} - p < \epsilon\right) > 1 - \frac{4}{\epsilon}\sqrt{\left(\frac{pq}{n}\right)}$$

provided that n is large.

Thus, for example, if $p = \frac{1}{4}$, then the probability that the average number of heads in 100 000 000 trials will be within .01 of the probability $p = \frac{1}{4}$ of obtaining heads in one trial is at least .98. This result should be considered in conjunction with the remarks concerning the law of large numbers made in §9.42.

9.47 Cauchy distribution

A continuous random variable X with the probability density function

$$f(x) = \frac{1}{\pi} \frac{1}{1 + x^2}$$

is said to have the Cauchy distribution. We mention this distribution only to point out that it *does not have a mean*. As pointed out in example 9.29, the integral

$$\frac{1}{\pi} \int_{-\infty}^{\infty} \frac{x}{1 + x^2} dx$$

diverges. For random variables which share this property with X, the law of large numbers *does not hold*.

10

Multiple integrals

This chapter is chiefly concerned with three techniques — i.e. evaluating repeated integrals, changing the order of integration in repeated integrals and changed variables in multiple integrals. The important feature of each technique is the necessity of ensuring that the correct limits of integration are used. This is fairly straightforward in the two variable cases provided one is not tempted to hurry things along by such short cuts as dispensing with a diagram.

The theoretical explanation of the formula for changing variables given in §10.7 and §10.8 should be omitted if this is found troublesome. The next section §10.9 can be read independently. However, it is worth taking note in passing of the geometric interpretation of a determinant as described in §10.7 if this has not been met before.

Techniques similar to those which work for multiple integrals also work for multiple sums except that things are usually rather easier in the latter case. Some discussion of multiple sums and series has been given in §10.15 but this should certainly be omitted if §9.35 has not been studied.

10.1 Introduction

Suppose that $f: \mathbb{R}^2 \to \mathbb{R}$ and that D is some region in $\mathbb{R}^2$. Then the *double integral*

$$\iint_D f(x, y)\,dx\,dy$$

may be interpreted as the volume beneath the surface $z = f(x, y)$ and above the region D.

$$z = f(x, y)$$

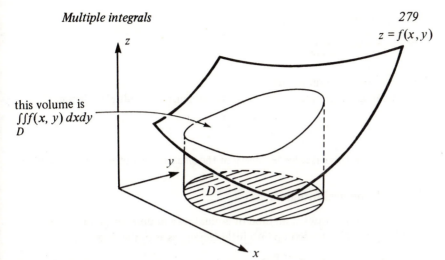

this volume is
$\iint_D f(x, y)\, dxdy$

As in the case of simple integrals, volume beneath the (x, y) plane is counted negative (see §9.2).

In the following diagram, the region D has been split up into a large number of small regions. The shaded region has area A_i and contains the point (x_i, y_i). Thus

$$f(x_i, y_i) A_i$$

is an approximation to the volume of the 'obelisk' drawn and hence

$$\sum_i f(x_i, y_i) A_i$$

is an approximation to the volume under the surface.

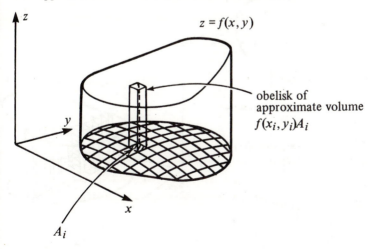

$$z = f(x, y)$$

obelisk of
approximate volume
$f(x_i, y_i) A_i$

As the number of regions increases and the area of each individual region shrinks to zero, each of these approximating sums approaches.

$$\iint_D f(x,y)dxdy$$

provided that f is continuous.

Similar considerations apply to higher dimensional multiple integrals

$$\iint \ldots \int_D f(x_1, x_2, \ldots, x_n)dx_1 dx_2 \ldots dx_n$$

except that in this case the geometry of the situation is not so easily visualised.

10.2   Repeated integrals

The most straightforward method of evaluating a double integral is to think of the region D as divided up into little rectangles as in the diagram.

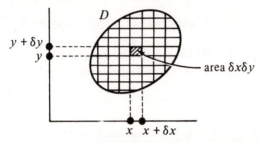

We then begin by considering each row of little rectangles separately.

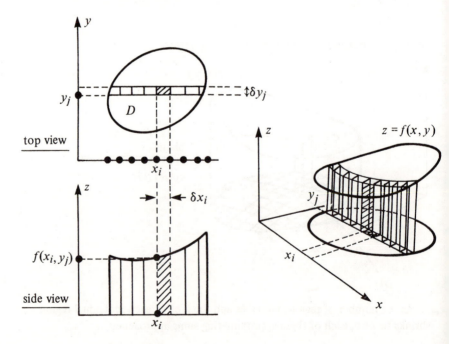

The volume above the row of little rectangles in D between y_j and $y_j + \delta y_j$ is approximately

$$\sum_i f(x_i, y_j)\delta x_i \delta y_j = \delta y_j \sum_i f(x_i, y_j)\delta x_i.$$

This in turn is approximately

$$\delta y_j \int_{a(y_j)}^{b(y_j)} f(x, y_j)dx$$

where the limits of integration are as indicated below:

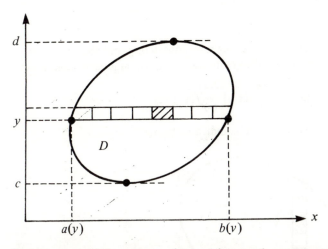

Having estimated the volumes obtained from each separate row, we can now estimate the total volume by adding these up. An approximation to the volume under the surface is therefore

$$\sum_j \delta y_j \int_{a(y_j)}^{b(y_j)} f(x, y_j)dx.$$

This in turn is approximately equal to

$$\int_c^d \left\{ \int_{a(y)}^{b(y)} f(x, y)dx \right\} dy$$

which it is more convenient to write as

$$\int_c^d dy \int_{a(y)}^{b(y)} f(x, y)dx.$$

All of these approximations improve as the number of little rectangles increases and the area of each individual rectangle shrinks to zero. It follows that

$$\iint_D f(x,y)dxdy = \int_c^d dy \int_{a(y)}^{b(y)} f(x,y)dx.$$

Thus the double integral may be evaluated by regarding it as a repeated integral. We first compute

$$I(y) = \int_{a(y)}^{b(y)} f(x,y)dx$$

for each y ($c \le y \le d$) and then evaluate

$$\int_c^d I(y)dy.$$

One can equally well, of course, begin by considering contributions from each column separately (instead of each row). A similar argument then shows that

$$\iint_D f(x,y)dxdy = \int_a^b dx \int_{c(x)}^{d(x)} f(x,y)dy$$

where the limits of integration are as indicated below:

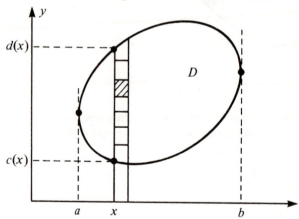

This means, in particular, that

$$\int_c^d dy \int_{a(y)}^{b(y)} f(x,y)dx = \int_a^b dx \int_{c(x)}^{d(x)} f(x,y)dy.$$

Example 10.3.

Calculate the mass of the flat rectangular plate indicated below given that its density at the point $(x,y)^T$ is $\rho(x,y) = x^2y^3 + 3x$.

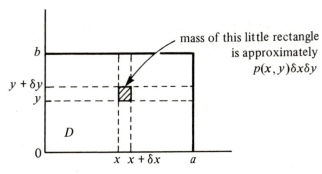

The problem reduces to evaluating

$$M = \iint\limits_{D} \rho(x,y)dxdy = \iint\limits_{D} (x^2y^3 + 3x)dxdy.$$

We can arrange this double integral as a repeated integral in two ways. Only one of these repeated integrals need be calculated. However, we give both calculations to check that the answers are equal.

(i) By rows   (ii) By columns

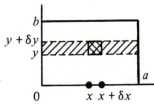

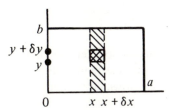

$$M = \int_0^b dy \int_0^a (x^2y^3 + 3x)dx \qquad M = \int_0^a dx \int_0^b (x^2y^3 + 3x)dy$$

$$= \int_0^b dy \left[\frac{1}{3}x^3y^3 + \frac{3}{2}x^2\right]_0^a \qquad = \int_0^a dx \left[\frac{1}{4}x^2y^4 + 3xy\right]_0^b$$

$$= \int_0^b \left(\frac{1}{3}a^3y^3 + \frac{3}{2}a^2\right) dy \qquad = \int_0^a \left(\frac{1}{4}x^2b^4 + 3xb\right) dx$$

$$= \left[\frac{1}{12}a^3y^4 + \frac{3}{2}a^2y\right]_0^b \qquad = \left[\frac{1}{12}x^3b^4 + \frac{3}{2}x^2b\right]_0^a$$

$$= \frac{1}{12}a^3b^4 + \frac{3}{2}a^2b \qquad = \frac{1}{12}a^3b^4 + \frac{3}{2}a^2b$$

Example 10.4.

The case when D is a rectangle is particularly easy. We therefore next consider the problem of calculating the area of the triangle illustrated below:

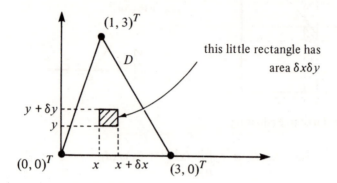

To make the problem interesting we shall feign ignorance of the formula '$\frac{1}{2}$ (base × height)' and instead evaluate the double integral

$$A = \iint_D dxdy$$

Again we consider both repeated integrals. Note that, as in example 10.3, we begin by drawing diagrams. Without the aid of such diagrams it is easy to make a mistake when determining the limits of integration in the repeated integrals.

(i) <u>By rows</u>   (ii) <u>By columns</u>

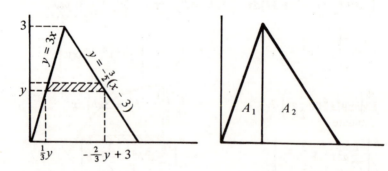

In the second case it is best to proceed by splitting A into two areas A_1 and A_2 as indicated above.

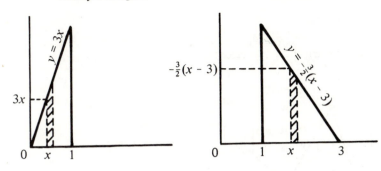

$$A = \int_0^3 dy \int_{y/3}^{-2y/3 + 3} dx \qquad\qquad A = A_1 + A_2$$

$$= \int_0^3 {}^{-2y/3+3}_{\ \ y/3}[\, x\,]\; dy \qquad\qquad = \int_0^1 dx \int_1^{3x} dy + \int_1^3 dx \int_0^{-3(x-3)/2} dy$$

$$= \int_0^3 \left(-\frac{2}{3}y + 3 - \frac{1}{3}y\right) dy \qquad = \int_0^1 dx \,{}^{3x}_{\ 0}[\,y\,] + \int_1^3 dx \,{}^{-3(x-3)/2}_{\qquad 0}[\quad y\,]$$

$$= \int_0^3 (3-y) dy \qquad\qquad = \int_0^1 3x\,dx + \int_1^3 -\frac{3}{2}(x-3) dx$$

$$= {}^3_0\!\left[3y - \frac{1}{2}y^2\right] \qquad\qquad = {}^1_0\!\left[\frac{3}{2}x^2\right] + {}^3_1\!\left[-\frac{3}{2}\!\left(\frac{1}{2}x^2 - 3x\right)\right]$$

$$= 9 - \frac{9}{2} = 4\tfrac{1}{2}. \qquad\qquad = \frac{3}{2}\left(1 + \frac{9}{2} - \frac{5}{2}\right) = 4\tfrac{1}{2}.$$

Example 10.5.
Evaluate

$$I = \iint_D xy\,dxdy$$

when D is the region
indicated in the diagram.

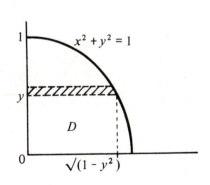

We have that

$$I = \int_0^1 dy \int_0^{\sqrt{(1-y^2)}} xy\,dx$$

$$= \int_0^1 ydy \int_0^{\sqrt{(1-y^2)}} x\,dx$$

$$= \int_0^1 ydy \left[\frac{1}{2}x^2\right]_0^{\sqrt{(1-y^2)}}$$

$$= \int_0^1 \frac{1}{2}y(1-y^2)\,dy$$

$$= \left[\frac{1}{4}y^2 - \frac{1}{8}y^4\right]_0^1 = \frac{1}{4} - \frac{1}{8} = \frac{1}{8}.$$

Example 10.6.
Change the order of integration in the repeated integral

$$J = \int_0^a dx \int_0^{a-\sqrt{(a^2-x^2)}} \frac{xe^y}{(y-a)^2}\,dy$$

where $a > 0$ and hence evaluate it. (Note that it is quite intractable in its present form.)

The first and vital step is to determine the region D over which we are integrating. For a fixed value of x, y ranges between 0 and $a - \sqrt{(a^2 - x^2)}$. This observation leads us to the left-hand diagram below. What we have to do is to rearrange the integral as indicated in the right-hand diagram.

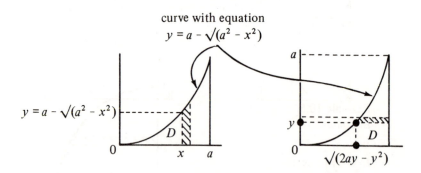

curve with equation
$y = a - \sqrt{(a^2 - x^2)}$

Solving the equation $y = a - \sqrt{(a^2 - x^2)}$ for x yields that $x = \sqrt{(2ay - y^2)}$. Thus when we reverse the order of integration we obtain that

$$J = \int_0^a dy \int_{\sqrt{(2ay-y^2)}}^a \frac{xe^y}{(y-a)^2} dx$$

$$= \int_0^a \frac{e^y}{(y-a)^2} dy \int_{\sqrt{(2ay-y^2)}}^a x\,dx$$

$$= \int_0^a \frac{e^y}{(y-a)^2} dy \left[\frac{1}{2} x^2 \right]_{\sqrt{(2ay-y^2)}}^a$$

$$= \int_0^a \frac{e^y}{(y-a)^2} \left(\frac{1}{2} a^2 - \frac{1}{2}(2ay - y^2) \right) dy$$

$$= \int_0^a \frac{e^y}{(y-a)^2} \frac{1}{2}(y-a)^2 dy = \frac{1}{2} \int_0^a e^y dy = \frac{1}{2}(e^a - 1).$$

10.7 Determinants and area†

Let M be a 2×2 non-singular matrix and consider the linear function $L : \mathbb{R}^2 \to \mathbb{R}^2$ defined by $y = L(x) = Mx$.

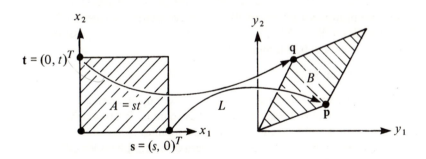

What is the relation between the areas A and B in the above diagram? We begin by calculating B using the fact that the area of the triangle in the following diagram is equal to $\frac{1}{2}ab \sin \theta$.

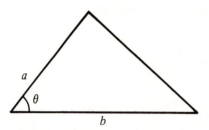

Thus
$$B^2 = \|\mathbf{p}\|^2 \|\mathbf{q}\|^2 \sin^2\theta = \|\mathbf{p}\|^2 \|\mathbf{q}\|^2 (1 - \cos^2\theta)$$
and so
$$B^2 = \|\mathbf{p}\|^2 \|\mathbf{q}\|^2 - \langle \mathbf{p}, \mathbf{q} \rangle^2.$$

But $\mathbf{p} = M\mathbf{s}$ and $\mathbf{q} = M\mathbf{t}$. Therefore

$$
\left. \begin{aligned}
p_1 &= m_{11}s \\
p_2 &= m_{21}s
\end{aligned} \right\}
\qquad
\left. \begin{aligned}
q_1 &= m_{12}t \\
q_2 &= m_{22}t
\end{aligned} \right\}
$$

It follows that

$$\|\mathbf{p}\|^2 = (m_{11}^2 + m_{21}^2)s^2; \qquad \|\mathbf{q}\|^2 = (m_{12}^2 + m_{22}^2)t^2$$

$$\langle \mathbf{p}, \mathbf{q} \rangle = (m_{11}m_{12} + m_{21}m_{22})st.$$

Thus
$$B^2 = \{(m_{11}^2 + m_{21}^2)(m_{12}^2 + m_{22}^2) - (m_{11}m_{12} + m_{21}m_{22})^2\}s^2 t^2$$

$$= (\det M)^2 A^2.$$

(To check this last step, expand $(\det M)^2 = (m_{11}m_{22} - m_{12}m_{21})^2$.)

We have shown that $B^2 = (\det M)^2 A^2$ and so either $B = (\det M)A$ or else $B = -(\det M)A$. Since $B > 0$ we require the alternative with a positive right-hand side. Since $A > 0$, this means that

$$B = |\det M| A.$$

This useful geometric interpretation of a determinant extends to the 3×3 case provided that A and B are interpreted as volumes. It also extends to the $n \times n$ case with A and B interpreted as 'hypervolumes'.

When applying the result in the next section, we shall be using affine functions rather than linear functions. If M is a 2×2 non-singular matrix and the affine function $\alpha : \mathbb{R}^2 \to \mathbb{R}^2$ is defined by

$$\mathbf{y} = \alpha(\mathbf{x}) = M(\mathbf{x} - \boldsymbol{\xi}) + \boldsymbol{\eta},$$

then the equation $B = |\det M| A$ still holds with A and B equal to the areas indicated below.

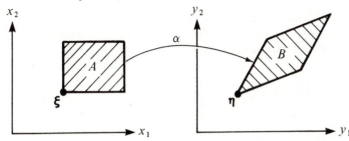

(To see this, simply take $\mathbf{X} = \mathbf{x} - \boldsymbol{\xi}$ and $\mathbf{Y} = \mathbf{y} - \boldsymbol{\eta}$ and consider the equation $\mathbf{Y} = M\mathbf{X}$.) A similar result holds, of course, in the case when M is an $n \times n$ matrix.

10.8   Change of variable in multiple integrals†

In §6.16 we considered the question of changing variables in $\mathbb{R}^2$. Similar considerations are valid in $\mathbb{R}^n$. In §6.19 we explained how such changes of variable transform differential operators. In this section we shall examine how such a change of variable transforms a multiple integral. Recalling the success which this technique enjoys for the evaluation of one-dimensional integrals, we should anticipate that this will prove a profitable enterprise.

Suppose that it is proposed to introduce the change of variable defined by

(1)
$$\left.\begin{array}{l} u = \phi_1(x,y) \\ v = \phi_2(x,y) \end{array}\right\}$$

into the double integral

(2)
$$\iint_D f(x,y)\,dx\,dy.$$

It is useful to consider the function ϕ defined by

$$\begin{pmatrix} u \\ v \end{pmatrix} = \phi(x,y) = \begin{pmatrix} \phi_1(x,y) \\ \phi_2(x,y) \end{pmatrix}$$

Suppose that this function maps the region D onto the region Δ.

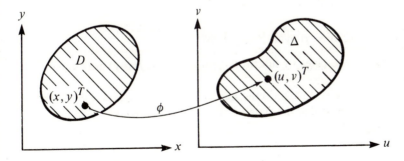

We shall restrict our attention to the case when $\phi : D \to \Delta$ admits an inverse function $\psi : \Delta \to D$. The reasons why this condition is necessary when changing co-ordinate systems in $\mathbb{R}^n$ are explained in §6.16. In the formula

$$\begin{pmatrix} x \\ y \end{pmatrix} = \psi(u, v) = \begin{pmatrix} \psi_1(u, v) \\ \psi_2(u, v) \end{pmatrix},$$

the equations

$$(3) \quad \begin{array}{c} x = \psi_1(u, v) \\ y = \psi_2(u, v) \end{array} \Bigg\}$$

are obtained by solving equations (1) for x and y in terms of u and v.

We now divide the region Δ into a large number of small rectangles. The lines $u = u_i$ and $v = v_i$ in the $(u, v)^T$ plane correspond to the curves $\phi_1(x, y) = u_i$ and $\phi_2(x, y) = v_i$ in the $(x, y)^T$ plane. Our subdivision of Δ therefore induces a corresponding subdivision of D (see §6.16 again).

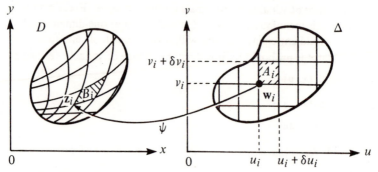

We shall write $\mathbf{z} = (x, y)^T$ and $\mathbf{w} = (u, v)^T$. The shaded area with a corner at $\mathbf{z}_i = (x_i, y_i)^T$ will be denoted by B_i and the corresponding shaded area with a corner at $\mathbf{w}_i = (u_i, v_i)^T$ will be denoted by A_i.

An approximating sum to the double integral (2) is

$$\sum_i f(x_i, y_i)B_i = \sum_i F(u_i, v_i)B_i$$

where

$$F(u, v) = f(\psi_1(u, v), \psi_2(u, v))$$

i.e. $F(u, v)$ is obtained by substituting for x and y in $f(x, y)$ using equations (3).

The next step is to express the area B_i in terms of the area $A_i = \delta u_i \delta v_i$. It is for this purpose that the material of §10.7 is required. We begin by observing that

$$\alpha(\mathbf{w}) = \psi'(\mathbf{w}_i)(\mathbf{w} - \mathbf{w}_i) + \psi(\mathbf{w}_i)$$

is a good approximation to $\psi(\mathbf{w})$ for values of $\mathbf{w}$ close to $\mathbf{w}_i$. The reason is that $\mathbf{z} = \alpha(\mathbf{w})$ is *tangent* to $\mathbf{z} = \psi(\mathbf{w})$ at the point $\mathbf{w}_i$ (see §4.4). It follows that C_i in the diagram below is a good approximation to B_i provided that δu_i and δv_i are small.

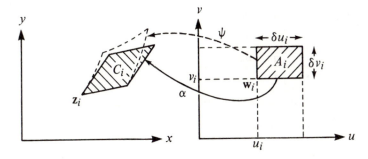

But, since α is an affine function, it follows from §10.7 that

$$B_i \doteq C_i = |\det \psi'(u_i, v_i)| A_i$$
$$= |\det \psi'(u_i, v_i)| \delta u_i \delta v_i.$$

Returning to our approximating sum, we obtain that

$$\sum_i f(x_i, y_i) B_i \doteq \sum_i F(u_i, v_i) |\det \psi'(u_i, v_i)| \delta u_i \delta v_i.$$

But this final sum is an approximating sum for the double integral,

$$(4) \qquad \iint_\Delta F(u, v) |\det \psi'(u, v)| \, du dv.$$

We conclude that (2) is equal to (4). We give some special attention to the term $|\det \psi'(u, v)|$ in the next section.

10.9 Jacobians

Given a differentiable function $\phi: \mathbb{R}^n \to \mathbb{R}^n$, the number

$$\det \phi'(\mathbf{z})$$

is called the *Jacobian* of the function ϕ at the point $\mathbf{z}$. A point at which the Jacobian is zero is called a *critical point* for the function. As we know from §6.16, it is essential for the argument given in §10.9 that there be *no* critical points for ϕ inside the region D.

In the case when

$$\begin{pmatrix} u \\ v \end{pmatrix} = \phi(x, y)$$

as in §10.8 we have that

$$\det \phi'(x, y) = \begin{vmatrix} \dfrac{\partial u}{\partial x} & \dfrac{\partial u}{\partial y} \\[2mm] \dfrac{\partial v}{\partial x} & \dfrac{\partial v}{\partial y} \end{vmatrix}$$

which leads naturally to the standard notation for a Jacobian, namely

$$\frac{\partial(u, v)}{\partial(x, y)} = \begin{vmatrix} \dfrac{\partial u}{\partial x} & \dfrac{\partial u}{\partial y} \\[2mm] \dfrac{\partial v}{\partial x} & \dfrac{\partial v}{\partial y} \end{vmatrix}.$$

In general we have that

$$\frac{\partial(u_1, u_2, \ldots, u_n)}{\partial(x_1, x_2, \ldots, x_n)} = \begin{vmatrix} \dfrac{\partial u_1}{\partial x_1} & \dfrac{\partial u_1}{\partial x_2} & \cdots & \dfrac{\partial u_1}{\partial x_n} \\[2mm] \dfrac{\partial u_2}{\partial x_1} & \dfrac{\partial u_2}{\partial x_2} & \cdots & \dfrac{\partial u_2}{\partial x_n} \\ \cdot & & & \\ \cdot & & & \\ \cdot & & & \\ \dfrac{\partial u_n}{\partial x_1} & \dfrac{\partial u_n}{\partial x_2} & \cdots & \dfrac{\partial u_n}{\partial x_n} \end{vmatrix}$$

Jacobians have a number of convenient properties which follow from the fact that $\det(AB) = (\det A)(\det B)$ for any $n \times n$ matrices A and B. For example, we know that

$$\frac{d\mathbf{z}}{d\mathbf{x}} = \frac{d\mathbf{z}}{d\mathbf{y}}\frac{d\mathbf{y}}{d\mathbf{x}}$$

and hence

$$\det\left(\frac{d\mathbf{z}}{d\mathbf{x}}\right) = \det\left(\frac{d\mathbf{z}}{d\mathbf{y}}\right)\det\left(\frac{d\mathbf{y}}{d\mathbf{x}}\right)$$

provided that $\mathbf{x}, \mathbf{y}$ and $\mathbf{z}$ are $n \times 1$ column vectors. It follows that

$$\frac{\partial(z_1,\ldots,z_n)}{\partial(x_1,\ldots,x_n)} = \frac{\partial(z_1,\ldots,z_n)}{\partial(y_1,\ldots,y_n)}\frac{(y_1,\ldots,y_n)}{\partial(x_1,\ldots,x_n)}.$$

In the context of changing the variable in multiple integrals, however, the really useful formula is

$$(5) \quad \frac{\partial(x_1,x_2,\ldots,x_n)}{\partial(u_1,u_2,\ldots,u_n)} = \left\{\frac{\partial(u_1,u_2,\ldots,u_n)}{\partial(x_1,x_2,\ldots,x_n)}\right\}^{-1}$$

which follows from the fact that

$$\frac{dx}{du} = \left(\frac{du}{dx}\right)^{-1}.$$

In §10.8 we obtained the formula

$$\iint_D f(x,y)\,dxdy = \iint_\Delta F(u,v)\left|\frac{\partial(x,y)}{\partial(u,v)}\right|\,dudv$$

The general version of this is

$$\iint_D \ldots \int f(x_1,\ldots,x_n)dx_1,\ldots,dx_n =$$

$$\iint_\Delta \ldots \int F(u_1,\ldots,u_n)\left|\frac{\partial(x_1,\ldots,x_n)}{\partial(u_1,\ldots,u_n)}\right|du_1,\ldots,du_n.$$

The significance of formula (5) is that it allows us to calculate the necessary Jacobian without first solving equations (1) of §10.8. This sometimes saves a lot of work.

Example 10.10.
We return to example 10.5 and consider the double integral

$$I = \iint_D xy\,dxdy.$$

Since part of the boundary of D is circular, it is natural to think of using polar co-ordinates.

We have that

$$x = r\cos\theta$$

$$y = r\sin\theta$$

from which it follows that the lines $x = 0$ and $y = 0$ correspond to $\theta = \pi/2$ and $\theta = 0$ respectively. Moreover, since $r^2 = x^2 + y^2$, the circle $x^2 + y^2 = 1$ corresponds to $r = 1$. These observations make it clear that Δ is the region indicated below:

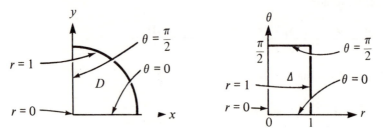

The Jacobian is given by

$$\frac{\partial(x, y)}{\partial(r, \theta)} = \begin{vmatrix} \dfrac{\partial x}{\partial r} & \dfrac{\partial x}{\partial \theta} \\[2mm] \dfrac{\partial y}{\partial r} & \dfrac{\partial y}{\partial \theta} \end{vmatrix} = \begin{vmatrix} \cos\theta & -r\sin\theta \\[1mm] \sin\theta & r\cos\theta \end{vmatrix} = r(\cos^2\theta + \sin^2\theta) = r.$$

It follows that

$$\iint_D xy\,dxdy = \iint_\Delta (r\cos\theta)(r\sin\theta)|r|\,drd\theta$$

$$= \iint_\Delta r^3\cos\theta\,\sin\theta\,drd\theta.$$

Because Δ is rectangular, it is easy to evaluate this double integral. We obtain that

$$\iint_D xy\,dxdy = \int_0^1 r^3\,dr \int_0^{\pi/2} \cos\theta\,\sin\theta\,d\theta$$

$$= \int_0^1 r^3\,dr \int_0^{\pi/2} \frac{\sin 2\theta}{2}\,d\theta$$

$$= \int_0^1 r^3\,dr \left[-\frac{\cos 2\theta}{4}\right]_0^{\pi/2}$$

$$= \frac{1}{2}\int_0^1 r^3\,dr = \frac{1}{2}\left[\frac{1}{4}r^4\right]_0^1 = \frac{1}{8}.$$

(Note that there is a critical point at $(x, y)^T = (0, 0)^T$ but this is not inside the region D.)

Example 10.11.

We give a second example using polar co-ordinates because of the inportance of the result in statistics. The problem is to evaluate

$$I = \int_{-\infty}^{\infty} e^{-x^2/2} \, dx.$$

We have that

$$I^2 = \left\{ \int_{-\infty}^{\infty} e^{-x^2/2} \, dx \right\} \left\{ \int_{-\infty}^{\infty} e^{-y^2/2} \, dy \right\}$$

$$= \int_{-\infty}^{\infty} \int_{-\infty}^{\infty} e^{-(x^2+y^2)/2} \, dxdy.$$

The region Δ in the $(r, \theta)^T$ plane which corresponds to D in the case when D is the whole plane $\mathbb{R}^2$ is the set of those $(r, \theta)^T$ which satisfy $r \geqslant 0$ and $-\pi < \theta \leqslant \pi$ (see §6.18). Hence

$$I^2 = \iint_{\Delta} e^{-r^2/2} \left| \frac{\partial(x, y)}{\partial(r, \theta)} \right| drd\theta$$

$$= \iint_{\Delta} re^{-r^2/2} \, drd\theta$$

$$= \int_{-\pi}^{\pi} d\theta \int_{0}^{\infty} re^{-r^2/2} \, dr$$

$$= \int_{-\pi}^{\pi} d\theta \left[-e^{-r^2/2} \right]_{0}^{\infty} = \int_{-\pi}^{\pi} d\theta = 2\pi$$

It follows that

$$I = \sqrt{(2\pi)}.$$

Example 10.12.

Let D be the set of those $(x, y)^T$ in the first quadrant (i.e. where x and y are both non-negative) which lie above the curve $yx^2 = a$ and between the curves $y = bx^2$ and $y = cx^2$ (where $a > 0$ and $c > b > 0$). Evaluate

$$\iint_{D} \frac{dxdy}{y^2x^3}$$

by means of the change of variable

$$(6) \qquad \begin{aligned} u &= yx^2 \\ v &= \frac{y}{x^2}. \end{aligned} \Bigg\}$$

The first task is to sketch the region D and to determine the corresponding region Δ in the $(u, v)^T$ plane.

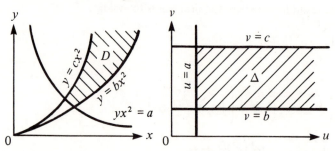

The region Δ can be identified by observing that the curves $yx^2 = a$, $y = bx^2$ and $y = cx^2$ map onto $u = a$, $v = b$ and $v = c$ respectively.

The Jacobian is

$$\frac{\partial(x, y)}{\partial(u, v)} = \left(\frac{\partial(u, v)}{\partial(x, y)} \right)^{-1} = \begin{vmatrix} 2xy & x^2 \\ -\dfrac{2y}{x^3} & \dfrac{1}{x^2} \end{vmatrix}^{-1}$$

$$= \left(\frac{2y}{x} + \frac{2y}{x} \right)^{-1} = \frac{x}{4y}$$

and so

$$\iint_D \frac{dx\,dy}{y^2 x^3} = \iint_\Delta \frac{1}{y^2 x^3} \left| \frac{\partial(x, y)}{\partial(u, v)} \right| du\,dv$$

$$= \iint_\Delta \frac{1}{y^2 x^3} \frac{x}{4y} du\,dv$$

$$= \frac{1}{4} \iint_\Delta \frac{1}{y^2 x^4} \frac{x^2}{y} du\,dv$$

$$= \frac{1}{4} \iint_\Delta \frac{1}{u^2} \frac{1}{v} du\,dv$$

$$= \frac{1}{4} \int_a^\infty \frac{du}{u^2} \int_b^c \frac{dv}{v}$$

$$= \frac{1}{4} \int_a^\infty \frac{du}{u^2} \left[\log v \right]_b^c$$

$$= \frac{1}{4} \log \frac{c}{b} \left[-\frac{1}{u} \right]_a^\infty = \frac{1}{4a} \log \frac{c}{b}.$$

(Note the delay in substituting for x and y in terms of u and v in case it should prove possible to avoid solving equations (6).)

Example 10.13.
Make the change of variable

$$u = x + y$$

$$v = \frac{y}{x}$$

in the integral

$$\int_0^1 dy \int_y^1 \frac{(x+y)}{x^2} e^{(x+y)} dx$$

and hence evaluate it.

It is first necessary, as in example 10.6, to identify the region D of integration. For a fixed value of y, x ranges between y and 1 and so the region D is as indicated in the diagram on the left. The corresponding region Δ is indicated on the right.

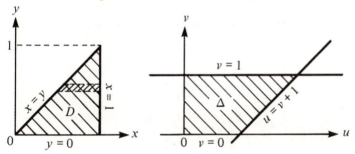

In determining the region Δ, one may begin by observing that the line $y = 0$ is mapped into $v = 0$ and the line $x = y$ into $v = 1$. When $x = 1$, we obtain that $u = y + 1$ and $v = y$. Thus $x = 1$ becomes $u = v + 1$. Normally this information would suffice to specify Δ but in this case one has to be a little careful since, although there is no critical point *inside D*, there is a problem about what happens when $x = 0$. This problem is most easily resolved by observing that, for each α with $0 \leqslant \alpha \leqslant 1$, the line segment $y = \alpha x \ (0 < x \leqslant 1)$ is mapped to $v = \alpha \ (0 < u \leqslant \alpha + 1)$.

The Jacobian is

$$\frac{\partial(x,y)}{\partial(u,v)} = \left|\frac{\partial(u,v)}{\partial(x,y)}\right|^{-1} = \begin{vmatrix} 1 & 1 \\ -\dfrac{y}{x^2} & \dfrac{1}{x} \end{vmatrix}^{-1} = \left(\frac{1}{x} + \frac{y}{x^2}\right)^{-1}.$$

It follows that

$$\int_0^1 dy \int_y^1 \frac{(x+y)}{x^2} e^{(x+y)} dx = \iint_D \frac{(x+y)}{x^2} e^{(x+y)} dx dy$$

$$= \iint_\Delta \frac{(x+y)}{x^2} e^{(x+y)} \left| \frac{x^2}{x+y} \right| du dv$$

$$= \iint_\Delta e^u \, du dv.$$

This may be evaluated as a repeated integral.

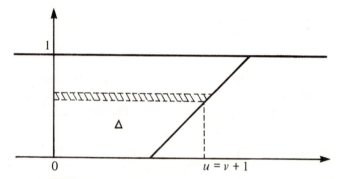

We obtain that

$$\iint_\Delta e^u \, du dv = \int_0^1 dv \int_0^{v+1} e^u \, du$$

$$= \int_0^1 dv [e^u]_0^{v+1} = \int_0^1 (e^{v+1} - 1) dv$$

$$= [e^{v+1} - v]_0^1 = e^2 - e - 1.$$

10.14   Unbounded regions of integration†

In example 10.12 and elsewhere we have manipulated a double integral over an unbounded region D quite freely, although our account of the theory given in §10.2 and §10.8 was restricted to the case when the integrand f is continuous and the region D is bounded. Such manipulations in the general case are legitimate, provided that the convergence of the double integral is dominated (see §9.30). This is guaranteed in the case of a convergent double integral of a *non-negative* integrand. However, applied mathematicians tend to give little attention to this question since it is only in exceptional cases that things can go wrong.

10.15   Multiple sums and series†

Multiple sums and series can be manipulated in much the same way as multiple integrals. We give some examples below.

Example 10.16.†
We have that, for $-1 < x < 1$,

$$(1-x)^{-1} = 1 + x + x^2 + x^3 + \ldots$$

$$= \sum_{n=0}^{\infty} x^n.$$

In example 9.36, we differentiated and obtained the result

$$(1-x)^{-2} = 1 + 2x + 3x^2 + 4x^3 + \ldots$$

$$= \sum_{n=1}^{\infty} nx^{n-1}$$

provided $-1 < x < 1$. The same result can also be obtained as follows. Note that, for $-1 < x < 1$,

$$(1-x)^{-2} = (1-x)^{-1}(1-x)^{-1}$$

$$= \left\{ \sum_{m=0}^{\infty} x^m \right\} \left\{ \sum_{n=0}^{\infty} x^n \right\}$$

$$= \sum_{m=0}^{\infty} \sum_{n=0}^{\infty} x^{m+n}.$$

We have that

$$\sum_{n=0}^{\infty} x^{m+n} = \sum_{l=m}^{\infty} x^l.$$

This is obtained by writing $l = m + n$ and observing that $l = m$ when $n = 0$. Hence

$$(1) \quad (1-x)^{-2} = \sum_{m=0}^{\infty} \sum_{l=m}^{\infty} x^l$$

We now propose to change the order of summation. As in the case of integration, a diagram is usually useful:

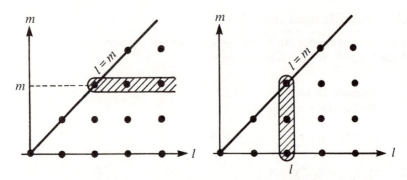

In the repeated series (1), we begin by summing over rows of points (as ·indicated in the left-hand diagram). If we first sum over columns (as in the right-hand diagram) we obtain that

$$\sum_{m=0}^{\infty} \sum_{l=m}^{\infty} x^l = \sum_{l=0}^{\infty} \sum_{m=0}^{l} x^l$$

$$= \sum_{l=0}^{\infty} x^l \sum_{m=0}^{l} 1$$

$$= \sum_{l=0}^{\infty} x^l (l+1)$$

Finally, we may write $k = l + 1$ and obtain

$$(1-x)^{-2} = \sum_{k=1}^{\infty} kx^{k-1}$$

$$= 1 + 2x + 3x^2 + \ldots$$

provided $-1 < x < 1$.

Example 10.17.[†]

When dealing with double series it is sometimes a good idea to begin neither with rows nor with columns but with *diagonals*. In the case of power series, this can be a particularly helpful technique. We have that

$$\left(\sum_{m=0}^{\infty} a_m x^m \right) \left(\sum_{n=0}^{\infty} b_n x^n \right) = \sum_{m=0}^{\infty} \sum_{n=0}^{\infty} a_m b_n x^{m+n}.$$

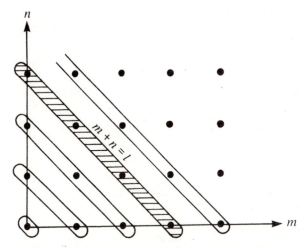

If we begin by summing along diagonals as indicated, we obtain that

$$\left(\sum_{m=0}^{\infty} a_m x^m \right) \left(\sum_{n=0}^{\infty} b_n x_n \right) = \sum_{l=0}^{\infty} \sum_{m=0}^{l} a_m b_{l-m} x^l$$

$$= \sum_{l=0}^{\infty} x^l \left\{ \sum_{m=0}^{l} a_m b_{l-m} \right\}$$

provided x lies inside the interval of convergence of the two power series.

Applying this result to the problem considered in the previous example, we obtain the following alternative argument:

$$(1-x)^{-2} = (1-x)^{-1}(1-x)^{-1} = \left(\sum_{m=0}^{\infty} x^m \right) \left(\sum_{n=0}^{\infty} x^n \right)$$

$$= \sum_{l=0}^{\infty} x^l \sum_{m=0}^{l} 1$$

$$= \sum_{l=0}^{\infty} (l+1) x^l$$

provided $-1 < x < 1$.

Exercises 10.18

1. By changing the order of integration in the repeated integral, prove that

$$\int_0^{\pi/2} dx \int_x^{\pi/2} \frac{\sin y}{y} dy = 1.$$

2.* Change the order of integration in the repeated integral

$$\int_0^1 dx \int_x^{2-x} \frac{x}{y} dy$$

and evaluate the result.

3. If $a > 0$, prove that

$$\int_0^a y dy \int_{y^2/a}^y f(x,y) dx = \int_0^a dx \int_x^{\sqrt{(ax)}} f(x,y) dy.$$

Deduce that

$$\int_0^a y dy \int_{y^2/a}^y \frac{dx}{(a-x)\sqrt{(ax-y^2)}} = \frac{1}{2}\pi a.$$

4.* Evaluate the double integral

$$\iint_D e^{(ax+by)} dxdy$$

where D is the triangle enclosed by the lines $x = 0, y = 0$ and $ax + by = 1$.

5.† The identity

$$I(a,b) = \int_0^\infty (\sin bx) e^{-ax} dx = \frac{b}{a^2 + b^2}$$

was given in exercise 9.37 (2). By considering

$$\frac{\partial}{\partial b} \int_a^\infty I(\alpha, b) d\alpha$$

deduce the identity

$$J(a,b) = \int_0^\infty (\cos bx) e^{-ax} dx = \frac{a}{a^2 + b^2}.$$

6.*† Deduce the identity given in the previous question for $I(a, b)$ from the identity given for $J(a, b)$ by considering

$$\frac{\partial}{\partial a} \int_0^b J(a, \beta) d\beta.$$

7. The region D is the part of the positive quadrant enclosed by the curves $x^2 + 2y^2 = 1, x^2 + 2y^2 = 4, y = 2x$ and $y = 5x$. Sketch D and evaluate the double integral

$$\iint_D \frac{y}{x} dxdy$$

by introducing the change of variable

$$u = x^2 + 2y^2$$
$$v = \frac{y}{x}.$$

8.* Evaluate

$$\iint_D x^2 y^2 (y^2 - x^2) dx dy$$

where D is the region in the first quadrant enclosed by the lines $y = x + 1$, $y = x + 3$ and the curves $xy = 1, xy = 4$ by making the change of variable

$$u = xy$$
$$v = x - y.$$

9. Consider the function $f: \mathbb{R}^2 \to \mathbb{R}^2$ defined by

$$\begin{pmatrix} u \\ v \end{pmatrix} = f(x, y) = \begin{pmatrix} xe^y \\ ye^x \end{pmatrix}.$$

Sketch the region in the $(x, y)^T$ plane which is mapped onto the square in the $(u, v)^T$ plane enclosed by the lines $u = 1, u = e, v = 1, v = e$. At what points does the Jacobian of f vanish and what significance does this have?
Use the function given above to change variables in the double integral

$$\iint_D (x^2 y^3 - x^3 y^4) dx dy$$

where D is the region enclosed by the curves $y + \log x = 0, y + \log x = 1$, $x + \log y = 0$ and $x + \log y = 1$. Hence evaluate the integral.
10.* Sketch the region D in the first quadrant which lies above the curve $y^2 x^3 = a$ and between the curves $y^2 = bx^3, y^2 = cx^3$ where $a > 0$ and $0 < b < c$. By making the change of variable

$$u = y^2 x^3$$
$$v = y^2 x^{-3},$$

evaluate the double integral

$$\iint_D \frac{dx dy}{x^4 y^7}.$$

11. Introduce the change of variable defined by

$$x = r \cos \theta \sin \phi$$
$$y = r \sin \theta \sin \phi$$
$$z = r \cos \phi$$

(see §6.18) into the triple integral

$$\iiint_D dxdydz$$

where D is the inside of a sphere with radius r and centre $(0, 0, 0)^T$. Hence calculate the volume of a sphere.

12.* Evaluate

$$\iiint_D \left\{ \frac{x^2}{a^2} + \frac{y^2}{b^2} + \frac{z^2}{c^2} \right\} dxdydz$$

where D is the inside of the ellipsoid

$$\frac{x^2}{a^2} + \frac{y^2}{b^2} + \frac{z^2}{c^2} = 1.$$

[Hint: Begin with the change of variable $X = x/a$, $Y = y/b$, $Z = z/c$ and then proceed as in the previous question.]

13.† Prove that $e^{x+y} = e^x e^y$ by considering

$$\sum_{n=0}^{\infty} \frac{(x+y)^n}{n!} = \sum_{n=0}^{\infty} \frac{1}{n!} \sum_{k=0}^{n} \binom{n}{k} x^k y^{n-k}.$$

14.*† Explain why

$$\sum_{j=1}^{\infty} \sum_{\substack{k=1 \\ j \neq k}}^{\infty} \frac{1}{j^2 - k^2} \neq \sum_{k=1}^{\infty} \sum_{\substack{j=1 \\ j \neq k}}^{\infty} \frac{1}{j^2 - k^2}.$$

SOME APPLICATIONS (OPTIONAL)

10.19 Joint probability distributions

Suppose that $f: \mathbb{R}^2 \to \mathbb{R}$ has the property that for all regions D in $\mathbb{R}^2$

$$\iint_D f(x, y) dxdy$$

represents the probability that the pair $(X, Y)^T$ of random variables lies in the region D. We then say that f is the *joint probability density function* for the continuous random variables X and Y. Similar considerations apply in the discrete case except that the double integral is replaced by a double sum.

To qualify as a joint probability density function in the continuous case, a function $f: \mathbb{R}^2 \to \mathbb{R}$ must satisfy $f(x, y) \geqslant 0$ for all $(x, y)^T$. Moreover

$$\iint_{\mathbb{R}^2} f(x, y) dx dy = 1.$$

Similarly in the discrete case.

10.20 Marginal probability distributions

Suppose that $f: \mathbb{R}^2 \to \mathbb{R}$ is the joint probability density function for the continuous random variables X and Y. Then

$$\text{Prob} \, (a \leqslant X \leqslant b) = \iint_D f(x, y) dx dy$$

where D is the strip in $\mathbb{R}^2$ indicated below:

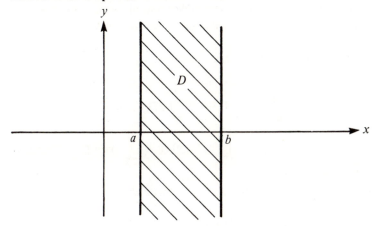

A point $(x, y)^T$ lies in this strip D provided $a \leqslant x \leqslant b$ regardless of the value of y. Observe that

$$\iint_D f(x, y) dx dy = \int_a^b dx \int_{-\infty}^{\infty} f(x, y) dy$$

and hence

$$\text{Prob} \, (a \leqslant X \leqslant b) = \int_a^b g(x) dx$$

where $g: \mathbb{R} \to \mathbb{R}$ is defined by

$$g(x) = \int_{-\infty}^{\infty} f(x, y)dy.$$

Thus g is the probability density function for X. We say that g is a *marginal* density function derived from the joint density function f.

Similarly, the probability density function for Y is given by

$$h(y) = \int_{-\infty}^{\infty} f(x, y)dx.$$

In the case of discrete random variables, the integrals must be replaced by sums.

10.21   Expectation, variance and covariance

An immediate consequence of the above discussion is that

$$\mathscr{E}(X + Y) = \mathscr{E}(X) + \mathscr{E}(Y)$$

which we used frequently in discussing applications of the results of the previous chapter. In the case of continuous random variables

$$
\begin{aligned}
\mathscr{E}(X + Y) &= \iint_{R^2} (x + y)f(x, y)\, dxdy \\
&= \iint_{R^2} xf(x, y)dxdy + \iint_{R^2} yf(x, y)dxdy \\
&= \int_{-\infty}^{\infty} xdx \int_{-\infty}^{\infty} f(x, y)dy + \int_{-\infty}^{\infty} ydy \int_{-\infty}^{\infty} f(x, y)dx \\
&= \int_{-\infty}^{\infty} xg(x)dx + \int_{-\infty}^{\infty} yh(y)dy \\
&= \mathscr{E}(X) + \mathscr{E}(Y).
\end{aligned}
$$

Write $\mathscr{E}(X) = \mu$ and $\mathscr{E}(Y) = \nu$. Then

$$
\begin{aligned}
\text{Var } (X + Y) &= \mathscr{E}\{(X + Y) - (\mu + \nu)\}^2 \\
&= \mathscr{E}(X - \mu)^2 + 2\mathscr{E}(X - \mu)(Y - \nu) + \mathscr{E}(Y - \nu)^2.
\end{aligned}
$$

The quantity $\mathscr{E}(X - \mu)(Y - \nu)$ is called the covariance of X and Y and denoted by

$$\text{Cov } (X, Y) = \mathscr{E}(X - \mu)(Y - \nu).$$

Thus

$$\text{Var } (X + Y) = \text{Var } (X) + 2\text{ Cov } (X, Y) + \text{Var } (Y)$$

The number

$$\rho(X, Y) = \frac{\text{Cov}(X, Y)}{\sqrt{((\text{Var } X)(\text{Var } Y))}}$$

is called the *correlation coefficient* for X and Y. It is a measure of the tendency of the two random variables to vary together. Its largest possible value is 1 and this is achieved when $X = Y$. Its smallest possible value is zero and this is achieved when X and Y are independent. In the latter case, $\mathscr{E}(XY) = (\mathscr{E}X)(\mathscr{E}Y)$ and so $\text{Cov}(X, Y) = 0$.

10.22 Independent random variables

Two random variables X and Y are independent if it is always the case that

$$\text{Prob }(\xi \leqslant X \leqslant x \text{ and } \eta \leqslant Y \leqslant y)$$

$$= \text{Prob }(\xi \leqslant X \leqslant x) \ \text{Prob }(\eta \leqslant Y \leqslant y).$$

In the continuous case, this means that

$$\int_\xi^x ds \int_\eta^y f(s, t)dt = \left(\int_\xi^x g(s)ds\right)\left(\int_\eta^y h(t)dt\right).$$

Differentiating both sides successively with respect to x and y, we obtain that

$$f(x, y) = g(x)h(y).$$

If the continuous random variables X and Y are independent, it follows that

$$\mathscr{E}(XY) = \iint_{\mathbf{R}^2} xyf(x, y)dxdy$$

$$= \iint_{\mathbf{R}^2} xyg(x)h(y)dxdy$$

$$= \left(\int_{-\infty}^\infty xg(x)dx\right)\left(\int_{-\infty}^\infty yh(y)dy\right)$$

$$= \mathscr{E}(X)\mathscr{E}(Y).$$

It is often useful to know the probability density function k for the random variable $Z = X + Y$ in the case when X and Y are independent. We have that

$$\text{Prob }(a \leqslant Z \leqslant b) = \iint_D f(x, y)dxdy$$

where D is the region indicated in the diagram below:

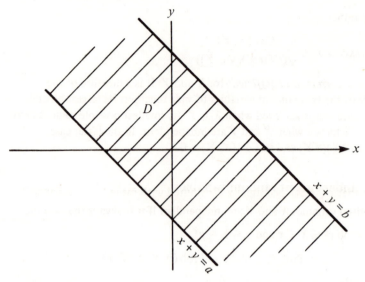

with the change of variable

$$t = x \atop z = x+y \Big\},$$

we obtain that

$$\iint\limits_{D} f(x, y)dxdy = \int_{a}^{b} dz \int_{-\infty}^{\infty} f(t, z - t)dt$$

$$= \int_{a}^{b} dz \int_{-\infty}^{\infty} g(t)h(z - t)dt.$$

Hence

$$k(z) = \int_{-\infty}^{\infty} g(t)h(z - t)dt.$$

As an example, consider the case when X_1 and X_2 are independent standardized normal random variables. Then the probability density function for $X_1 + X_2$ is

$$k(z) = \frac{1}{2\pi} \int_{-\infty}^{\infty} e^{-t^2/2} \, e^{-(z-t)^2/2} \, dt$$

$$= \frac{1}{2\pi} \int_{-\infty}^{\infty} e^{-(2t^2 - 2zt - z^2)/2} \, dt$$

$$= \frac{1}{2\pi} e^{-z^2/4} \int_{-\infty}^{\infty} e^{-(t-z/2)^2} \, dt$$

where we have 'completed the square' in the exponent (see §9.13). In the final integral, make the change of variable $t - z/2 = u/\sqrt{2}$. Then

$$k(z) = \frac{1}{2\pi}e^{-z^2/4}\frac{1}{\sqrt{2}}\int_{-\infty}^{\infty}e^{-u^2/2}\,du$$

$$= \frac{1}{2\pi}e^{-z^2/4}\frac{1}{\sqrt{2}}\sqrt{(2\pi)}$$

(example 10.11)

$$= \frac{1}{\sqrt{(4\pi)}}e^{-(z/\sqrt{2})^2/2}$$

It follows that $X_1 + X_2$ is normally distributed with mean 0 and variance 2. In the case when X_1 is normally distributed with mean μ_1 and variance σ_1^2 and X_2 is independent and normally distributed with mean μ_2 and variance σ_2^2, then $X_1 + X_2$ is normally distributed with mean $\mu = \mu_1 + \mu_2$ and variance $\sigma^2 = \sigma_1^2 + \sigma_2^2$. A more complicated version of the argument given above suffices to prove this. (See also §12.16.)

10.23 Generating functions

Suppose that X_1 is a discrete random variable with probability density function g_1 and that X_2 is an independent discrete random variable with probability density function g_2. We shall suppose that X_1 and X_2 do not take negative values so that $g_1(j) = g_2(j) = 0$ when $j = -1, -2, -3, \ldots$.

The functions G_1 and G_2 defined by the power series

$$G_1(x) = \sum_{j=0}^{\infty} g_1(j)x^j; \quad G_2(x) = \sum_{k=0}^{\infty} g_2(k)x^k$$

are called *generating functions* for g_1 and g_2. They are useful because, by example 10.17,

$$G_1(x)G_2(x) = \sum_{l=0}^{\infty} h(l)x^l,$$

where

$$h(l) = \sum_{j=0}^{l} g_1(j)g_2(l-j) = \sum_{j=-\infty}^{\infty} g_1(j)g_2(l-j),$$

since $g_1(j) = 0$ for $j < 0$ and $g_2(l-j) = 0$ for $j > l$. It follows from §10.22 that h is the probability density function of the random variable $X_1 + X_2$.

As an example, consider the case in which X_1 and X_2 are independent Poisson random variables with parameters λ_1 and λ_2 respectively.

Then

$$G_1(x) = \sum_{j=0}^{\infty} \frac{\lambda_1^j e^{-\lambda_1}}{j!} x^j = e^{-\lambda_1} \sum_{j=0}^{\infty} \frac{(\lambda_1 x)^j}{j!} = e^{\lambda_1(x-1)}.$$

Similarly, $G_2(x) = e^{\lambda_2(x-1)}$. Thus

$$\begin{aligned}
G_1(x)G_2(x) &= e^{\lambda_1(x-1)}e^{\lambda_2(x-1)} \\
&= e^{(\lambda_1 + \lambda_2)(x-1)} \\
&= e^{\lambda(x-1)}
\end{aligned}$$

where $\lambda = \lambda_1 + \lambda_2$. It follows that the random variable $X_1 + X_2$ is Poisson with parameter $\lambda = \lambda_1 + \lambda_2$.

10.24   Multivariate normal distributions

We conclude this chapter by considering the multivariate normal distribution. This will give us the opportunity to review a number of useful techniques.

The continuous random variables $X_1, X_2, \ldots, X_n$ are said to have the *multivariate normal distribution* provided that their joint probability density function $f: \mathbb{R}^n \to \mathbb{R}$ is of the form

$$f(x_1, x_2, \ldots, x_n) = \frac{\sqrt{(\det A)}}{(\sqrt{(2\pi)})^n} e^{-(x-\mu)^T A(x-\mu)/2}$$

where μ is an $n \times 1$ column vector and A is a positive definite $n \times n$ matrix (see §5.6). Since A is positive definite if and only if its eigenvalues are positive, it is legitimate to denote the eigenvalues by $s_1^2, s_2^2, \ldots, s_n^2$. Note that the determinant of a square matrix is equal to the product of the eigenvalues (§5.6) and so $\sqrt{(\det A)} = s_1 s_2 \ldots s_n$.

The first thing to check is that f qualifies as a probability density function. We require that $f(x_1, x_2, \ldots, x_n) \geqslant 0$ for all $(x_1, x_2, \ldots, x_n)^T$ and that

$$\iint \ldots \int_{\mathbb{R}^n} f(x_1, x_2, \ldots, x_n) dx_1 dx_2 \ldots dx_n = 1.$$

In proving the latter condition, it is useful to recall from §5.4 that there exists an orthogonal matrix E (i.e. $E^T E = I$) such that $E^T A E = D$ where D is the diagonal matrix whose diagonal entries are $s_1^2, s_2^2, \ldots, s_n^2$. Write $D = S^T S = SS$ where

$$S = \begin{pmatrix} s_1 & 0 \ldots 0 \\ 0 & s_2 \ldots 0 \\ \vdots & \\ 0 & 0 \ldots s_n \end{pmatrix}.$$

Then $A = EDE^T = ES^T SE^T = (SE^T)^T (SE^T)$. This representation makes it natural to introduce the change of variable defined by

$$SE^T(\mathbf{x} - \mathbf{\mu}) = \mathbf{y}$$

because then

$$(\mathbf{x} - \mathbf{\mu})^T A (\mathbf{x} - \mathbf{\mu}) = \mathbf{y}^T \mathbf{y}.$$

The appropriate Jacobian is

$$\det\left(\frac{d\mathbf{x}}{d\mathbf{y}}\right) = \det(ES^{-1}) = (\det A)^{-1/2}.$$

It follows that

$$\int\int \ldots \int_{\mathbf{R}^n} f(x_1, \ldots, x_n) dx_1 \ldots dx_n$$

$$= \frac{1}{(\sqrt{(2\pi)})^n} \int\int \ldots \int_{\mathbf{R}^n} e^{-\mathbf{y}^T \mathbf{y}/2} \, dy_1 \ldots dy_n$$

$$= \frac{1}{(\sqrt{(2\pi)})^n} \int\int \ldots \int_{\mathbf{R}^n} e^{-(y_1^2 + y_2^2 + \ldots + y_n^2)/2} \, dy_1 \ldots dy_n$$

$$= \left\{ \frac{1}{\sqrt{(2\pi)}} \int_{-\infty}^{\infty} e^{-y_1^2/2} \, dy_1 \right\} \ldots \left\{ \frac{1}{\sqrt{(2\pi)}} \int_{-\infty}^{\infty} e^{-y_n^2/2} \, dy_n \right\}$$

$$= 1.$$

We next show that $\mathscr{E}(\mathbf{X}) = \mathbf{\mu}$. This simply means that $\mathscr{E}(X_1) = \mu_1$, $\mathscr{E}(X_2) = \mu_2, \ldots, \mathscr{E}(X_n) = \mu_n$. Since $\mathbf{x} = \mathbf{\mu} + ES^{-1}\mathbf{y}$, the above argument gives

$$\mathscr{E}(\mathbf{X}) = \int\int \ldots \int_{\mathbf{R}^n} \mathbf{x} f(x_1, \ldots, x_n) dx_1 \ldots dx_n$$

$$= \mathbf{\mu} + \frac{1}{(\sqrt{(2\pi)})^n} ES^{-1} \int\int \ldots \int_{\mathbf{R}^n} \mathbf{y} e^{-\mathbf{y}^T \mathbf{y}/2} \, dy_1 \ldots dy_n.$$

But the final term is the zero vector. This follows from the fact that, for each $j = 1, 2, \ldots, n$,

$$\int\int \ldots \int_{\mathbf{R}^n} y_j e^{-\mathbf{y}^T \mathbf{y}/2} \, dy_1 \ldots dy_n$$

$$= \left\{ \int_{-\infty}^{\infty} e^{-y_1^2/2} \, dy_1 \right\} \ldots \left\{ \int_{-\infty}^{\infty} y_j e^{-y_j^2/2} \, dy_j \right\} \ldots \left\{ \int_{-\infty}^{\infty} e^{-y_n^2/2} \, dy_n \right\}$$

and the jth term is zero.

It remains to interpret the matrix A in terms of the random variables $X_1, X_2, \ldots, X_n$. It turns out that A is the inverse of their *covariance matrix*

$$
\begin{pmatrix}
\text{Var}\,(X_1) & \text{Cov}\,(X_1, X_2) \ldots \text{Cov}\,(X_1, X_n) \\
\text{Cov}\,(X_2, X_1) & \text{Var}\,(X_2) \ \ldots \text{Cov}\,(X_2, X_n) \\
\vdots & \\
\text{Cov}\,(X_n, X_1) & \text{Cov}\,(X_n, X_2) \ldots \ \ \ \text{Var}\,(X_n)
\end{pmatrix}
$$

This matrix may be regarded as the expectation of the $n \times n$ matrix $(\mathbf{X} - \boldsymbol{\mu})(\mathbf{X} - \boldsymbol{\mu})^T$. But

$$
\mathscr{E}(\mathbf{X} - \boldsymbol{\mu})(\mathbf{X} - \boldsymbol{\mu})^T
$$

$$
= \frac{1}{(\sqrt{(2\pi)})^n} \iint \ldots \int_{\mathbb{R}^n} ES^{-1}\mathbf{y}\mathbf{y}^T S^{-1} E^T e^{-\mathbf{y}^T\mathbf{y}/2}\, dy_1 \ldots dy_n
$$

$$
= \frac{1}{(\sqrt{(2\pi)})^n} ES^{-1} \left\{ \iint \ldots \int_{\mathbb{R}^n} \mathbf{y}\mathbf{y}^T e^{-\mathbf{y}^T\mathbf{y}/2}\, dy_1 \ldots dy_n \right\} S^{-1} E^T.
$$

But

$$
\iint \ldots \int_{\mathbb{R}^n} y_i y_j e^{-\mathbf{y}^T\mathbf{y}/2}\, dy_1 \ldots dy_n = \begin{cases} (\sqrt{(2\pi)})^n, & i = j \\ 0, & i \neq j \end{cases}
$$

(see §9.45) and it follows that

$$
\iint \ldots \int_{\mathbb{R}^n} \mathbf{y}\mathbf{y}^T e^{-\mathbf{y}^T\mathbf{y}/2}\, dy_1 \ldots dy_n = (\sqrt{(2\pi)})^n T.
$$

Thus

$$
\mathscr{E}(\mathbf{X} - \boldsymbol{\mu})(\mathbf{X} - \boldsymbol{\mu})^T = ES^{-1}S^{-1}E^T
$$

$$
= ED^{-1}E^T
$$

$$
= (EDE^T)^{-1} = A^{-1}.
$$

11

Differential and difference equations (I)

Differential equations are of immense importance in many different areas. In this chapter we provide only the briefest of introductions to the most elementary techniques. None of this material is difficult and all of it is essential.

The section on difference equations is even more abbreviated than those on differential equations. It is intended only to indicate that every technique from the theory of differential equations has an analogue for difference equations. Further material on differential and difference equations will be found in chapter 13.

11.1 Differential equations

A *differential equation* is an equation which contains at least one derivative of an unknown function. Some examples of differential equations are given below.

(1) $\dfrac{dy}{dx} = \cos x$

(2) $\dfrac{d^2 y}{dx^2} + k^2 y = 0$

(3) $\left(\dfrac{d^2 w}{dx^2}\right)^3 - x\dfrac{dw}{dx} + w = 0$

(4) $\dfrac{d^2 u}{dt^2} + 7\left(\dfrac{du}{dt}\right)^4 - 8u = 0$

(5) $\dfrac{\partial^2 V}{\partial x^2} + \dfrac{\partial^2 V}{\partial y^2} = 0$

(6) $x\dfrac{\partial f}{\partial x} + y\dfrac{\partial f}{\partial y} = nf.$

A *solution* of a differential equation is a relation between the variables involved in the equation which is free from derivatives and which is consistent with the differential equation. For example, $y = \sin x$ is a solution of (1) and $f = x^n + y^n$ is a solution of (6).

For obvious reasons, equations (5) and (6) above are called *partial* differential equations. The other equations are called *ordinary* differential equations.

The *order* of a differential equation is the order of the derivative of highest order appearing in the equation. The *degree* of the equation is the algebraic degree with which the derivative of highest order appears in the equation.

In the examples given above, (1) and (6) are first order equations. The rest are second order equations. All but (3) are of first degree. Equation (3) is of degree 3.

We shall concentrate almost entirely on ordinary differential equations (with only fleeting references to partial differential equations in §11.18). In this chapter, the ordinary differential equations will be of order one and we shall devote no attention at all to equations of degree two or more. To say that our treatment is incomplete would therefore be an understatement of monumental proportions.

11.2   General solutions of ordinary equations

Consider the expression

(1) $(x - c)^2 + y^2 = c^2$

where c is a constant. This may be rewritten as

(2) $x^2 - 2xc + y^2 = 0$

We differentiate implicitly with respect to x and obtain

$$2x - 2c + 2y\frac{dy}{dx} = 0.$$

Hence

$$c = x + y\frac{dy}{dx}.$$

Substituting this result in (2) yields the differential equation

(3) $x^2 - 2\left(x + y\dfrac{dy}{dx}\right)x + y^2 = 0.$

This is a first order, ordinary differential equation. Observe that (1) is a solution of this differential equation whatever the value of the constant c. Thus the differential equation (3) does *not* have a *unique* solution. Each possible value for c generates a different solution for (3).

It is fairly evident that the same argument can be employed in general with any equation which, like (1), connects x, y and an 'arbitrary constant' c. The arbitrary constant c may be eliminated, leaving as the result an ordinary differential equation of the first order. The original equation is then a solution of the resulting ordinary differential equation for all values of the arbitrary constant c.

A reversal of this reasoning leads us to the notion of a *general solution* of an ordinary differential equation of the *first order*. Such a general solution is a solution involving *one arbitrary constant*. For example,

$$(x - c)^2 + y^2 = c^2$$

is a general solution of

$$x^2 - 2\left(x + y\frac{dy}{dx}\right)x + y^2 = 0.$$

Similar reasoning leads us to define a *general solution* of an ordinary differential equation of the *nth order* to be a solution containing *n arbitrary constants*. For example,

$$(x - c)^2 + y^2 = r^2$$

where c and r are arbitrary constants is a general solution of the second order, ordinary differential equation

$$1 + \left(\frac{dy}{dx}\right)^2 + y\left(\frac{d^2y}{dx^2}\right) = 0.$$

This can be checked by differentiating the equation $(x - c)^2 + y^2 = r^2$ twice with respect to x.

11.3 Boundary conditions

The rate of increase with time t of a population P of a colony of bacteria is proportional to the population P at time t – i.e.

$$\frac{dP}{dt} = kP$$

where k is a constant. It is easy to check that the general solution of this differential equation is

$$P(t) = ce^{kt}$$

where c is an arbitrary constant.

Suppose that one needs to calculate the population at time $t = 100$ in the case $k = 2$. Then

$$P(100) = ce^{200}.$$

To proceed any further, it is necessary to assign a particular value to the arbitrary constant c. For this, further information is necessary. The conditions which supply this information are usually called the *boundary conditions* for the problem.

In our particular problem, the boundary condition might take the form $P(5) = 60$ – i.e. it is known that the population was 60 at time $t = 5$. It then follows that

$$60 = P(5) = ce^{2 \times 5} = ce^{10}.$$

and so

$$c = 60e^{-10}.$$

Thus

$$P(t) = 60e^{-10}e^{2t} = 60e^{2t-10}$$

and, in particular, $P(100) = 60e^{190}$.

Alternatively, it may be that the boundary condition takes the form $P'(1) = 3$ – i.e. it is known that the rate of increase of the population was 3 at time 1. Since

$$P'(t) = cke^{kt},$$

it follows that

$$3 = P'(1) = c2e^{2 \times 1} = 2ce^2.$$

Thus

$$c = \tfrac{3}{2}e^{-2}$$

and so

$$P(t) = \tfrac{3}{2}e^{-2}e^{2t} = \tfrac{3}{2}e^{2(t-1)}.$$

In particular, $P(100) = \tfrac{3}{2}e^{198}$.

Note that, without an appropriate boundary conditon, the problem of finding $P(100)$ is *indeterminate*. In brief, to select the particular solution of a differential equation which is relevant to the problem in hand one requires *boundary conditions*.

11.4   Separating variables

We now begin to study some systematic methods for finding the general solution of first order, ordinary differential equations.

Consider, to begin with, the differential equation

$$\frac{dy}{dx} = \cos x.$$

From the fundamental theorem of calculus (see §9.3 and §9.5), we have that

$$y = \int \frac{dy}{dx}\, dx + c,$$

where c is the arbitrary constant of integration. Thus, integrating both sides of our differential equation, we obtain that

$$y = \int \frac{dy}{dx}\, dx + c = \int (\cos x)\, dx + c = \sin x + c.$$

The required general solution is therefore

$$y = \sin x + c$$

where c is an arbitrary constant (which must never be forgotten).

We usually express this argument somewhat more concisely. We begin by rewriting the differential equation in the form

(1) $dy = (\cos x)\, dx.$

Integration then yields

$$y = \int dy = \int (\cos x)\, dx = \sin x + c$$

as before.

The step which leads to (1) is called *separating the variables*. In the examples below we use the same technique with some slightly more complicated differential equations.

Example 11.5.
Find the general solution of the differential equation

$$xy \frac{dy}{dx} = 2(y + 3).$$

We begin by writing the equation in the form

$$xy\, dy = 2(y + 3)\, dx.$$

Next, separate the variables. This gives

$$\frac{y\, dy}{y + 3} = \frac{2\, dx}{x}$$

and so

$$\int \frac{y \, dy}{y+3} = 2 \int \frac{dx}{x} + c$$

where c is an arbitrary constant. Observe that

$$\int \frac{y \, dy}{y+3} = \int \frac{y+3-3}{y+3} \, dy = \int \left(1 - \frac{3}{y+3}\right) dy$$

$$= y - 3 \log (y+3).$$

Hence the general solution is

$$y - 3 \log (y+3) = 2 \log x + c.$$

This can be expressed more neatly. We have that

$$y = \log (x^2) + \log (y+3)^3 + c$$
$$e^y = x^2 (y+3)^3 e^c.$$

Put $a = e^c$. If c is an arbitrary constant, then so is a. Hence the general solution takes the form

$$e^y = ax^2 (y+3)^3.$$

Example 11.6.
Find the general solution of the differential equation

$$(4+x)\frac{dy}{dx} = y^3.$$

We have that

$$(4+x)dy = y^3 dx$$

$$\frac{dy}{y^3} = \frac{dx}{4+x}$$

$$\int \frac{dy}{y^3} = \int \frac{dx}{4+x} + c$$

$$-\tfrac{1}{2} y^{-2} = \log (4+x) + c$$

where c is an arbitrary constant.

Example 11.7.
Ignoring the effect of the atmosphere, with what velocity must a particle be projected radially from the surface of the earth if it is not to fall back again? – i.e. what is the *escape velocity*?

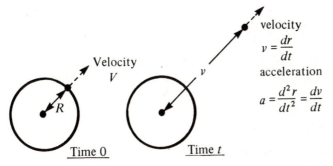

velocity
$$v = \frac{dr}{dt}$$
acceleration
$$a = \frac{d^2 r}{dt^2} = \frac{dv}{dt}$$

Velocity
V

R

Time 0   Time *t*

Newton's law of gravitation asserts that the (outward) acceleration *a* satisfies

$$a = -\frac{k}{r^2}$$

where the positive constant *k* is given by $k = gR^2$. (The well-known constant *g* is approximately 981 when distances are measured in centimetres.) Observe that

$$a = \frac{dv}{dt} = \frac{dv}{dr}\frac{dr}{dt} = v\frac{dv}{dr}.$$

We therefore obtain the differential equation

$$v\frac{dv}{dr} = -\frac{k}{r^2}.$$

We separate the variables. Then

$$v dv = -\frac{k}{r^2}dr$$

$$\int v dv = -k \int \frac{dr}{r^2} + c$$

$$\frac{1}{2}v^2 = \frac{k}{r} + c.$$

This is the *general solution*. We require the particular solution which satisfies the *boundary condition* $v = V$ when $r = R$. If this is satisfied, then

$$\frac{1}{2}V^2 = \frac{k}{R} + c$$

and so

$$c = \frac{1}{2}V^2 - \frac{k}{R}.$$

It follows that

$$\frac{1}{2}v^2 = k\left(\frac{1}{r} - \frac{1}{R}\right) + \frac{1}{2}V^2.$$

The condition that the particle never falls back to the Earth may be expressed by requiring that, for all v, $v > 0$. We therefore require that

$$k\left(\frac{1}{r} - \frac{1}{R}\right) + \frac{1}{2}V^2 > 0$$

for *all* $r \geq R$. But this is true if and only if

$$\frac{1}{2}V^2 - \frac{k}{R} \geq 0$$

i.e.

$$V^2 \geq \frac{2k}{k} = 2gR.$$

It follows that the escape velocity is $\sqrt{(2gR)}$.

11.8   Exact equations

Consider the equation

$$f(x, y) = c$$

where c is a constant. Introducing differentials, we obtain that

$$df = \frac{\partial f}{\partial x}dx + \frac{\partial f}{\partial y}dy = 0.$$

Thus, if an ordinary differential equation of the first order

$$M(x, y)dx + N(x, y)dy = 0$$

has the property that

$$(1) \quad M = \frac{\partial f}{\partial x} \quad \text{and} \quad N = \frac{\partial f}{\partial y}$$

for some function $f: \mathbb{R}^2 \to \mathbb{R}$, then the general solution takes the form

$$f(x, y) = c$$

where c is an arbitrary constant. Such a differential equation is said to be *exact*.

If equations (1) are satisfied, then

$$\frac{\partial M}{\partial y} = \frac{\partial}{\partial y}\left(\frac{\partial f}{\partial x}\right) = \frac{\partial^2 f}{\partial y \partial x} = \frac{\partial^2 f}{\partial x \partial y} = \frac{\partial}{\partial x}\left(\frac{\partial f}{\partial y}\right) = \frac{\partial N}{\partial x}.$$

Thus, for the equation $Mdx + Ndy = 0$ to be exact, we require that

(2) $\qquad \dfrac{\partial M}{\partial y} = \dfrac{\partial N}{\partial x}.$

Examples 11.9.

Some simple exact equations together with their solutions are given below. Note that (2) holds in each case.

(i) $\qquad ydx + xdy = 0.$

We have that $d(xy) = xdy + ydx$ and so the general solution is $xy = c$.

(ii) $\qquad \dfrac{dx}{y} - \dfrac{xdy}{y^2} = 0.$

We have that

$$d\left(\frac{x}{y}\right) = \frac{ydx - xdy}{y^2}$$

and so the general solution is $x = yc$.

(iii) $\qquad \left(1 - \dfrac{y^2}{x^2}\right)dx + 2\dfrac{y}{x}dy = 0.$

We have that

$$d\left(\frac{x^2 + y^2}{x}\right) = \left(1 - \frac{y^2}{x^2}\right)dx + 2\frac{y}{x}dy$$

and so the general solution is $x^2 + y^2 = cx$.

In general, of course, it is rather unlikely that a given equation

$$M(x,y)dx + N(x,y)dy = 0$$

will be *exact*. Sometimes, however, it is possible to find a function $\mu(x,y)$ such that

$$\mu Mdx + \mu Ndy = 0$$

is exact. Such a function μ is called an *integrating factor* for the equation. This technique is particularly important for the class of differential equations considered next.

11.10 Linear equations of order 1

A *linear* differential equation of the first order has the form

(1) $\dfrac{dy}{dx} + P(x)y = Q(x).$

(The word 'linear' is only meant to signify that the equation is of the form $au + bv = c$ where the 'variables' u and v are y and its derivative respectively.)

A suitable integrating factor for such a differential equation is given by

$$\mu = e^{\int P(x)dx}.$$

This is easily proved directly. Note first that

$$\frac{d\mu}{dx} = Pe^{\int Pd_2} = P\mu.$$

Now multiply through (1) by μ. Then

$$\mu\frac{dy}{dx} + \mu Py = \mu Q$$

$$\frac{d}{dx}(\mu y) = \mu Q$$

It follows that the general solution is

$$\mu y = \int \mu Q dx + c.$$

Example 11.11.

Find the general solution of

$$\frac{dy}{dx} + \frac{2y}{x} = 8x.$$

This equation is linear with $P = 2/x$ and $Q = 8x$. The integrating factor is

$$\mu = e^{\int Pdx} = e^{\int (2/x)dx} = e^{2\log x} = x^2.$$

(For the last step, see §6.7). We therefore consider

$$x^2\frac{dy}{dx} + 2xy = 8x^3$$

i.e.

$$\frac{d}{dx}(x^2 y) = 8x^3.$$

The general solution is therefore

$$x^2 y = \int 8x^3 dx + c$$

i.e.

$$x^2 y = 2x^4 + c.$$

Example 11.12.

Find the general solution of

$$x(x+1)\frac{dy}{dx} + y = x + 1.$$

We first put this into the standard form of §11.10 i.e.

$$\frac{dy}{dx} + \frac{1}{x(x+1)}y = \frac{1}{x}.$$

This is linear with $P = 1/x(x+1)$ and $Q = 1/x$. Since

$$\int \frac{dx}{x(x+1)} = \int \left(\frac{1}{x} - \frac{1}{x+1}\right)dx = \log x - \log (x+1)$$

$$= \log\left(\frac{x}{x+1}\right),$$

it follows that the appropriate integrating factor is

$$\mu = e^{\log(x/x+1)} = \frac{x}{x+1}.$$

We therefore consider

$$\frac{x}{x+1}\frac{dy}{dx} + \frac{1}{(x+1)^2}y = \frac{1}{x+1}$$

i.e.

$$\frac{d}{dx}\left(\frac{xy}{x+1}\right) = \frac{1}{x+1}.$$

The general solution is therefore

$$\frac{xy}{x+1} = \log(x+1) + c.$$

Example 11.13.

Find the general solution of

$$(3x - xy + 2)\frac{dy}{dx} + y = 0.$$

There is no way that this can be put into the form of §11.10 because of the 'non-linear' term

$$y\frac{dy}{dx}.$$

But we may reverse the roles of x and y by writing

$$\frac{dx}{dy} = \frac{3x - xy + 2}{-y}$$

i.e.

$$\frac{dx}{dy} + \left(\frac{3}{y} - 1\right)x = -\frac{2}{y}.$$

This equation is linear with $P(y) = 3y^{-1} - 1$ and $Q(y) = -2y^{-1}$. The appropriate integrating factor is

$$\mu(y) = e^{\int(3/y - 1)dy} = e^{3 \log y - y} = y^3 e^{-y}$$

and so we consider

$$y^3 e^{-y}\frac{dx}{dy} + \left(\frac{3}{y} - 1\right)y^3 e^{-y}x = -2y^2 e^{-y}$$

i.e.

$$\frac{d}{dy}(y^3 e^{-y}x) = -2y^2 e^{-y}.$$

The general solution is therefore

$$y^3 e^{-y}x = -2 \int y^2 e^{-y}dy + c.$$

The integral may be evaluated by parts. Recalling the remarks made about notation in §9.5, we consider

$$\int_0^y t^2 e^{-t}dt = [t^2(-e^{-t})]_0^y - \int_0^y 2t(-e^{-t})dt$$

$$= -y^2 e^{-y} + [2t(-e^{-t})]_0^y - \int_0^y 2(-e^{-t})dt$$

$$= -y^2 e^{-y} - 2ye^{-y} + [2(-e^{-t})]_0^y$$

$$= -e^{-y}(y^2 + 2y + 2) + 2.$$

Returning to the general solution, we have that

$$y^3 e^{-y}x = 2e^{-y}(y^2 + 2y + 2) + a$$

$$x = \frac{2}{y} + \frac{4}{y^2} + \frac{4}{y^3} + a\frac{1}{y^3}e^y$$

where $a = c + 2$ is arbitrary because the same is true of c.

11.14   Changing variables

Occasionally an unpleasant differential equation can be converted to something manageable by making a change of variable. Unfortunately it is seldom easy to think of such a change of variable.

Example 11.15.
Find the general solution of

$$\frac{dy}{dx} = (9x + 4y + 1)^2.$$

The change of variable $z = 9x + 4y + 1$ looks as though it might be helpful. We then have that

$$\frac{dz}{dx} = 9 + 4\frac{dy}{dx}$$

and so the differential equation becomes

$$\frac{dz}{dx} = 9 + 4z^2$$

$$\frac{dz}{9 + 4z^2} = dx$$

$$\int \frac{dz}{9 + 4z^2} = \int dx + c$$

$$\frac{1}{6}\arctan\left(\frac{2z}{3}\right) = x + c$$

$$2z = 3\tan(6x + a)$$

$$2(9x + 4y + 1) = 3\tan(6x + a).$$

Example 11.16.
Find the general solution of

$$\sin x \frac{dy}{dx} - 2(4\sin^2 x - y)\cos x = 0.$$

The change of variable $z = \sin x$ reduces this to

$$z\frac{dy}{dz} - 2(4z^2 - y) = 0$$

which we have already solved in example 11.11. The general solution is

i.e.

$$z^2 y = 2z^4 + c$$

$$(\sin^2 x)y = 2(\sin x)^4 + c.$$

Example 11.17.

Find the general solution of

$$(x^2 - xy + y^2)dx - xydy = 0.$$

The substitution $y = vx$ always works when the functions M and N in the equation $Mdx + Ndy = 0$ are homogeneous of the same degree (see exercise 2.22 (13)). We have that $dy = vdx + xdv$ and hence the differential equation becomes

$$(x^2 - x^2 v + v^2 x^2)dx - x^2 v(vdx + xdv) = 0$$

$$(1 - v)dx = xvdv$$

$$\frac{dx}{x} = \frac{vdv}{1 - v}$$

$$\int \frac{dx}{x} = \int \frac{vdv}{1 - v} + c$$

$$\log x = -\int \frac{(1 - v) - 1}{1 - v}dv + c$$

$$= -v - \log (1 - v) + c$$

$$\log (x(1 - v)) = c - v$$

i.e.

$$x(1 - v) = ae^{-v}$$

$$x\left(1 - \frac{y}{x}\right) = ae^{-y/x}$$

$$(x - y)e^{y/x} = a.$$

11.18   Partial differential equations

The general solution of a partial differential equation of order n requires n arbitrary *functions.*

Suppose, for example, that

(1) $f(x, y) = \phi(y)x + y.$

Then

$$\phi(y) = \frac{f - y}{x}.$$

and hence

$$0 = \frac{\partial}{\partial x}\left(\frac{f-y}{x}\right)$$

$$= \frac{1}{x}\frac{\partial f}{\partial x} - \frac{1}{x^2}(f-y).$$

Thus (1) is the general solution of the partial differential equation

$$x\frac{\partial f}{\partial x} = f + y.$$

Example 11.19.

Find the general solution of

$$\frac{\partial f}{\partial x} = 3y^2 + x.$$

When the partial derivative was calculated, we kept y constant. To recover f we must therefore integrate with respect to x *keeping y constant*. The constant of integration obtained will then really be constant only if y is held constant – i.e. the 'constant of integration' is actually an arbitrary function of y.

In the case of the current problem, we obtain that

$$f(x,y) = \int (3y^2 + x)dx + \phi(y)$$

$$= 3y^2 x + \tfrac{1}{2}x^2 + \phi(y)$$

where ϕ is an arbitrary function. This result can be checked by differentiating partially with respect to x.

Example 11.20.

Find the general solution of

$$\frac{\partial^2 f}{\partial x \partial y} = 3y^2 + x.$$

We have that

$$\frac{\partial f}{\partial y} = \int (3y^2 + x)dx + \phi(y)$$

$$\frac{\partial f}{\partial y} = (3y^2 x + \tfrac{1}{2}x^2) + \phi(y)$$

$$f = \int (3y^2 x + \tfrac{1}{2}x^2)dy + \int \phi(y)dy + \psi(x)$$

$$f = y^3 x + \tfrac{1}{2}x^2 y + \Phi(y) + \psi(x)$$

where Φ and ψ are arbitrary functions.

Example 11.21.
Find the general solution of

$$\frac{\partial f}{\partial x} = \frac{\partial f}{\partial y}.$$

The change of variable

$$\left. \begin{aligned} x &= u + v \\ y &= u - v \end{aligned} \right\}$$

leads to

$$\left. \begin{aligned} \frac{\partial}{\partial u} &= \frac{\partial}{\partial x} + \frac{\partial}{\partial y} \\[2mm] \frac{\partial}{\partial v} &= \frac{\partial}{\partial x} - \frac{\partial}{\partial y} \end{aligned} \right\}$$

Hence

$$\frac{\partial f}{\partial x} - \frac{\partial f}{\partial y} = \frac{\partial f}{\partial v} = 0.$$

The general solution is therefore

$$f = \phi(u) = \phi_1(x + y)$$

where ϕ_1 is an arbitrary function.

Example 11.22.
In a two-dimensional universe, gravitational potential $f(x, y)$ would satisfy the two-dimensional wave equation

$$\frac{\partial^2 f}{\partial x^2} + \frac{\partial^2 f}{\partial y^2} = 0.$$

If, like Newton, we are interested in the gravitational potential induced by a single point mass, it seems a good idea to introduce polar co-ordinates $(r, \theta)^T$. The reason is that we know that the gravitational potential at a point will depend *only* on its distance r from the gravitating mass. Thus

$$\frac{\partial f}{\partial \theta} = 0.$$

From example 6.22 we have that

$$\frac{\partial^2 f}{\partial x^2} + \frac{\partial^2 f}{\partial y^2} = \frac{\partial^2 f}{\partial r^2} + \frac{1}{r}\frac{\partial f}{\partial r} + \frac{1}{r^2}\frac{\partial^2 f}{\partial \theta^2}.$$

But since f is only a function of r, the wave equation becomes

$$\frac{d^2 f}{dr^2} + \frac{1}{r}\frac{df}{dr} = 0.$$

If we write

$$g = \frac{df}{dr},$$

this differential equation reduces to

$$\frac{dg}{dr} + \frac{g}{r} = 0$$

i.e.

$$\frac{dg}{g} = -\frac{dr}{r}$$

$$\log g = -\log r + c$$

$$g = \frac{a}{r}$$

where a is an arbitrary constant. Thus

$$\frac{df}{dr} = \frac{a}{r}$$

and so

$$f = -\frac{a}{r^2} + b$$

where b is an arbitrary constant.

Exercises 11.23

1. Determine the order and the degree of the following ordinary differential equations:

(i) $\dfrac{d^3 y}{dx^3} + 2xy\dfrac{dy}{dx} = \left(\dfrac{d^2 y}{dx^2}\right)^4$

(ii) $\left(\dfrac{d^2 y}{dx^2}\right)^2 - 2\left(\dfrac{dy}{dx}\right)^3 + yx = 0.$

2.* Eliminate the arbitrary constants a and b from the following equations and hence find differential equations which they satisfy.

(i) $x^2y = 1 + ax$ (ii) $y = ae^{-x} + be^{2x}$

(iii) $y^2 = 4ax$ (iv) $y = a\cos(2x + b)$.

3. Separate the variables and hence find the general solutions of the following differential equations.

(i) $(4 + x)\dfrac{dy}{dx} = y^3$ (ii) $(xy + y)\dfrac{dy}{dx} = (x - xy)$

(iii) $x^2y\dfrac{dy}{dx} = e^y$ (iv) $\dfrac{1}{y}\dfrac{dy}{dx} = (\log x)(\log y)$.

4.* Separate the variables and hence find the general solutions of the following differential equations.

(i) $2x\dfrac{dy}{dx} = 3y$ (ii) $\dfrac{dy}{dx} = \dfrac{(y + 1)(y + 2)}{(x + 1)(x + 2)}$

(iii) $\dfrac{dy}{dx} = xy^2$ (iv) $\dfrac{dy}{dx} = x(1 + x^2)y$

(v) $\sin x\dfrac{dy}{dx} = \sin y$ (see exercise 9.24 (6)).

5. Which of the following equations are exact?

(i) $3x(xy - 2)dx + (x^3 + 2y)dy = 0$

(ii) $(x^5 + 3y)dx - xdy = 0$.

6.* Which of the following equations are exact?

(i) $(1 + y^2)dx + (x^3y + y)dy = 0$

(ii) $(2x^3 - xy^2 - 2y + 3)dx - (x^2y + 2x)dy = 0$

(iii) $(2xy + x^2 + x^4)dx - (1 + x^2)dy = 0$.

7. Determine which of the equations of question 5 are linear. Find the general solution of such equations using an integrating factor.

8.* Determine which of the equations of question 6 are linear. Find the general solution of such equations using an integrating factor.

9. Find the general solution of the following linear differential equation:

$$(1 + x^2)\dfrac{dy}{dx} = 2xy + x^2 + x^4.$$

10.* Find the general solution of the following linear differential equations:

(i) $\dfrac{dy}{dx} = x - 2y$ (ii) $\dfrac{dy}{dx} + 2xy = 2x^3$.

11. Make the change of variable

$$\left. \begin{array}{l} u = \tan x \\ v = \cos y \end{array} \right\}$$

and hence find the general solution of

$$(3 \tan x - 2 \cos y) \sec^2 x\, dx + \tan x \sin y\, dy = 0.$$

12.* Make the changes of variable indicated and hence find the general solutions of the given differential equations.

(i) $\dfrac{dy}{dx} = 1 + 6xe^{x-y}$

$x = u + y$

(ii) $2(2x^2 + y^2)dx - xy\, dy = 0$

$y = vx$

(iii) $ydx + xdy = -\dfrac{x}{y^2} dy + \dfrac{dx}{y}$

$$\left. \begin{array}{l} u = xy \\ v = \dfrac{x}{y}. \end{array} \right\}$$

13. Make the change of variable

$$\left. \begin{array}{l} u = x^2 + y^2 \\ v = x^2 - y^2 \end{array} \right\}$$

and hence find the general solution of

$$\frac{1}{x} \frac{\partial f}{\partial x} + \frac{1}{y} \frac{\partial f}{\partial y} = 1$$

(see example 6.20).

14.* Make the change of variable

$$\left. \begin{array}{l} x = u + v \\ y = u - v \end{array} \right\}$$

and hence find the general solution of

$$\frac{\partial^2 f}{\partial x^2} - \frac{\partial^2 f}{\partial y^2} = x.$$

15.† Show that the equation

$$(3x^2 y^2 + 2yx)dx + (2x^3 y + x^2)dy = 0$$

is exact. Find the general solutions of the partial differential equations

$$\left. \begin{aligned} \frac{\partial f}{\partial x} &= 3x^2 y^2 + 2yx \\[2mm] \frac{\partial f}{\partial y} &= 2x^3 y + x^2. \end{aligned} \right\}$$

By comparing these solutions, solve the original ordinary differential equation.

16.*† Show that the equation

$$3x(xy - 2)dx + (x^3 + 2y)dy = 0$$

is exact. Find the general solutions for

$$\left. \begin{aligned} \frac{\partial f}{\partial x} &= 3x(xy - 2) \\[2mm] \frac{\partial f}{\partial y} &= x^3 + 2y. \end{aligned} \right\}$$

By comparing these solutions, solve the original ordinary differential equation.

11.24   Difference equations

Difference equations involve an unknown *sequence* y_x instead of an unknown function as in differential equations. The variable x is therefore to be understood to take only the values

$$x = 0, 1, 2, 3, \ldots .$$

Some examples of sequences are

$$y_x = 1 \qquad\qquad (x = 0, 1, 2, \ldots)$$

$$y_x = x + 1 \qquad\qquad (x = 0, 1, 2, \ldots)$$

$$y_x = (-1)^x 2^x \qquad\qquad (x = 0, 1, 2, \ldots).$$

For the last of these sequences, we have that $y_0 = 1, y_1 = -2, y_2 = 4, y_3 = -8. y_4 = 16, \ldots .$

Some typical difference equations are listed below. The first three of these equations are of order one. Equation (4) is of second order and equation (5) is of fourth order.

(1) $y_{x+1} - y_x = 0$

(2) $y_{x+1} - 2y_x = 1$

(3) $y_{x+1} + x^2 y_x = 2^x$

(4) $y_{x+2} + 2y_{x+1} + y_x = 0$

(5) $y_{x+4}^2 + y_x y_{x+1} + x = 3.$

To find the order of a difference equation it is only necessary to locate the term y_{x+n} with largest subscript. The order is then simply n.

As with differential equations, the general solution of a difference equation of order n contains n arbitrary constants.

Consider, for example, equation (4). We can choose y_0 and y_1 in any way we like. Thus y_0 and y_1 will serve as our arbitrary constants. But the subsequent terms of the sequence must satisfy

$$y_2 = -2y_1 - y_0$$

$$y_3 = -2y_2 - y_1 = -2(-2y_1 - y_0) - y_1 = 3y_1 + 2y_0$$

$$y_4 = -2y_3 - y_2 = -2(3y_1 + 2y_0) - (-2y_1 - y_0)$$

$$= -4y_1 - 3y_0$$

and so on. Thus the subsequent terms are all determined by the choice of y_0 and y_1.

11.25 Difference operator

The *shift operator* E is defined by

$$Ey_x = y_{x+1}.$$

It is natural to write $E^2 y_x = E(Ey_x) = Ey_{x+1} = y_{x+2}$ and $E^3 y_x = E(E^2 y_x) = Ey_{x+2} = y_{x+3}$. In general,

$$E^n y_x = y_{x+n}.$$

Difference equations can conveniently be expressed in terms of the shift operator. For example, the difference equation $y_{x+2} + 2y_{x+1} + y_x = 0$ can be expressed in the form $E^2 y_x + 2Ey_x + y_x = 0$ which we usually write as

$$(E^2 + 2E + 1)y_x = 0.$$

In this chapter, we seek to stress the similarities between differential and difference equations. These are most apparent after the introduction of the *difference operator* Δ (more properly called the forward difference operator). This is defined by

$$\Delta y_x = y_{x+1} - y_x.$$

We also define $\Delta^2 y_x = \Delta(\Delta y_x)$, $\Delta^3 y_x = \Delta(\Delta^2 y_x)$ and so on. Observe that

$$Ey_x = y_{x+1} = y_{x+1} - y_x + y_x = \Delta y_x + y_x = (\Delta + 1)y_x.$$

We express this result by simply writing $E = \Delta + 1$. More useful is the generalization

$$E^n = (\Delta + 1)^n.$$

In the case $n = 2$, for example, we have that

$$E^2 y_x = (\Delta + 1)^2 y_x = (\Delta^2 + 2\Delta + 1)y_x = \Delta^2 y_x + 2\Delta y_x + y_x.$$

These results allow us to express any difference equation in terms of the difference operator Δ. In the case of the equation $y_{x+2} + 2y_{x+1} + y_x = 0$, for example, we begin by introducing the shift operator and obtain $(E^2 + 2E + 1)y_x = 0$. Using the fact that $E = (\Delta + 1)$ we then obtain the equation in the form

$$((\Delta + 1)^2 + 2(\Delta + 1) + 1)y_x = 0$$

$$(\Delta^2 + 4\Delta + 4)y_x = 0.$$

The difference operator Δ has strong similarities to the differential operator D. In particular, a version of the fundamental theorem of calculus holds with Δ replacing D and Σ replacing $\int$. We have that

$$\sum_{x=0}^{X-1} \Delta y_x = (y_1 - y_0) + (y_2 - y_1) + (y_3 - y_2)$$

$$+ \ldots + (y_X - y_{X-1}) = y_X - y_0.$$

Thus, if F_x is a 'primitive' for f_x — i.e. $\Delta F_x = f_x$, the difference equation

$$\Delta y_x = f_x$$

has general solution

$$y_x = F_x + c$$

where c is an arbitrary constant. To see this, it is only necessary to observe that

$$0 = \sum_{x=0}^{X-1} (\Delta y_x - f_x) = \sum_{x=0}^{X-1} \Delta(y_x - F_x)$$

$$= (y_X - F_X) - (y_0 - F_0)$$

$$= y_X - F_X - c.$$

The following table lists some simple sequences f_x with their primitives F_x (i.e. $\Delta F_x = f_x$).

$f_x = 0$	$F_x = c$
$f_x = b$	$F_x = bx + c$
$f_x = ax$	$F_x = \frac{1}{2}ax(x-1) + c$
$f_x = r^x$	$F_x = (r-1)^{-1} r^x + c \qquad (r \neq 1)$

Example 11.26.
The difference equation

$$y_{x+1} = y_x + 3x + 1$$

may be written in the form

$$\Delta y_x = 3x + 1.$$

Taking $a = 3$ and $b = 1$ in the above table, we obtain the general solution

$$y_x = \frac{3}{2}x(x-1) + x + c$$
$$y_x = \frac{3}{2}x^2 - \frac{1}{2}x + c.$$

Example 11.27.
The difference equation

$$y_{x+2} - 2y_{x+1} + y_x = 0$$

may be written in the form

$$(E^2 - 2E + 1)y_x = 0$$
$$(E-1)^2 y_x = 0$$
$$\Delta^2 y_x = 0.$$

From this we deduce that

$$\Delta y_x = c$$

$$y_x = cx + d$$

where c and d are arbitrary constants.

Example 11.28.

A linear difference equation of order one may be expressed in the form $\Delta y_x + P_x y_x = Q_x$. We shall consider only the special case in which the difference equation takes the form

$$y_{x+1} - ry_x = Q_x.$$

Multiplying through by the 'summation factor' $\mu_x = r^{-x-1}$, we obtain

$$r^{-x-1} y_{x+1} - r^{-x} y_x = r^{-x-1} Q_x$$

i.e.

$$\Delta(r^{-x} y_x) = r^{-x-1} Q_x.$$

If a 'primitive' for the right-hand side is known, then the difference equation can be solved.

A significant special case arises in the case of capital accumulation under compound interest. Suppose that $1000 is invested at 20% interest per year. How much will have accumulated after 40 years? The appropriate difference equation is

$$y_{x+1} = y_x + \frac{20}{100} y_x \quad (x = 0, 1, 2 \dots)$$

which we rewrite as

$$y_{x+1} - ry_x = 0$$

where $r = 1.2$. From the analysis given above, we know that this difference equation may be expressed in the form

$$\Delta(r^{-x} y_x) = 0$$

which has general solution $r^{-x} y_x = c$ i.e.

$$y_x = cr^x$$

where c is an arbitrary constant. To decide on an appropriate value of c for the problem in hand, we appeal to the *boundary condition* $y_0 = 1000$. This gives $y_0 = 1000 = cr^0$ and so

$$y_x = 1000(1.2)^x$$

In particular $y_{40} = \$1\,469\,771.6$.

Suppose that a further $1000 is invested each year. What then will be the capital accumulation after 40 years? In this case the appropriate difference equation is

$$y_{x+1} = y_x + \frac{20}{100}y_x + 1000$$

which we rewrite as

(1) $y_{x+1} - ry_x = Q$

where $r = 1.2$ and $Q = 1000$. From the analysis given above, this difference equation may be expressed in the form

$$\Delta(r^{-x}y_x) = r^{-x-1}Q.$$

A 'primitive' for r^{-x} is $(r^{-1} - 1)^{-1}r^{-x}$ and so the general solution is

$$r^{-x}y_x = r^{-1}Q(r^{-1} - 1)^{-1}r^{-x} + c$$

$$y_x = \frac{Q}{1-r} + cr^x.$$

To decide on the proper value of c, we use the boundary condition $y_0 = 1000 = Q$. Then

$$Q = y_0 = \frac{Q}{1-r} + cr^0$$

$$c = Q\left(1 - \frac{1}{1-r}\right) = \frac{rQ}{r-1} = 6000.$$

Thus
$$y_x = -5000 + 6000(1.2)^x.$$

In particular, $y_{40} = \$8\,813\,629.5$.

Example 11.29.
The difference equation

$$y_{x+1} = \frac{y_x - 1}{y_x + 3}$$

may be solved by introducing the change of variable

$$y_x = \frac{1}{z_x} - 1.$$

The equation reduces to

$$\frac{1 - z_{x+1}}{z_{x+1}} = \frac{1 - 2z_x}{1 + 2z_x}$$

$$1 + 2z_x - z_{x+1} - 2z_x z_{x+1} = z_{x+1} - 2z_x z_{x+1}$$

$$z_{x+1} - z_x = \tfrac{1}{2}$$

$$\Delta z_x = \tfrac{1}{2}$$

which has general solution

$$z_x = \tfrac{1}{2}x + c.$$

The general solution of the original equation is therefore

$$\frac{1}{1+y_x} = \frac{1}{2}x + c.$$

Exercises 11.30

1. Determine the order of each of the following difference equations:

(i) $y_{x+1} + y_x = 0$ (ii) $y_{x+1} - y_x = x$

(iii) $y_{x+1} y_x = 1$ (iv) $y_{x+2}^2 y_x^3 = 1$

(v) $y_{x+2} + y_{x+1} + y_x = 0$ (vi) $x^2 y_{x+3} + 2^x y_x^4 = x^3$.

2.* Prove the following results

(i) $\Delta(\tfrac{1}{2}ax(x-1) + bx + c) = ax + b$

(ii) $\Delta((r-1)^{-1}r^x + c) = r^x$ $(r \neq 1)$

(iii) $\Delta((r-1)^{-1}xr^x - (r-1)^{-2}r^{x+1}) = xr^x$ $(r \neq 1)$.

3. Find general solutions for the equations

(i) $y_{x+1} - y_x = 2x$ (ii) $y_{x+1} - y_x = 2^x$

(iii) $y_{x+1} + 2y_x = x$ (iv) $y_{x+1} - 2y_x = 3^x$.

4.* Find general solutions for the equations

(i) $y_{x+1} - y_x = 1 + x$ (ii) $y_{x+1} - y_x = x2^x$

(iii) $3y_{x+1} - y_x = 1$ (iv) $2y_{x+1} + 3y_x = x$.

(In the case of (ii) and (iv) use the result of question 2(iii).)

5. Make the change of variable $y_x = 2^{z_x}$ and hence find the general solution of the difference equation

$$y_{x+1} y_x = 2.$$

6.* Make the change of variable $z_x = y_x/x$ and hence find the general solution of the difference equation

$$xy_{x+1} + (x+1)y_x = 0.$$

7. A sum of money y_0 is invested at an interest rate of 10% per year. Translate the following questions into problems concerning difference equations with suitable boundary conditions and hence solve them.

(i) How much was originally invested if the amount which has accumulated after 100 years is \$1 000 000?

(ii) Suppose that $y_0 = \$10$ and the average amount held over the first X years is \$15.93. How much is held at the beginning of the $(X + 1)$th year?

8.* An investor has \$1 000 000 invested at an interest rate of 10% per year. He transfers \$1000 from this account to a second account each year for ten consecutive years. In the second account, the interest rate is 20%. What will his capital be at the beginning of the eighth year.

SOME APPLICATIONS (OPTIONAL)

11.31   Cobweb model

We shall be concerned with the price P_x, the demand D_x and the supply S_x of a commodity at successive time periods $x = 0, 1, 2, \ldots$. These quantities are assumed to be related by the equations

$$
\left.
\begin{array}{ll}
(1) & D_x = a - bP_x \\[2mm]
(2) & S_x = D_x \\[2mm]
(3) & S_{x+1} = c + dP_x
\end{array}
\right\} \quad (x = 0, 1, 2, \ldots)
$$

where a, b, c and d are constants with $b > 0$ and $d > 0$. The story that accompanies these equations is the following. Having produced S_x of the commodity, the supplier decides on a price P_x. This determines the demand D_x as described in equation (1). (Since $b > 0$, demand falls as price increases.) The supplier chooses the price to ensure that all of the amount S_x that he has produced is sold in period x. This explains equation (2). The supplier then decides how much to produce for the *next* time period. This will depend on his production costs and his predictions about what price he will be able to obtain for the amount he chooses to supply. It is assumed that the supplier does not understand the way that the market works but simply makes a prediction about the price P_{x+1} for the next period which is based solely on the current price P_x. The amount S_{x+1} supplied in the next period will then be a function of the single variable P_x. Equation (3) is the simplest form such a function can take. (Since $d > 0$, supply in the next period rises as current price increases.)

The diagram below makes it clear why this model is called the *cobweb model.*

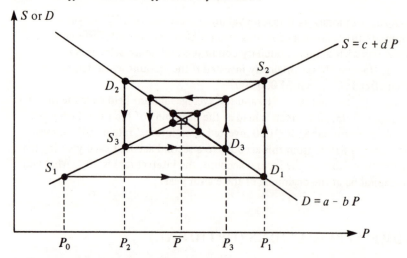

Suppose that P_0 is given. Then S_1 may be determined from the equation $S_1 = c + dP_0$. But then D_1 may be found because $D_1 = S_1$. Thus P_1 can be determined from the equation $D_1 = a - bP_1$. But then S_2 can be calculated from the equation $S_2 = c + dP_1$ and so on.

The quantity $\bar{P}$ in the diagram is called the *equilibrium price*. It is chosen so that
$$a - b\bar{P} = c + d\bar{P}$$

i.e. $\bar{P} = (a - c)/(b + d)$. If $P_0 = \bar{P}$, it is clear that the price, demand and supply will all remain constant for all time periods with $P_x = \bar{P}$ and $S_x = D_x = a - b\bar{P} = c + d\bar{P}$.

The diagrammatic method is adequate for many purposes but often an analytic solution for this type of problem is necessary. From equations (1), (2) and (3), we obtain the *difference equation*

$$a - bP_{x+1} = c + dP_x$$

i.e.

$$P_{x+1} + \left(\frac{d}{b}\right)P_x = \frac{a - c}{b}$$

$$\left(-\frac{d}{b}\right)^{-x-1} P_{x+1} - \left(-\frac{d}{b}\right)^{-x} P_x = \left(\frac{a-c}{b}\right)\left(-\frac{d}{b}\right)^{-x-1}$$

$$\Delta\left(-\frac{b}{d}\right)^{x} P_x = -\left(\frac{a-c}{d}\right)\left(-\frac{b}{d}\right)^{x}$$

$$\left(-\frac{b}{d}\right)^{x} P_x = -\left(\frac{a-c}{d}\right)\left(-\frac{b}{d}\right)^{x}\frac{1}{-\frac{b}{d}-1} + k$$

where k is an arbitrary constant. Hence

$$P_x = \frac{a-c}{b+d} + k\left(-\frac{d}{b}\right)^x.$$

This same result can be obtained with less trouble by introducing the change of variable $Q_x = P_x - \bar{P}$. The difference equation reduces to

$$-bQ_{x+1} = dQ_x$$

i.e.

$$Q_{x+1} + \left(\frac{d}{b}\right)Q_x = 0$$

$$\Delta\left(-\frac{b}{d}\right)^x Q_x = 0$$

$$P_x - \bar{P} = Q_x = k\left(-\frac{d}{b}\right)^x.$$

Observe that P_x is alternately larger and smaller than $\bar{P}$. Thus P_x 'oscillates' about $\bar{P}$ (and so D_x and S_x oscillate about $\bar{D}$ and $\bar{S}$). If $d < b$, the magnitude of these oscillations decreases as x grows larger and P_x approaches closer and closer to $\bar{P}$. We say that, in this case, the oscillations are *damped* and that P_x *converges* to $\bar{P}$. (The diagram above illustrates this case.)

If $d > b$, the oscillations become wilder and wilder and we say that P_x *oscillates infinitely*. If $d = b$, then P_x hops back and forward between $\bar{P} - k$ and $\bar{P} + k$. In this case, we say that P_x *oscillates finitely*. In neither of these cases does P_x converge. We say that P_x *diverges*.

The manner in which the suppliers predict future prices in the above model is not very enlightened. They never notice for example that the current price P_x is always on the wrong side of $\bar{P}$ relative to P_{x+1}. This does not matter too much in the case when $d < b$ since then we obtain convergence to the equilibrium price $\bar{P}$ in spite of the obtusity of the suppliers. But when $d > b$, it is clear that the system will suffer a breakdown before very long.

In a more sophisticated economy, the suppliers might base their estimates of the coming price on the average of the prices in the two previous time periods. We would then obtain the equations

(1) $D_x = a - bP_x$

(2) $S_x = D_x$

(3) $S_{x+2} = c + d \dfrac{P_x + P_{x+1}}{2}$

from which we extract the second order difference equation

$$2bQ_{x+2} + dQ_{x+1} + dQ_x = 0$$

in which $Q_x = P_x - \bar{P}$. This leads to a more satisfactory situation in that P_x converges to $\bar{P}$ whenever $d < 2b$. However, we shall leave the analysis of this difference equation until chapter 13 and concentrate instead on an even more sophisticated economy in which difference equations are replaced by differential equations.

The equations for this version of the cobweb model are

(1) $D(x) = a - bP(x)$

(2) $S(x) = D(x)$

(3) $S(x) = c + d \displaystyle\int_{-\infty}^{x} e^{y-x} P(y) dy.$

The only change in the first two equations is that x now represents a continuous variable which may take any real value instead of being a discrete variable restricted to the values $0, 1, 2, \ldots$. The third equation is more complicated. In this equation

$$\int_{-\infty}^{x} e^{y-x} P(y) dy$$

may be regarded as a prediction of the price at time x calculated as a 'weighted average' over prices at all previous times. Note that the prices in the recent past are given a much larger weight than prices in remote times. The weight e^{y-x} is sometimes called a 'discounting factor'.

The three equations together yield the *integral equation*

$$a - bP(x) = c + d \int_{-\infty}^{x} e^{y-x} P(y) dy.$$

At equilibrium, the price $P(x)$ will be equal to a constant $\bar{P}$. This constant satisfies

$$a - b\bar{P} = c + d \int_{-\infty}^{x} e^{y-x} \bar{P} dy = c + d\bar{P}$$

and so $\bar{P} = (a - c)/(b + d)$ as before. The change of variable $Q(x) = P(x) - \bar{P}$ reduces the integral equation to the simpler form

$$-bQ(x) = d \int_{-\infty}^{x} e^{y-x}Q(y)dy$$

$$= de^{-x} \int_{-\infty}^{x} e^{y}Q(y)dy.$$

Integral equations can often be solved by converting them into differential equations. (Equally, differential equations can sometimes be solved by converting them into integral equations.) Using the formula for differentiating a product in conjunction with the fundamental theorem of calculus (§9.3), we obtain that

$$-b\frac{dQ}{dx} = -de^{-x} \int_{-\infty}^{x} e^{y}Q(y)dy + de^{-x}e^{x}Q(x)$$

$$= -(-bQ) + dQ$$

$$= (b+d)Q.$$

Separating the variables, we obtain that

$$\frac{dQ}{Q} = -\left(\frac{b+d}{b}\right)dx$$

$$\int \frac{dQ}{Q} = -\left(\frac{b+d}{b}\right) \int dx + k$$

$$\log Q = -\left(\frac{b+d}{b}\right)x + k$$

$$P(x) - \bar{P} = Q(x) = Ke^{-x(b+d)/b}.$$

This is a very much more satisfactory result than that obtained in the original version of the cobweb model. This is not too surprising since the prediction method is much less naive. As long as

$$\frac{b+d}{b} > 0,$$

which will always be the case if $b > 0$ and $d > 0$, we have that

$$e^{-x(b+d)/b} \to 0 \quad \text{as} \quad x \to \infty$$

and so $P(x)$ converges smoothly to $\bar{P}$.

It may be instructive to go through the above analysis *without* making the change of variable $Q(x) = P(x) - \bar{P}$. The differential equation then obtained is linear and may be solved by the method of §11.10.

Note, incidentally, that a more natural continuous cobweb model is given by

$$
\left.
\begin{aligned}
&(1) \quad D(x) = a - bP(x) \\
&(2) \quad S(x) = D(x) \\
&(3) \quad S(x) = c + d\frac{1}{x}\int_0^x P(y)dy
\end{aligned}
\right\} \quad (x > 0).
$$

It may be of interest to solve this problem and to compare the result with that obtained above. Also of interest is the case in which (3) is replaced by

$$
S(x) = c + d \int_0^x \frac{2y}{x^2}P(y)dy.
$$

12

Complex numbers

The next chapter on differential and difference equations of order greater than one requires the use of complex numbers. Only a little knowledge of complex numbers is required for this purpose but it is essential that this small amount of information be properly understood. The relevant material is given in §12.1–§12.9 inclusive. Although readers may have met some or all of these ideas before, it is suggested that a review of this material would be advisable.

The application of complex numbers to the solution of differential equations is just one of a vast number of their uses. Some applications in statistics, for example, are given at the end of the chapter. More important, however, than all their many uses in applied work is the light an understanding of complex numbers casts on theoretical issues. For example, in §6.7 we considered the exponential and logarithmic functions and in §6.8 we considered the trigonometric functions. On the face of it, the two sections seem to be entirely unrelated. It is not until the complex versions of these functions are studied that it becomes apparent that this is far from being the case. Consider, for example, the delightful formula $e^{\pi i} = -1$ which contains all four of the mysterious entities of elementary mathematics (i.e. -1, e, π and i). A very brief discussion on this subject appears in §12.13. This is not difficult but no further use is made of the material in the main body of the text.

12.1 Quadratic equations

In §9.13, we completed the square in the quadratic equation

$$ax^2 + bx + c = 0 \quad (a \neq 0)$$

and obtained the formula

$$x = \frac{-b \pm \sqrt{(b^2 - 4ac)}}{2a}.$$

If

$$\alpha = \frac{-b + \sqrt{(b^2 - 4ac)}}{2a} \quad \text{and} \quad \beta = \frac{-b - \sqrt{(b^2 - 4ac)}}{2a}$$

it is easy to check that

$$ax^2 + bx + c = a(x - \alpha)(x - \beta)$$

for all values of x. We call α and β the *roots* of the quadratic $ax^2 + bx + c$. (If $\alpha = \beta$, the roots are said to be co-incident.)

When a, b and c are real and $\Delta = b^2 - 4ac \geqslant 0$, complex numbers have no role to play in the above analysis. If $\Delta > 0$, we know that $z^2 = \Delta$ has two *real* solutions, one positive and the other negative. We denote the positive solution by $\sqrt{\Delta}$ and the negative solution by $-\sqrt{\Delta}$. If $\Delta = 0$, $\sqrt{\Delta}$ and $-\sqrt{\Delta}$ are both zero (and so $\alpha = \beta = -b/2a$). However, if $\Delta < 0$ complex numbers are necessary to make sense of the formulae.

12.2   Complex numbers

It is easiest to think of a complex number z as a pair (x, y) of real numbers. It may then be plotted on a diagram as below:

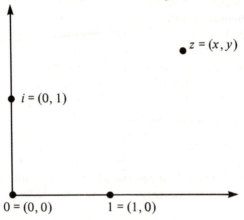

The real number x is identified with the complex number $(x, 0)$. In particular, we write $0 = (0, 0)$ and $1 = (1, 0)$. The complex numbers $(0, y)$ are called purely *imaginary*. Of particular importance is the imaginary number

$$i = (0, 1).$$

Addition and multiplication of complex numbers are defined by the rules

(1) $\quad z_1 + z_2 = (x_1, y_1) + (x_2, y_2) = (x_1 + x_2, y_1 + y_2)$

(2) $\quad z_1 z_2 = (x_1, y_1)(x_2, y_2) = (x_1 x_2 - y_1 y_2, x_1 y_2 + x_2 y_1).$

If z_1 and z_2 are both real (i.e. $y_1 = y_2 = 0$), these formulae reduce to the usual laws of addition and multiplication for real numbers. If α is a real number and z is complex, then (2) says that

$$\alpha z = \alpha(x, y) = (\alpha x, \alpha y).$$

Thus, given any complex number z, we have that

$$z = (x, y) = (x, 0) + (0, y) = x(1, 0) + y(0, 1)$$

i.e.

$$\boxed{z = x + iy}$$

This representation leads us to call x the *real part* of z and to call y the *imaginary part* of z. We write

$$x = \mathscr{R}z \quad \text{and} \quad y = \mathscr{I}\!\!\mathit{m}z.$$

(Note that $\mathscr{I}\!\!\mathit{m}z \neq iy$.)

Next observe that

$$i^2 = (0, 1)(0, 1) = (0.0 - 1.1, 0.1 + 1.0) = (-1, 0) = -1$$

i.e.

$$\boxed{i^2 = -1}$$

The rules for addition and multiplication of complex numbers are cunningly chosen so that *complex numbers satisfy all the usual laws of arithmetic.* This information, together with the boxed results above is all that is necessary for the manipulation of complex quantities.

Examples 12.3

(i) $(1 + 2i) + (3 + 4i) = (1 + 3) + (2 + 4)i = 4 + 6i$

(ii) $(1 + 2i) - (3 + 4i) = (1 - 3) + (2 - 4)i = -2 - 2i$

(iii) $(1 + 2i)(3 + 4i) = 1(3 + 4i) + 2i(3 + 4i)$

$$= 3 + 4i + 6i + 8i^2$$

$$= 3 + 10i - 8 = -5 + 10i$$

(iv) $\dfrac{(1 + 2i)}{(3 + 4i)} = \dfrac{(1 + 2i)(3 - 4i)}{(3 + 4i)(3 - 4i)} = \dfrac{3 - 4i + 6i - 8i^2}{9 - 12i + 12i - 16i^2}$

$$= \dfrac{11 + 2i}{25} = \dfrac{11}{25} + \dfrac{2}{25}i.$$

Note that the latter trick always works. We have that

$$\frac{a+ib}{c+id} = \frac{(a+ib)(c-id)}{(c+id)(c-id)} = \frac{(ac+bd)+i(bc-ad)}{c^2+d^2}$$

provided that c and d are not both zero. (Division by zero is *never* valid.)

12.4   Modulus and argument

If is often useful to express a complex number in terms of polar co-ordinates. If $z = x + iy$, we write

(1)
$$\left.\begin{array}{l} x = r\cos\theta \\ y = r\sin\theta \end{array}\right\}$$

and obtain

(2) $z = r(\cos\theta + i\sin\theta)$.

The number $r \geqslant 0$ in this expression is called the *modulus* of z. We write

$$r = |z| = \{x^2 + y^2\}^{1/2}.$$

The number θ is called the *argument* of z. We write

$$\theta = \arg z$$

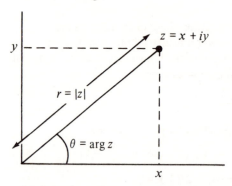

Note that the above remarks do not define θ unambiguously. If θ satisfies equations (1), then so does $\theta + 2n\pi$ for any integer $n = 0, \pm 1, \pm 2, \ldots$. Usually (but not always), we select the value of θ which lies in the range $-\pi < \theta \leqslant \pi$. Note also that the formula

$$\theta = \arctan\frac{y}{x}$$

is valid only for $-\pi/2 < \theta < \pi/2$.

The importance of the identity (2) lies in the fact that
$\cos \theta + i \sin \theta = e^{i\theta}$. We shall comment on the reasons for this in §12.13.
For the moment, we shall only observe that (2) may be rewritten in the form

$$z = re^{i\theta}$$

Examples 12.5

(i) $1 + \sqrt{3}i = 2e^{i\pi/3}$ (ii) $e^{2n\pi i} = 1$ $(n = 0, \pm 1, \pm 2 \ldots)$

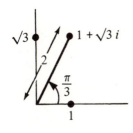
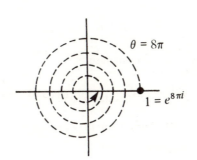

We have that

$$|1 + \sqrt{3}i| = \{1^2 + (\sqrt{3})^2\}^{1/2} = 2$$

$$\arg (1 + \sqrt{3}i) = \arctan \frac{\sqrt{3}}{1} = \frac{\pi}{3}$$

(iii) $i = e^{\pi i/2}$ (iv) $e^{\pi i} = -1$.

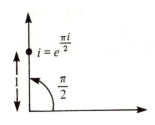
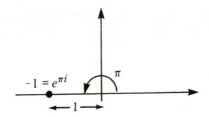

The equation $e^{\pi i} = -1$ is particularly pleasing since it incorporates all the 'mysterious' entities of elementary mathematics — i.e. $-1, \pi, e$ and i.

Exercises 12.6

1. Find the real and imaginary parts of the complex numbers

$$\text{(i)} \quad (z_1 + z_2)z_3 \quad \text{(ii)} \quad (z_1 z_2) - z_3 \quad \text{(iii)} \quad \frac{z_1 - z_3}{z_2} \quad \text{(iv)} \quad \frac{z_1 z_3}{z_2}$$

in the case when $z_1 = 1 - 2i, z_2 = -2 + 3i, z_3 = -3 - 4i$. Find the modulus and the argument of z_1, z_2 and z_3.

2.* Repeat the previous question with $z_1 = 2 - i, z_2 = -3 + 2i$, $z_3 = -4 - 3i$.

3. Prove the following results.

$$\text{(i)} \quad \frac{1}{i} = -i \quad \text{(ii)} \quad i^4 = 1 \quad \text{(iii)} \quad i^3 = -i \quad \text{(iv)} \quad 1 + i = \sqrt{2}e^{i\pi/4}.$$

4.* Prove the following results

$$\text{(i)} \quad e^{-i\theta} = \cos\theta - i\sin\theta \quad \text{(ii)} \quad \cos\theta = \frac{e^{i\theta} + e^{-i\theta}}{2} \quad \text{(iii)} \quad \sin\theta = \frac{e^{i\theta} - e^{-i\theta}}{2i}$$

5. Prove DeMoivre's theorem — i.e.

$$(\cos\theta + i\sin\theta)^n = \cos n\theta + i\sin n\theta \quad (n = 1, 2, 3 \ldots).$$

By considering the case $n = 3$, show that

$$\cos 3\theta = 4\cos^3\theta - 3\cos\theta.$$

6.* Prove that

$$\cos(\theta + \phi) + i\sin(\theta + \phi) = (\cos\theta + i\sin\theta)(\cos\phi + i\sin\phi).$$

7.† The *complex conjugate* $\bar{z}$ of a complex number $z = x + iy$ is given by

$$\bar{z} = x - iy.$$

Prove that

$$\text{(i)} \quad \overline{z_1 + z_2} = \bar{z}_1 + \bar{z}_2 \quad \text{(ii)} \quad \overline{z_1 z_2} = \bar{z}_1 \bar{z}_2 \quad \text{(iii)} \quad \mathcal{R}z = \frac{z + \bar{z}}{2}$$

$$\text{(iv)} \quad \overline{(\bar{z})} = z \quad \text{(v)} \quad \mathcal{R}\left(\frac{z}{w}\right) = \frac{1}{|w|^2}\mathcal{R}(z\bar{w}).$$

8.*† With the notation of the previous question, prove that

(i) $\overline{z_1 - z_2} = \overline{z}_1 - \overline{z}_2$ (ii) $\overline{\left(\dfrac{z_1}{z_2}\right)} = \dfrac{\overline{z}_1}{\overline{z}_2}$ (iii) $\mathscr{I}\!\mathit{m}\, z = \dfrac{z - \overline{z}}{2i}$

(iv) $|z|^2 = z\overline{z}$ (v) $\mathscr{I}\!\mathit{m}\left(\dfrac{z}{w}\right) = \dfrac{1}{|w|^2}\mathscr{I}\!\mathit{m}(z\overline{w}).$

12.7   Complex roots

We now return to the quadratic equation of §12.1 since we can now deal with the case $\Delta = b^2 - 4ac < 0$. In this case, the roots α and β are the complex numbers

$$\alpha = \frac{-b + i\sqrt{(4ac - b^2)}}{2a} \quad \text{and} \quad \beta = \frac{-b - i\sqrt{(4ac - b^2)}}{2a}.$$

This follows from the fact that the equation $z^2 = -1$ has precisely two complex roots, namely $z = i$ and $z = -i$. In general, for any complex number $w \neq 0$, the equation

$$z^n = w$$

has precisely n distant complex solutions for z. We call these solutions the *nth roots* of w. They can be evaluated in the following way. Write $w = Re^{i\phi}$ and write $z = re^{i\theta}$. Then

$$(re^{i\theta})^n = Re^{i\phi}$$

i.e.

$$r^n e^{i\theta n} = Re^{i\phi}$$

(see exercise 12.6 (5) or §12.13). We deduce that $r^n = R$ and so $r = R^{1/n}$ (i.e. r is the positive real number whose nth power is R). Also,

$$\theta n = \phi + 2k\pi$$

for some $k = 0, \pm 1, \pm 2, \ldots$. (Note that we cannot deduce simply that $\theta n = \phi$ for the reasons explained in §12.4.) The possible values of θ are therefore

$$\theta = \frac{\phi}{n} + \frac{2k\pi}{n} \quad (k = 0, \pm 1, \pm 2, \ldots).$$

Note that only n of these possibilities give rise to distinct values of $z = re^{i\theta}$. The reasons will be apparent from the following example.

Example 12.8.
Find the complex fourth roots of -4. We begin by writing

$$-4 = 4e^{i\pi}.$$

The complex fourth roots are therefore of the form $z = re^{i\theta}$ where

$$\begin{cases} r = 4^{1/4} = \sqrt{2} \\ \theta = \dfrac{\pi}{4} + \dfrac{2k\pi}{4} \quad (k = 0, \pm 1, \pm 2, \ldots). \end{cases}$$

Only four of these possibilities give rise to distinct values of z and the required roots are

$$\begin{cases} z_0 = \sqrt{2}e^{i\pi/4} \\ z_1 = \sqrt{2}e^{3i\pi/4} \\ z_2 = \sqrt{2}e^{5i\pi/4} \\ z_3 = \sqrt{2}e^{7i\pi/4} \end{cases}$$

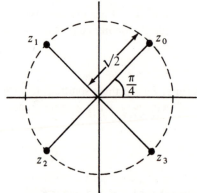

Note that

$$z_4 = \sqrt{2}e^{9i\pi/4} = \sqrt{2}e^{i\pi/4}e^{2\pi i} = \sqrt{2}e^{i\pi/4} = z_0,$$
$$z_{-1} = \sqrt{2}e^{-i\pi/4} = \sqrt{2}e^{7i\pi/4}e^{-2\pi i} = \sqrt{2}e^{7i\pi/4} = z_3.$$

12.9   Polynomials

A *polynomial* is an expression of the form

$$P(z) = a_n z^n + a_{n-1} z^{n-1} + \ldots + a_1 z + a_0$$

in which the coefficients $a_n, a_{n-1}, \ldots, a_1, a_0$ are real or complex numbers. If $a_n \neq 0$, the polynomial is of *degree n*.

An important theorem asserts that every polynomial of degree n has n complex *roots* – i.e. $P(z)$ may be expressed uniquely in the form

$$P(z) = a_n(z - \zeta_1)(z - \zeta_2) \ldots (z - \zeta_n).$$

As in the quadratic case some of the roots $\zeta_1, \zeta_2, \ldots, \zeta_n$ may be co-incident. If a particular root is repeated k times, we say that this root is of *multiplicity k*.

Examples 12.10

(i) The quadratic polynomial $P(x) = x^2 + 1$ has no *real* roots as the diagram indicates. But $(x + i)(x - i) = x^2 - i^2 = x^2 + 1$ and hence

$$x^2 + 1 = (x + i)(x - i).$$

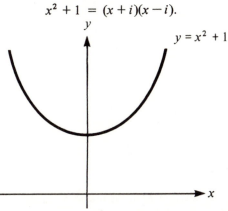

(ii) Let $P(z) = z^4 + 4$. From example 12.8 we have that $P(z_0) = P(z_1) = P(z_2) = P(z_3) = 0$ and so

$$z^4 + 4 = (z - \sqrt{2}e^{i\pi/4})(z - \sqrt{2}e^{3i\pi/4})(z - \sqrt{2}e^{5i\pi/4})(z - \sqrt{2}e^{7i\pi/4}).$$

It is often important to determine the roots of a given polynomial. For example, the eigenvalues of an $n \times n$ matrix are the roots of a polynomial of degree n (see §5.4). Further examples appear in the next chapter. However, it is usually not at all easy to find the roots of a polynomial even when n is reasonably small. We shall just describe a trial and error method which is applicable only in very simple cases.

The first step is to guess a root ζ. Having made such a guess, it is next necessary to check that the guess is accurate by confirming that $P(\zeta) = 0$. One may then write

$$P(z) = (z - \zeta)Q(z)$$

where $Q(z)$ is a polynomial of degree one less than $P(z)$ and repeat the exercise.

Example 12.11.

Consider the polynomial

$$P(z) = z^3 - 7z^2 + 16z - 12.$$

Try $z = 1$. Then $P(1) = 1 - 7 + 16 - 12 = -2 \neq 0$ and so $z = 1$ is *not* a root. Try $z = 2$. Then $P(2) = 8 - 28 + 32 - 12 = 40 - 40 = 0$ and so $z = 2$ *is* a root. Thus

$$P(z) = (z - 2)Q(z).$$

Next we require a formula for $Q(z)$. This can be obtained by 'long division' as below.

$$
\begin{array}{r}
z^2 - 5z + 6 \\
z - 2 \overline{\smash{\big)}\, z^3 - 7z^2 + 16z - 12} \\
\underline{z^3 - 2z^2} \\
-5z^2 + 16z - 12 \\
\underline{-5z^2 + 10z} \\
6z - 12 \\
\underline{6z - 12}
\end{array}
$$

We obtain that $Q(z) = z^2 - 5z + 6 = (z - 2)(z - 3)$ and so

$$P(z) = (z - 2)(z - 2)(z - 3) = (z - 2)^2(z - 3).$$

Observe that $z = 2$ is a root of multiplicity 2.

As an alternative to the use of 'long division' as in the previous example, one may observe that $P(z) = (z - \zeta)Q(z)$ if and only if

$$p_n z^n + p_{n-1} z^{n-1} + \ldots + p_1 z + p_0$$
$$= (z - \zeta)(q_{n-1} z^{n-1} + \ldots + q_1 z + q_0)$$
$$= q_{n-1} z^n + (q_{n-2} - \zeta q_{n-1})z^{n-1} + \ldots + (q_0 - \zeta q_1)z - \zeta q_0$$

from which we obtain that

$$p_k = q_{k-1} - \zeta q_k$$

i.e.

(1) $q_{k-1} = p_k + \zeta q_k$ $(k = 0, 1, 2, \ldots, n)$

where it is understood that $q_n = q_{-1} = 0$.

Example 12.12.

To calculate $Q(z)$ from the formula $P(z) = (z - 2)Q(z)$ of example 12.11, we use the table below:

n	3	2	1	0	-1
p_n	1	-7	16	-12	0
q_n	0	1	-5	6	0

The coefficients of $Q(x) = q_2 x^2 + q_1 x + q_0$ occupy the shaded boxes. We calculate these successively from left to right using formula (1) above — i.e.

$$c = a + \zeta b.$$

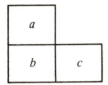

(It is worth checking the calculation by confirming that $q_{-1} = 0$.)

12.13 Elementary functions[†]

In §9.35, we considered real power series. One can equally well consider complex power series. Of particular importance are the series

$$1 + \frac{z}{1!} + \frac{z^2}{2!} + \frac{z^3}{3!} + \dots$$

$$1 - \frac{z^2}{2!} + \frac{z^4}{4!} - \frac{z^6}{6!} + \dots$$

$$z - \frac{z^3}{3!} + \frac{z^5}{5!} - \frac{z^7}{7!} + \dots.$$

These power series converge for all complex numbers z and we use them to define e^z, $\cos z$ and $\sin z$ respectively for *complex* values of z. By and large, these complex functions retain all of the properties of their real counterparts which can be mathematically expressed without the use of the symbols $>$ or $<$. For example,

$$e^{z+w} = e^z e^w$$

because

$$e^{z+w} = \sum_{n=0}^{\infty} \frac{(z+w)^n}{n!} = \sum_{n=0}^{\infty} \frac{1}{n!} \sum_{k=0}^{n} \binom{n}{k} z^k w^{n-k}$$

$$= \sum_{n=0}^{\infty} \sum_{k=0}^{n} \left(\frac{z^k}{k!} \right) \left(\frac{w^{n-k}}{(n-k)!} \right)$$

$$= \sum_{k=0}^{\infty} \frac{z^k}{k!} \sum_{n=k}^{\infty} \frac{w^{n-k}}{(n-k)!}$$

$$= \sum_{k=0}^{\infty} \frac{z^k}{k!} \sum_{l=0}^{\infty} \frac{w^l}{l!} = e^z e^w.$$

Similarly, we have that

$$(e^z)^n = e^{nz}.$$

But the complex functions also have properties which the real functions do not. In particular, the next result shows that the exponential and trigonometric functions are intimately related. If θ is real,

$$e^{i\theta} = 1 + \frac{(i\theta)}{1!} + \frac{(i\theta)^2}{2!} + \frac{(i\theta)^3}{3!} + \frac{(i\theta)^4}{4!} + \ldots$$

$$= 1 + i\theta - \frac{\theta^2}{2!} - \frac{i\theta^3}{3!} + \frac{\theta^4}{4!} - \ldots$$

$$= \left(1 - \frac{\theta^2}{2!} + \frac{\theta^4}{4!} - \ldots\right) + i\left(\theta - \frac{\theta^3}{3!} + \frac{\theta^5}{5!} - \ldots\right)$$

$$= \cos\theta + i\sin\theta.$$

and we therefore obtain the formula

$$\boxed{e^{i\theta} = \cos\theta + i\sin\theta}$$

introduced in §12.14. Note in particular that De Moivre's theorem (exercise 12.6 (5)) is then simply the observation that

$$(e^{i\theta})^n = e^{i\theta n}.$$

Example 12.14.[†]
In exercise 9.37 (2), we considered the integrals

$$I(a, b) = \int_0^\infty (\sin bx)e^{-ax}dx, \quad J(a, b) = \int_0^\infty (\cos bx)e^{-ax}dx.$$

The same integrals were considered again in exercises 10.18 (5) and (6). The integrals may be evaluated more simply by considering

$$\int_0^\infty e^{ibx}e^{-ax}dx = \left[\frac{e^{(ib-a)x}}{(ib-a)}\right]_0^\infty$$

$$= \frac{-1}{(ib-a)} = \frac{a+ib}{a^2+b^2}.$$

Thus

$$I(a, b) = \mathcal{I}m \int_0^\infty e^{ibx}e^{-ax}dx = \frac{b}{a^2+b^2}$$

$$J(a, b) = \mathscr{R} \int_0^\infty e^{ibx} e^{-ax} dx = \frac{a}{a^2 + b^2}.$$

The complex logarithm is less easy to deal with because the equation

$$w = e^z$$

has many solutions for z. If $w = Re^{i\phi}$, these solutions are of the form

$$z = \log R + i\phi$$

Thus the logarithm of w is of the form

(1) $\log w = \log R + i\phi$

but, since the value of ϕ is ambiguous, so is the value of $\log R + i\phi$. To make use of formula (1), it is therefore necessary to say which value of ϕ is to be selected. Different possible selections for ϕ determine different possible 'branches' of the logarithm. We shall just observe that the branch of the logarithm for which

$$\log (1 + w) = w - \frac{w^2}{2} + \frac{w^3}{3} - \ldots \qquad (|w| < 1)$$

is that determined by the restriction $-\pi < \arg (1 + w) \leqslant \pi$.

These remarks are made only to indicate that the complex logarithm must be treated very carefully.

Exercises 12.15

1. Find all complex roots of the polynomials

    (i) $z^2 - i$ (ii) $z^5 + 1$.

2.* Find all complex roots of the polynomials

    (i) $z^3 + i$ (ii) $z^4 - 16$.

3. Factorize the polynomials

    (i) $z^3 - 3z^2 - 10z$ (ii) $z^4 + z^3 - 3z^2 - 5z - 2$.

4.*Factorize the polynomials

    (i) $z^3 + 6z^2 + 11z + 6$ (ii) $z^4 + 2z^2 + 1$.

5.† Let $P(z)$ be a polynomial of degree n with *real* coefficients. Prove that

$$\overline{P(z)} = P(\bar{z})$$

(see exercise 12.6 (7)). Deduce that if ζ is a root of $P(z)$, then so is $\bar{\zeta}$. If n is odd, show that $P(z)$ has at least one real root.

6.*† Continuing the previous question, show that

$$P(x) = \pm (x - \xi_1)(x - \xi_2) \ldots (x - \xi_k)|Q(x)|^2$$

where $\xi_1, \xi_2, \ldots, \xi_k$ are the real roots of $P(x)$ and $Q(x)$ is a polynomial with no real roots (see exercise 12.6 (7(iv))).

SOME APPLICATIONS (OPTIONAL)

12.16 Characteristic functions

In §10.23 we discussed generating functions for the probability density functions of discrete random variables. The notion of a *characteristic function* or *Fourier transform* is a much more powerful version of the same idea.

If X is a continuous random variable with probability density function f, we define its characteristic function ϕ by

$$\phi(t) = \int_{-\infty}^{\infty} e^{ixt} f(x) dx.$$

(This converges for all real t because $|e^{ixt}| = 1$. See §9.30.) If ψ is the characteristic function for an *independent* random variable Y with probability density function g, we have that

$$\phi(t)\psi(t) = \int_{-\infty}^{\infty} e^{ixt} f(x) dx \int_{-\infty}^{\infty} e^{iyt} g(y) dy$$

$$= \int_{-\infty}^{\infty} dx \int_{-\infty}^{\infty} e^{i(x+y)t} f(x) g(y) dy$$

$$= \int_{-\infty}^{\infty} dx \int_{-\infty}^{\infty} e^{izt} f(x) g(z-x) dz$$

$$= \int_{-\infty}^{\infty} e^{izt} dz \int_{-\infty}^{\infty} f(x) g(z-x) dx$$

$$= \int_{-\infty}^{\infty} e^{izt} h(z) dz$$

where h is the probability density function of the random variable $X + Y$ (see §10.22). It follows that the characteristic function for $X + Y$ is the *product* of the characteristic functions for X and Y.

In the case of a discrete random variable X we define the characteristic function ϕ by

$$\phi(t) = \sum_{n=-\infty}^{\infty} e^{itn} f(n).$$

Then $\phi(t) = G(e^{it})$ where G is the corresponding generating function. The result concerning the multiplication of characteristic functions is therefore the same result that we met in §10.23.

As an example, we shall calculate the characteristic function ϕ of a normally distributed random variable with mean μ and variance σ^2. The appropriate formula is

$$\phi(t) = \frac{1}{\sigma\sqrt{(2\pi)}} \int_{-\infty}^{\infty} e^{ixt} e^{-\frac{1}{2}\left(\frac{x-\mu}{\sigma}\right)^2} dx$$

$$= \frac{1}{\sqrt{(2\pi)}} \int_{-\infty}^{\infty} e^{i(\mu+\sigma y)t} e^{-y^2/2} dy$$

$$= e^{i\mu t}\phi_0(\sigma t)$$

where ϕ_0 is the characteristic function of the standardized normal distribution – i.e.

$$\phi_0(t) = \frac{1}{\sqrt{(2\pi)}} \int_{-\infty}^{\infty} e^{ity} e^{-y^2/2} dy.$$

But

$$\frac{d\phi_0}{dt} = \frac{1}{\sqrt{(2\pi)}} \int_{-\infty}^{\infty} iye^{ity} e^{-y^2/2} dy$$

$$= \frac{1}{\sqrt{(2\pi)}} [ie^{ity}(-e^{-y^2/2})]_{-\infty}^{\infty} - \frac{1}{\sqrt{(2\pi)}} \int_{-\infty}^{\infty} i(ite^{ity})(-e^{-y^2/2}) dy$$

$$= \frac{i^2 t}{\sqrt{(2\pi)}} \int_{-\infty}^{\infty} e^{ity} e^{-y^2/2} dy = -t\phi_0(t).$$

Thus

$$\frac{d\phi_0}{dt} = -t\phi_0$$

$$\int \frac{d\phi_0}{\phi_0} = -\int t dt + c$$

$$\log \phi_0 = -\tfrac{1}{2}t^2 + c$$

$$\phi_0(t) = ke^{-t^2/2}.$$

Since $\phi_0(0) = 1$, it follows that $k = 1$. We conclude that

$$\phi(t) = e^{i\mu t} e^{-\sigma^2 t^2/2}.$$

Now suppose that X_1 and X_2 are two *independent* normally distributed random variables with characteristic functions ϕ_1 and ϕ_2 respectively. Then

$$\phi_1(t)\phi_2(t) = e^{i(\mu_1 + \mu_2)} e^{-(\sigma_1^2 + \sigma_2^2)t^2/2}$$

which is the characteristic function of a normally distributed random variable with mean $\mu_1 + \mu_2$ and variance $\sigma_1^2 + \sigma_2^2$. It follows that $X_1 + X_2$ is such a random variable (see §10.22).

12.17   Central limit theorem

Recall from §9.46 that, if $X_1, X_2, X_3, \ldots$ is a sequence of *independent*, identically distributed random variables, each with mean μ and variance σ^2, then, under very general conditions, the distribution of

$$Y = \frac{X_1 + X_2 + \ldots + X_n - n\mu}{\sigma\sqrt{n}}$$

is approximately standard normal provided that n is sufficiently large. This result is called the *central limit theorem*. We give an outline 'proof' below. This is far from being complete but it does indicate how this remarkable result comes about.

Suppose that X_k has probability density function f and characteristic function ϕ. Then

$$\phi(0) = \int_{-\infty}^{\infty} e^{i0x}f(x)dx = \int_{-\infty}^{\infty} f(x)dx = 1$$

Also

$$\left. \begin{array}{l} \phi'(t) = \displaystyle\int_{-\infty}^{\infty} ixe^{ixt}f(x)dx \\[3mm] \phi''(t) = \displaystyle\int_{-\infty}^{\infty} -x^2 e^{ixt}f(x)dx \end{array} \right\}$$

and so

$$\phi'(0) = i\mu \quad \phi''(0) = -\sigma^2 - \mu^2.$$

From §12.16 we know that the characteristic function of $S = X_1 + X_2 + \ldots + X_n$ is

$$\Phi(t) = \{\phi(t)\}^n.$$

It follows that the characteristic function for $(S - a)/b$ is

$$e^{-iat/b}\Phi\left(\frac{t}{b}\right)$$

and, in particular, that Y has the characteristic function

$$\psi(t) = e^{-i\mu\sqrt{n}t/\sigma}\Phi\left(\frac{t}{\sigma\sqrt{n}}\right)$$

$$= e^{-i\mu\sqrt{nt}/\sigma}\left\{\phi\left(\frac{t}{\sigma\sqrt{n}}\right)\right\}^{n}$$

$$\log \psi(t) = \frac{-i\mu\sqrt{nt}}{\sigma} + n \log \phi\left(\frac{t}{\sigma\sqrt{n}}\right).$$

The Taylor series expansion for ϕ about the origin is given by

$$\phi(x) = \phi(0) + x\phi'(0) + \frac{x^2}{2!}\phi''(0) + \dots$$

$$= 1 + i\mu x - \left(\frac{\sigma^2 + \mu^2}{2}\right)x^2 + \dots.$$

We also require the formula

$$\log(1 + z) = z - \frac{z^2}{2} + \frac{z^3}{3} - \dots \qquad (|z| < 1).$$

Systematically ignoring third and higher order terms as they arise, we obtain the approximation

$$\log \phi\left(\frac{t}{\sigma\sqrt{n}}\right) = \log\left(1 + \frac{i\mu t}{\sigma\sqrt{n}} - \left(\frac{\sigma^2 + \mu^2}{2}\right)\frac{t^2}{\sigma^2 n}\right)$$

$$\doteq \frac{i\mu t}{\sigma\sqrt{n}} - \left(\frac{\sigma^2 + \mu^2}{2}\right)\frac{t^2}{\sigma^2 n} - \frac{1}{2}\left(\frac{i\mu t}{\sigma\sqrt{n}} - \left(\frac{\sigma^2 + \mu^2}{2}\right)\frac{t^2}{\sigma^2 n^2}\right)^2$$

$$\doteq \frac{i\mu t}{\sigma\sqrt{n}} - \left(\frac{\sigma^2 + \mu^2}{2}\right)\frac{t^2}{\sigma^2 n} - \frac{1}{2}\left(-\frac{\mu^2 t^2}{\sigma^2 n}\right)$$

$$= \frac{i\mu t}{\sigma\sqrt{n}} - \frac{1}{2}\frac{t^2}{n}.$$

It follows that

$$\frac{-i\mu\sqrt{nt}}{\sigma} + n \log \phi\left(\frac{t}{\sigma\sqrt{n}}\right) \doteq -\frac{1}{2}t^2$$

and so we obtain the approximation

$$\psi(t) \doteq e^{-t^2/2}$$

for the characteristic function of the random variable Y. But the right-hand side is the characteristic function of the standard normal variable.

13

Differential and difference equations (II)

This continues chapter 11. The techniques are easy provided some mastery of complex numbers has been garnered from the previous chapter. A brief theoretical justification for the techniques described appears in §13.1–§13.5 inclusive but readers impatient with theory will find it adequate to confine their attention in these sections to the manipulation of operators as described in examples 13.2 and 13.3 and to the assertions of §13.5 without delving into the reasons why these are correct.

It should be noted that the technique for finding particular solutions of non-homogeneous equations described in §13.14 is only one of several. We have chosen to present this technique because it involves no essentially new ideas. But it is often quicker to find particular solutions using operators in a more adventurous way than described here.

In §13.17 there is a brief discussion of stability questions which is continued in §13.20 in the context of linear systems of differential and difference equations. This material is not hard but some readers may prefer to leave these topics until they can be studied in a more comprehensive manner at a later stage.

13.1 Operators

If $y = \phi(x)$, a function ϕ is sometimes said to transform x into y. Alternatively, one may say that ϕ operates on x to produce y. This leads to functions being described as transformations or *operators* in certain contexts. The latter terminology is often used in particular when x and y are themselves functions. For example, the *differential operator D* is a function which transforms a differentiable function f into its derivative f' — i.e.

$$Df(x) = f'(x).$$

In §11.25, we also met the *shift operator E* and the *difference operator* Δ. These operate on sequences. For example, the shift operator is a function which transforms a sequence s into the sequence obtained by omitting the first term of s — i.e.

$$Es_x = s_{x+1}.$$

The *identity operator* I should also be mentioned. This transforms everything into itself.

If ϕ and ψ are operators, we define the operators $\alpha\phi + \beta\psi$ and $\phi\psi$ by

$$(\alpha\phi + \beta\psi)(x) = \alpha\phi(x) + \sigma\psi(x)$$

$$(\phi\psi)(x) = \phi(\psi(x))$$

provided the right-hand expressions are meaningful. Note in particular that $\phi\psi$ means 'first apply ψ and then apply ϕ'.

These are not unfamiliar conventions. Indeed, matrix algebra is based on their use. In particular, the definitions of scalar multiplication, matrix addition and matrix multiplication given in §1.2, §1.3 and §1.4 are chosen because, and only because, with these definitions,

$$(\alpha A + \beta B)x = \alpha A x + \beta B x$$

$$(AB)x = A(Bx)$$

whenever the matrices A and B and the column vector x are such that the right-hand expressions are meaningful.

The algebra of operators is very similar to the algebra of numbers. But it is *not* identical. When dealing with operators, division is always a problem (as we know from considering inverse matrices). We shall avoid this problem here by not using division with operators at all. Multiplication of operators is not straightforward either. If ϕ and ψ are operators, it need *not* be the case that $\phi\psi = \psi\phi$ (see exercise 1.15 (3)). We shall avoid this problem in this chapter by restricting our attention to operators which are 'polynomials' in D or E. Such polynomials can be manipulated just like ordinary polynomials.

Example 13.2.
Consider the differential equation

$$\frac{d^3f}{dx^3} + 3\frac{d^2f}{dx^2} + \frac{df}{dx} + 3f = 0.$$

This may be rewritten in the form

$$D^3f + 3D^2f + Df + 3f = 0$$

$$(D^3 + 3D^2 + D + 3I)f = 0.$$

The use of the identity operator I is somewhat pedantic here and we usually write 3 instead of $3I$. We then obtain

$$P(D)f = 0$$

where $P(D)$ is the polynomial

$$P(D) = D^3 + 3D^2 + D + 3.$$

It is often convenient to factorize such a polynomial. For this purpose one requires the roots of the *auxiliary equation*

$$z^3 + 3z^2 + z + 3 = 0.$$

Little ingenuity is required to see that $z = -3$ is a root. Using the method of §12.9, we then obtain that

$$z^3 + 3z^2 + z + 3 = (z + 3)(z^2 + 1)$$

and thus

$$P(D) = D^3 + 3D^2 + D + 3 = (D + 3)(D^2 + 1).$$

We shall check the observations of §13.1 concerning the algebra of polynomials in D by confirming that

$$(D^3 + 3D^2 + D + 3)f = (D + 3)(D^2 + 1)f = (D^2 + 1)(D + 3)f.$$

We have that

(1) $\begin{aligned}[t] (D + 3)(D^2 + 1)f &= (D + 3)(f'' + f) \\ &= (D + 3)f'' + (D + 3)f \\ &= (f''' + 3f'') + (f' + 3f) \\ &= (D^3 + 3D^2 + D + 3)f. \end{aligned}$

Also

(2) $\begin{aligned}[t] (D^2 + 1)(D + 3)f &= (D^2 + 1)(f' + 3f) \\ &= (D^2 + 1)f' + 3(D^2 + 1)f \\ &= (f''' + f') + (3f'' + 3f) \\ &= f''' + 3f'' + f' + 3f \\ &= (D^3 + 3D^2 + D + 3)f. \end{aligned}$

Example 13.3.

Consider the difference equation

$$y_{x+3} + y_{x+2} - y_{x+1} - y_x = 0 \quad (x = 0, 1, 2, \ldots)$$

which may be rewritten in the form

$$P(E)y_x = (E^3 + E^2 - E - 1)y_x = 0.$$

As in the previous example, it is usually convenient to factorize $P(E)$ and so we consider the auxiliary equation

$$z^3 + z^2 - z - 1 = 0$$

of which $z = 1$ is clearly a root. The method of §12.9 then yields that

$$z^3 + z^2 - z - 1 = (z - 1)(z + 1)^2$$

and thus

$$P(E) = E^3 + E^2 - E - 1 = (E - 1)(E + 1)^2 = (E^2 - 1)(E + 1).$$

Observe that

$$
\begin{aligned}
(3) \quad (E^2 - 1)(E + 1)y_x &= (E^2 - 1)(y_{x+1} + y_x) \\
&= (E^2 - 1)y_{x+1} + (E^2 - 1)y_x \\
&= (y_{x+3} - y_{x+1}) + (y_{x+2} - y_x) \\
&= y_{x+3} + y_{x+2} - y_{x+1} - y_x \\
&= (E^3 + E^2 - E - 1)y_x.
\end{aligned}
$$

Also

$$
\begin{aligned}
(4) \quad (E + 1)(E^2 - 1)y_x &= (E + 1)(y_{x+2} - y_x) \\
&= (E + 1)y_{x+2} - (E + 1)y_x \\
&= (y_{x+3} + y_{x+2}) - (y_{x+1} + y_x) \\
&= (E^3 + E^2 - E - 1)y_x.
\end{aligned}
$$

13.4 Linear operators

If A is an $m \times n$ matrix and $\mathbf{x}$ and $\mathbf{y}$ are $n \times 1$ column vectors, then

$$A(\alpha \mathbf{x} + \beta \mathbf{y}) = \alpha A \mathbf{x} + \beta A \mathbf{y}.$$

The differential operator D and the shift operator E share the same property

$$D(\alpha f + \beta g) = \alpha Df + \beta Dg$$

$$E(\alpha s_x + \beta t_x) = \alpha E s_x + \beta E t_x.$$

Operators with this property are said to be *linear*. Note that, not only are D and E linear, but so are any polynomials in D and E. (We have already used this fact in steps (1), (2), (3) and (4) of the preceding examples.)

The *kernel* or (null-space) of an $m \times n$ matrix A is the set of all $n \times 1$ column vectors $\mathbf{x}$ which satisfy

$$A\mathbf{x} = \mathbf{0}.$$

The kernel is a vector subspace of $\mathbb{R}^n$. (Its dimension is called the 'nullity' of A. The formula 'rank + nullity $= n$' may be familiar.)

We shall be interested in the kernels of operators of the form $P(D)$ and $P(E)$ where P is a polynomial of degree k. The reason is that a function f is in the kernel of $P(D)$ if and only if it is a solution of the kth order differential equation

$$P(D)f = 0.$$

Similarly, a sequence s is in the kernel of $P(E)$ if and only if it satisfies the kth order difference equation

$$P(E)s = 0$$

i.e.

(1) $p_n s_{x+n} + p_{n-1} s_{x+n-1} + \ldots + p_1 s_{x+1} + p_0 s_x = 0$

It is easy to see that the kernel of $P(E)$ has dimension k because once the values of $s_0, s_1, \ldots, s_{k-1}$ are known, then (1) determines all the remaining terms of a sequence s which satisfies $P(E)s = 0$. However $s_0, s_1, \ldots, s_{k-1}$ may be chosen quite freely. Each k-dimensional vector $(s_0, s_1, \ldots, s_{k-1})^T$ therefore determines a solution of $P(E)s = 0$ and vice versa. Hence the set of solutions has dimension k.

An essentially similar argument (but using Taylor series) establishes that the set of solutions of $P(D)f = 0$ also has dimension k.

We use these results in the next sections.

13.5   Homogeneous, linear, differential equations

A linear differential equation of order n has the form

$$p_n(x)\frac{d^n y}{dx^n} + p_{n-1}(x)\frac{d^{n-1} y}{dx^{n-1}} + \ldots + p_1(x)\frac{dy}{dx} + p_0(x)y = q(x).$$

In this chapter we shall consider *only* the case in which the coefficients p_n, $p_{n-1}, \ldots, p_1, p_0$ are *constant*. The differential equation can then be written in the form

$$P(D)y = q(x)$$

where P is a polynomial of degree n.

In this section we consider the *homogeneous* case when $q(x) = 0$. We know from §13.4 that the space of solutions of the homogeneous equation

(1) $P(D)y = 0$

has dimension n. Our problem therefore reduces to finding a *basis* for this solution space. If $y_1, y_2, \ldots, y_n$ is such a basis, then any other solution y may be expressed as a *linear combination* of $y_1, y_2, \ldots, y_n$ — i.e.

(2) $y = a_1 y_1 + a_2 y_2 + \ldots + a_n y_n$

where $a_1, a_2, \ldots, a_n$ are constants. Note that (2) involves n arbitrary constants and hence this result accords with our definition of a general solution of a differential equation of order n (see §11.2).

In seeking a basis for the solution space of (1), it is not enough to find n different solutions $y_1, y_2, \ldots, y_n$. We must find n *linearly independent* solutions. This means that no y_k must be expressible as a linear combination of the other ys. Fortunately, a set of n linearly independent solutions is easy to write down provided that we can factorize the polynomial $P(D)$.

Consider the equation

$$P(D)y = (D - \alpha_1)^{m_1} (D - \alpha_2)^{m_2} \ldots (D - \alpha_k)^{m_k} y = 0$$

in which $\alpha_1, \alpha_2, \ldots, \alpha_k$ are distinct real or complex numbers. Each factor $(D - \alpha)^m$ contributes m linearly independent solutions to $P(D)y = 0$ — namely

$$y = e^{\alpha x}, \; y = x e^{\alpha x}, \; y = x^2 e^{\alpha x}, \ldots, \; y = x^{m-1} e^{\alpha x}$$

Taking each factor in turn, we obtain $m_1 + m_2 + \ldots + m_k = n$ linearly independent solutions in all, where n is the degree of the polynomial $P(D)$.

As an example, consider the equation

$$P(D)y = (D - \alpha)(D - \beta)^3 (D - \gamma)^2 y = 0$$

in which α, β, and γ are distinct. This has *six* linearly independent solutions. The general solution is

$$y = (A_0 e^{\alpha x}) + (B_0 e^{\beta x} + B_1 x e^{\beta x} + B_2 x^2 e^{\beta x}) + (C_0 e^{\gamma x} + C_1 x e^{\gamma x})$$

where A_0, B_0, B_1, B_2, C_0 and C_1 are arbitrary constants.

It is instructive to see how this result may be justified. We begin with the observation that $y = e^{\alpha x}$ is a solution of $(D - \alpha)y = 0$. This can be checked by a direct substitution. Alternatively, one may rewrite $(D - \alpha)y = 0$ as

$$\frac{dy}{dx} = \alpha y$$

and separate the variables to obtain

$$\frac{dy}{y} = \alpha dx$$

$$y = A e^{\alpha x}.$$

The fact that $y = e^{\alpha x}$ is a solution of $(D - \alpha)y = 0$ implies that $y = e^{\alpha x}$ is also a solution of $P(D)y = 0$ because

$$
\begin{aligned}
P(D)e^{\alpha x} &= (D - \alpha)(D - \beta)^3 (D - \gamma)^2 e^{\alpha x} \\
&= (D - \beta)^3 (D - \gamma)^2 (D - \alpha)e^{\alpha x} \\
&= (D - \beta)^3 (D - \gamma)^2 0 = 0.
\end{aligned}
$$

We next consider the equation $(D - \beta)^3 y = 0$. One can check that $e^{\beta x}$, $xe^{\beta x}$ and $x^2 e^{\beta x}$ are solutions by direct substitution. An alternative method is to write $u = (D - \beta)^2 y$ and $v = (D - \beta)y$. Then

$$(D - \beta)u = 0$$

and so

$$u = be^{\beta x}$$

i.e.

$$(D - \beta)v = be^{\beta x}$$

$$\frac{dv}{dx} - \beta v = be^{\beta x}.$$

This is a first order linear equation (§11.10) for which a suitable integrating factor is $\mu = e^{-\beta x}$. Using this integrating factor, we obtain that

$$\frac{d}{dx}(e^{-\beta x}v) = e^{-\beta x}\frac{dv}{dx} - \beta e^{-\beta x}v = b$$

$$e^{-\beta x}v = bx + c$$

$$v = bxe^{\beta x} + ce^{\beta x}$$

i.e.

$$(D - \beta)y = bxe^{\beta x} + ce^{\beta x}$$

This is another first order linear equation for which the same integrating factor is appropriate. Arguing as before, we obtain that

$$y = B_2 x^2 e^{\beta x} + B_1 xe^{\beta x} + B_0 e^{\beta x},$$

where $B_2 = \frac{1}{2}b$ and $B_1 = c$. It follows in particular that $y = e^{\beta x}$, $y = xe^{\beta x}$ and $y = x^2 e^{\beta x}$ are solutions of $(D - \beta)^3 y = 0$. We may conclude that these three functions are also solutions of $P(D)y = 0$. For example,

$$
\begin{aligned}
P(D)xe^{\beta x} &= (D - \alpha)(D - \beta)^3 (D - \gamma)^2 xe^{\beta x} \\
&= (D - \alpha)(D - \gamma)^2 (D - \beta)^3 xe^{\beta x} \\
&= (D - \alpha)(D - \gamma)^2 0 = 0.
\end{aligned}
$$

It remains to show that $y = e^{\gamma x}$ and $y = xe^{\gamma x}$ are solutions of $(D - \gamma)^2 y = 0$. This we leave as an exercise.

Example 13.6.

Consider the equation

$$\frac{d^3y}{dx^3} + \frac{d^2y}{dx^2} - \frac{dy}{dx} - y = 0.$$

This may be written as

$$P(D)y = (D^3 + D^2 - D - 1)y = 0.$$

The auxiliary equation is $P(z) = z^3 + z^2 - z - 1 = 0$. A root is $z = 1$. Thus $P(z) = (z-1)Q(z)$.

n	3	2	1	0	-1
p_n	1	1	-1	-1	0
q_n	0	1	2	1	0

The table gives the coefficients of Q. We obtain that
$P(z) = (z-1)(z^2 + 2z + 1) = (z-1)(z+1)^2$. Thus

$$P(D)y = (D-1)(D+1)^2 y = 0$$

has general solution

$$y = A_0 e^x + B_0 e^{-x} + B_1 x e^{-x}$$

where A_0, B_0 and B_1 are arbitrary constants.

13.7 Complex roots of the auxiliary equation

In §13.5 we explained how the general solution of a differential equation $P(D)f = 0$ may be expressed in terms of the roots $\alpha_1, \alpha_2, \ldots, \alpha_n$ of the auxiliary equation $P(z) = 0$. If P is a polynomial with *real* coefficients, we shall usually only be interested in *real* solutions of the differential equation. But some of the numbers $\alpha_1, \alpha_2, \ldots, \alpha_n$ may be *complex*. How can the theory be adapted in this case?

The first observation that needs to be made is that, if $\alpha = \lambda + i\mu$ is a root of $P(z) = 0$, then so is $\bar{\alpha} = \lambda - i\mu$ provided that the coefficients of P are *real* (see exercises 12.6 (7) and 12.15 (5)). The non-real roots of $P(z) = 0$ therefore occur in *pairs* of the form $\alpha = \lambda + i\mu$ and $\bar{\alpha} = \lambda - i\mu$. The complex terms of the general solution of $P(D)y = 0$ therefore occur in pairs of the form

$$(1) \quad Ax^k e^{\alpha x} + Bx^k e^{\bar{\alpha} x}.$$

When we are interested only in *real* solutions this expression may be replaced by

(2) $ax^k e^{\lambda x}\cos \mu x + bx^k e^{\lambda x}\sin \mu x$

where a and b are arbitrary constants.

To justify this assertion, we need to recall that $e^{i\theta} = \cos \theta + i \sin \theta$ (and hence $e^{-i\theta} = \cos \theta - i \sin \theta$). Thus

$$e^{\alpha x} = e^{\lambda x}e^{i\mu x} = e^{\lambda x}(\cos \mu x + i \sin \mu x)$$

$$e^{\bar{\alpha} x} = e^{\lambda x}e^{-i\mu x} = e^{\lambda x}(\cos \mu x - i \sin \mu x).$$

Hence
$$Ax^k e^{\alpha x} + Bx^k e^{\bar{\alpha} x} = x^k e^{\lambda x}\{(A + B) \cos \mu x + i(A - B) \sin \mu x\}$$

$$= ax^k e^{\lambda x}\cos \mu x + bx^k e^{\lambda x}\sin \mu x$$

provided $a = A + B$ and $b = i(A - B)$. We shall, of course, wish only to consider *real* values of a and b. (This will usually require that A and B are complex.)

Occasionally it is useful to replace the arbitrary constants a and b by r and θ where $a = r \cos \theta$ and $b = r \sin \theta$. Making this substitution in (2), we obtain that (1) may be replaced by

(3) $rx^k e^{\lambda x}\cos (\mu x - \theta)$

where r and θ are arbitrary constants. Alternatively, the arbitrary constants a and b may be replaced by R and Θ where $a = R \sin \Theta$ and $b = R \cos \Theta$. We then obtain that (1) may be replaced by

(4) $Rx^k e^{\lambda x}\sin (\mu x + \Theta)$

where R and Θ are arbitrary constants.

Example 13.8.

Find the general solution of

$$\frac{d^2 y}{dx^2} + 4y = 0.$$

We write the equation in the form

$$P(D)y = (D^2 + 4)y = 0$$

i.e.
$$(D + 2i)(D - 2i)y = 0.$$

The general solution is therefore

$$y = Ae^{-2ix} + Be^{2ix}$$

which we may rewrite as

$$y = a \cos 2x + b \sin 2x$$

if we are interested only in real solutions.

Example 13.9.
Find the general solution of

$$\frac{d^3y}{dx^3} = y.$$

We write the equation in the form

$$P(D)y = (D^3 - 1)y = 0.$$

Observe that $P(z) = z^3 - 1 = (z - 1)(z^2 + z + 1)$ because of the formula for the sum of a geometric progression. The roots of $z^2 + z + 1 = 0$ are given by

$$\alpha = \frac{-1 + \sqrt{(1 - 4)}}{2} = \frac{-1 + i\sqrt{3}}{2}$$

$$\bar{\alpha} = \frac{-1 - \sqrt{(1 - 4)}}{2} = \frac{-1 - i\sqrt{3}}{2}.$$

Thus $P(z) = (z - 1)(z - \alpha)(z - \bar{\alpha})$. The general solution of $P(D)y = 0$ is therefore

$$y = Ae^x + Be^{(-1 + i\sqrt{3})x/2} + Ce^{(-1 - i\sqrt{3})x/2}$$

which we may rewrite in the form

$$y = Ae^x + be^{-x/2} \cos (\tfrac{1}{2}x\sqrt{3}) + ce^{-x/2} \sin (\tfrac{1}{2}x\sqrt{3}).$$

(Note that the roots of $z^3 = 1$ may also be found as in §12.7.)

13.10 Homogeneous, linear, difference equations

Instead of considering equations of the form $P(D)y = 0$ as in §13.5, we now consider equations of the form

$$P(E)y = 0.$$

The appropriate theory is very similar indeed to that studied in §13.5.
Suppose that

$$P(E)y = (E - \alpha_1)^{m_1} (E - \alpha_2)^{m_2} \ldots (E - \alpha_k)^{m_k}y = 0$$

where $\alpha_1, \alpha_2, \ldots, \alpha_k$ are distinct non-zero numbers. Each factor $(E - \alpha)^m$ contributes m linearly independent solutions to $P(E)y = 0$ – namely

$$y = \alpha^x, \ y = x\alpha^x, \ y = x^2\alpha^x, \ldots, \ y = x^{m-1}\alpha^x$$

Taking each factor in turn, we obtain $m_1 + m_2 + \ldots + m_k = n$ linearly independent solutions in all, where n is the degree of the polynomial $P(D)$.

As an example, consider the equation

$$P(E)y = (E - \alpha)(E - \beta)^3 (E - \gamma)^2 y = 0$$

where α, β, and γ are distinct non-zero numbers. This has *six* linearly independent solutions. The general solution is

$$y = (A_0\alpha^x) + (B_0\beta^x + B_1 x\beta^x + B_2 x^2\beta^x) + (C_0\gamma^x + C_1 x\gamma^x).$$

This result can be justified along the same lines as in §13.5 using the results about difference equations obtained in §11.25.

When seeking real solutions of $P(E)y = 0$ in the case when P is a polynomial with real coefficients, the necessary argument is slightly simpler than that of §13.7. If $\alpha = \lambda e^{i\mu}$ and $\bar\alpha = \lambda e^{-i\mu}$ are roots of the auxiliary equation $P(z) = 0$, then the expression

$$Ax^k\alpha^x + Bx^k\bar\alpha^x$$

in the general solution of $P(E)y = 0$ may be replaced by

$$y = ax^k\lambda^x \cos \mu x + bx^k\lambda^x \sin \mu x$$

where a and b are arbitrary constants. Occasionally, as in §13.7, it is useful to use one of the alternative expressions

$$y = rx^k\lambda^x \cos (\mu x - \theta)$$

or

$$y = Rx^k\lambda^x \sin (\mu x + \Theta).$$

Example 13.11.
Find the general solution of

$$y_{x+2} + 3y_{x+1} + 2y_x = 0.$$

We write this as

$$(E^2 + 3E + 2)y = 0$$

$$(E + 1)(E + 2)y = 0$$

and observe immediately that the general solution is

$$y = A(-1)^x + B(-2)^x.$$

Example 13.12.
Find the general solution of

$$(E^4 - 12E^3 + 56E^2 - 120E + 91)y = 0.$$

One may check that the polynomial factors so that

$$(E - 3 - i)^2 (E - 3 + i)^2 y = 0$$

(although to obtain this result from scratch would not be at all easy). The general solution is

$$y = \{A_0(3 + i)^x + A_1 x(3 + i)^x\} + \{B_0(3 - i)^x + B_1 x(3 - i)^x\}$$

$$= \{A_0(3 + i)^x + B_0(3 - i)^x\} + \{A_1 x(3 + i)^x + B_1 x(3 - i)^x\}.$$

If we are interested only in real solutions, we need to observe that

$$3 + i = \lambda e^{i\mu}; \quad 3 - i = \lambda e^{-i\mu}$$

where

$$|\lambda| = \{3^2 + 1^2\}^{1/2} = \sqrt{(10)}$$

and

$$\mu = \arctan \tfrac{1}{3}.$$

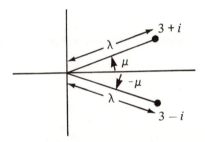

We may then write

$$y = \{a_0 10^{x/2} \cos \mu x + b_0 10^{x/2} \sin \mu x\}$$
$$+ \{a_1 x 10^{x/2} \cos \mu x + b_1 x 10^{x/2} \sin \mu x\}$$

and restrict our attention to *real* arbitrary constants a_0, b_0, a_1 and b_1.

Example 13.13.
Find the general solution of

$$y_{x+3} = 2y_{x+2} \quad (x = 0, 1, 2 \dots).$$

We may rewrite this as

$$P(E)y = (E^3 - 2E^2)y = 0$$

i.e.

$$E^2(E - 2)y = 0$$

but this is not particularly useful since, when $\alpha = 0$, the solutions given in
(1) of §13.10 all reduce to zero.

It is better to put $z = x + 2$ in which case the equation becomes

$$y_{z+1} = 2y_z \quad (z = 2, 3, 4, \dots).$$

Our equation is therefore 'really' the first order equation

$$(E - 2)y = 0$$

which has the general solution

$$y_z = C2^z \quad (z = 2, 3, 4 \dots).$$

This leaves us free to choose $y_0 = A$ and $y_1 = B$ where A and B are arbitrary.

13.14   Non-homogeneous equations

We begin by recalling some results about systems of linear equations. Suppose
that A is an $m \times n$ matrix and that $\mathbf{b}$ is an $m \times 1$ column vector. Let $\mathbf{x} = \boldsymbol{\xi}$ be
any particular solution of the system

$$A\mathbf{x} = \mathbf{b}.$$

Then any other $n \times 1$ column vector $\mathbf{u}$ is a solution of the system if and only
if

$$\mathbf{u} = \boldsymbol{\xi} + \mathbf{x}_0,$$

where $\mathbf{x}_0$ is a solution of the *homogeneous* system

$$A\mathbf{x} = \mathbf{0}$$

(i.e. $\mathbf{x}_0$ lies in the kernel of A).

The proof is quite easy. Suppose first that $\mathbf{u} = \boldsymbol{\xi} + \mathbf{x}_0$. Then.

$$A\mathbf{u} = A(\boldsymbol{\xi} + \mathbf{x}_0) = A\boldsymbol{\xi} + A\mathbf{x}_0 = \mathbf{b} + \mathbf{0} = \mathbf{b}$$

and so $\mathbf{u}$ is a solution of $A\mathbf{x} = \mathbf{b}$. Suppose, on the other hand, that $\mathbf{u}$ is a
solution of $A\mathbf{x} = \mathbf{b}$. Then

$$A(\mathbf{u} - \boldsymbol{\xi}) = A\mathbf{u} - A\boldsymbol{\xi} = \mathbf{b} - \mathbf{b} = \mathbf{0}$$

and so $\mathbf{x}_0 = \mathbf{u} - \boldsymbol{\xi}$ is a solution of $A\mathbf{x} = \mathbf{0}$.

The situation for a non-homogeneous differential or difference equation
is exactly the same. If η is a particular solution of the differential equation

$$P(D)y = q(x)$$

then the general solution is given by

$$y = \eta + z$$

where z is the general solution of the homogeneous equation

$$P(D)y = 0.$$

Exactly the same statement is true for difference equations with E replacing D.

The problem of determining the general solution of a non-homogeneous equation therefore reduces to two steps. One must locate any *particular solution* η of $P(D)y = q$ (or $P(E)y = q$) and one must find the general solution z of $P(D)y = 0$ (or $P(E)y = 0$). The latter quantity is called the *complementary function* (or sequence).

We shall provide a systematic method for locating a particular solution of a non-homogeneous equation only in the case when q is itself the solution of a homogeneous equation.

Suppose that we are seeking a particular solution η of $P(D)y = q$ where

$$Q(D)q = 0.$$

Then η must satisfy the homogeneous equation

$$Q(D)P(D)y = 0$$

because
$$Q(D)P(D)\eta = Q(D)q = 0.$$

If $Q(D)P(D)y = 0$ is an equation of order k with linearly independent solutions $y_1, y_2, \ldots, y_k$, then it follows that η may be expressed in the form

$$\eta = A_1 y_1 + A_2 y_2 + \ldots + A_k y_k.$$

One may then substitute this expression in the equation $P(D)\eta = q$ to obtain the appropriate values of the constants $A_1, A_2, \ldots, A_k$.

Precisely the same remarks are valid, of course, with D replaced everywhere by E.

Example 13.15.
Find the general solution of

$$(E-2)y = 3.$$

(This is equation (1) of example 11.28 with $r = 2$ and $Q = 3$.)
    We first find the complementary sequence – i.e. the general solution of

$$(E-2)y = 0.$$

This is given by

$$y = A2^x$$

where A is an arbitrary constant. Next we seek a particular solution η of $(E-2)y = 3$. The general solution of $(E-2)y = 3$ will then be of the form

$$y = \eta + A2^x.$$

Since $q = 3$ is a solution of $(E-1)q = 0$, we have that η is a solution of

$$(E-1)(E-2)y = 0$$

and hence may be expressed in the form

$$\eta = B + C2^x.$$

We already know that the constant C may take *any* value. The simplest particular solution is therefore obtained by taking $C = 0$. But what about the constant B?

To evaluate B, we substitute $\eta = B$ in the equation $(E-2)\eta = 3$ and obtain

$$(E-2)B = 3$$

i.e.

$$B - 2B = 3$$

$$B = -3$$

The general solution of $(E-2)y = 3$ is therefore

$$y = -3 + A2^x.$$

Example 13.16.
Find the general solution of

$$\frac{d^2y}{dx^2} + y = x \sin x.$$

We write this equation in the form

$$(D^2 + 1)y = x \sin x$$

i.e.

$$(D+i)(D-i)y = x \sin x.$$

The complementary function is found by determining the general solution of

$$(D+i)(D-i)y = 0$$

which is

$$y = Ae^{ix} + Be^{-ix}.$$

The general solution of $(D^2 + 1)y = x \sin x$ is therefore

$$y = \eta + Ae^{ix} + Be^{-ix}$$

where η is a particular solution of $(D^2 + 1)y = x \sin x$.

We next seek a homogeneous differential equation satisfied by $q = x \sin x$. Since

$$x \sin x = \frac{1}{2i}xe^{ix} - \frac{1}{2i}xe^{-ix}$$

(exercise 12.6 (4)), a suitable equation is

$$(D+i)^2(D-i)^2 q = 0.$$

We conclude that η is to be found among the solutions of

$$(D+i)^3(D-i)^3 y = 0$$

and so η may be expressed in the form

$$\eta = (a_0 e^{ix} + a_1 xe^{ix} + a_2 x^2 e^{ix}) + (b_0 e^{-ix} + b_1 xe^{-ix} + b_2 x^2 e^{-ix}).$$

It is already known that the constants a_0 and b_0 may take *any* values. The simplest particular solution is therefore obtained by taking $a_0 = 0$ and $b_0 = 0$. Then a_1, a_2, b_1 and b_2 may be evaluated by substituting in

$$(D^2 + 1)\eta = x \sin x = \frac{1}{2i}xe^{ix} - \frac{1}{2i}xe^{-ix}.$$

We have that

$$\begin{aligned}
D\eta &= a_1(e^{ix} + ixe^{ix}) + a_2(2xe^{ix} + ix^2 e^{ix}) + b_1(e^{-ix} - ixe^{-ix}) \\
&\quad + b_2(2xe^{-ix} - ix^2 e^{-ix}) \\
&= a_1 e^{ix} + (ia_1 + 2a_2)xe^{ix} + ia_2 x^2 e^{ix} + b_1 e^{-ix} \\
&\quad + (-ib_1 + 2b_2)xe^{-ix} - ib_2 x^2 e^{-ix}
\end{aligned}$$

and so

$$\begin{aligned}
D^2\eta &= ia_1 e^{ix} + (ia_1 + 2a_2)e^{ix} + (-a_1 + 2ia_2)xe^{ix} + 2ia_2 xe^{ix} \\
&\quad - a_2 x^2 e^{ix} - ib_1 e^{-ix} + (-ib_1 + 2b_2)e^{-ix} \\
&\quad + (-b_1 - 2ib_2)xe^{-ix} - 2ib_2 xe^{-ix} - b_2 x^2 e^{-ix}.
\end{aligned}$$

Thus

$$\begin{aligned}
\frac{1}{2i}xe^{ix} - \frac{1}{2i}xe^{-ix} &= (D^2 + 1)\eta \\
&= (2ia_1 + 2a_2)e^{ix} + 4ia_2 xe^{ix} \\
&\quad + (-2ib_1 + 2b_2)e^{-ix} - 4ib_2 xe^{-ix}.
\end{aligned}$$

It follows that

$$2ia_1 + 2a_2 = 0$$

$$4ia_2 = \frac{1}{2i}$$

$$-2ib_1 + 2b_2 = 0$$

$$-4ib_2 = -\frac{1}{2i}$$

and hence

$$a_2 = -\tfrac{1}{8}$$

$$b_2 = -\tfrac{1}{8}$$

$$a_1 = -\frac{a_2}{i} = \frac{1}{8i}$$

$$b_1 = \frac{b_2}{i} = -\frac{1}{8i}.$$

Thus

$$\eta = \left\{ \frac{1}{8i}xe^{ix} - \frac{1}{8}x^2 e^{ix} \right\} + \left\{ -\frac{1}{8i}xe^{-ix} - \frac{1}{8}x^2 e^{-ix} \right\}$$

$$= \frac{1}{4}x\frac{1}{2i}(e^{ix} - e^{-ix}) - \frac{1}{4}x^2 \frac{1}{2}(e^{ix} + e^{-ix})$$

$$= \frac{1}{4}x \sin x - \frac{1}{4}x^2 \cos x.$$

The general solution (in real form) is therefore

$$y = \tfrac{1}{4}x \sin x - \tfrac{1}{4}x^2 \cos x + a \cos x + b \sin x.$$

It is usually wise to substitute a result obtained after so much labour directly into $(D^2 + 1)y = x \sin x$ to check the validity of the calculations.

13.17 Convergence and divergence†

The behavior of solutions of differential and difference equations when x is very large is often of interest since it indicates what will happen 'in the long run'.

The diagrams below indicate the behavior of the functions which arise in the solution of the differential equations considered in this chapter.

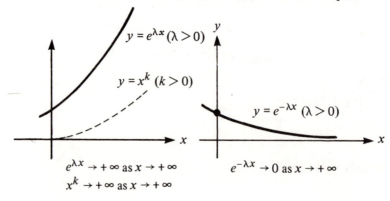

$$e^{\lambda x} \to +\infty \text{ as } x \to +\infty$$

$$x^k \to +\infty \text{ as } x \to +\infty$$

$$e^{-\lambda x} \to 0 \text{ as } x \to +\infty$$

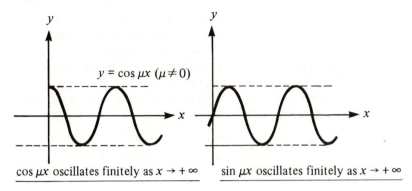

cos μx oscillates finitely as $x \to +\infty$ sin μx oscillates finitely as $x \to +\infty$

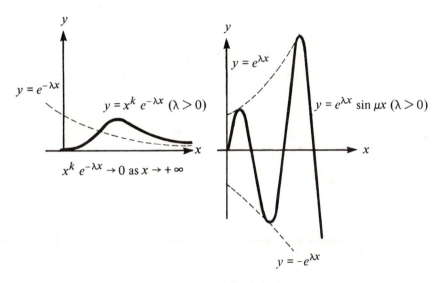

$x^k e^{-\lambda x} \to 0$ as $x \to +\infty$

$e^{\lambda x} \sin \mu x$ oscillates infinitely as $x \to +\infty$

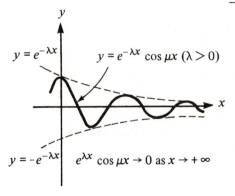

$e^{\lambda x} \cos \mu x \to 0$ as $x \to +\infty$

Note in particular that $x^k e^{-\lambda x} \to 0$ as $x \to +\infty$ no matter how large the constant k may be and no matter how small the positive constant λ may be. This is an example of the important fact that 'exponentials dominate powers'. The same phenomenon is also evident in the first diagram where the graph $y = e^{\lambda x}$ grows much faster than that of $y = x^k$.

These results are easy to confirm using the Taylor series expansion for the exponential function. We have that

$$e^{\lambda x} = 1 + \lambda x + \frac{(\lambda x)^2}{2!} + \ldots + \frac{(\lambda x)^{k+1}}{(k+1)!} + \ldots$$

and so

$$e^{\lambda x} > \frac{(\lambda x)^{k+1}}{(k+1)!} \quad (x \geqslant 0).$$

It follows that

$$x^k e^{-\lambda x} < \frac{(k+1)!}{\lambda^{k+1}} \cdot \frac{1}{x} \to 0 \quad \text{as} \quad x \to +\infty.$$

The situation for the solutions of difference equations is very similar except that terms of the form $e^{\lambda x}$ are replaced by terms of the form α^x.

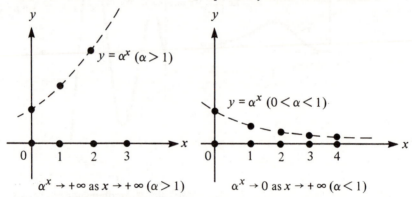

$$\alpha^x \to +\infty \text{ as } x \to +\infty \, (\alpha > 1) \qquad \alpha^x \to 0 \text{ as } x \to +\infty \, (\alpha < 1)$$

Example 13.18.[†]
In example 13.6 we found that the general solution of

$$\frac{d^3 y}{dx^3} + \frac{d^2 y}{dx^2} - \frac{dy}{dx} - y = 0$$

is given by

$$y = A_0 e^x + B_0 e^{-x} + B_1 x e^{-x}.$$

If $A_0 \neq 0$, the dominant term is the first. If $A_0 > 0$, we have that $y \to +\infty$ as $x \to +\infty$. If $A_0 < 0$, we have that $y \to -\infty$ as $x \to +\infty$.
If $A_0 = 0$, then $y \to 0$ as $x \to +\infty$.

Example 13.19.[†]
In example 13.11 we found that the general solution of

$$y_{x+2} + 3y_{x+1} + 2y_x = 0$$

is given by

$$y = A(-1)^x + B(-2)^x \quad (x = 0, 1, 2, \ldots).$$

If $B \neq 0$, the dominant term is the second and y oscillates *infinitely* as $x \to +\infty$. If $B = 0$, y *oscillates finitely* as $x \to +\infty$.

13.20 Systems of linear equations[†]

Consider the system

(1) $\quad Dy_1 = m_{11}y_1 + m_{12}y_2 + c_1$

$\quad\quad Dy_2 = m_{21}y_1 + m_{22}y_2 + c_2$

of simultaneous differential equations in which y_1 and y_2 are unknown functions of x and the other quantities are constant. We may rewrite this system in the form

$$D\mathbf{y} = M\mathbf{y} + \mathbf{c}.$$

If M is non-singular, the system has a unique solution in which y_1 and y_2 are both constant. This solution is given by $y_1 = \eta_1$ and $y_2 = \eta_2$ where the constant vector $\boldsymbol{\eta}$ satisfies

$$M\boldsymbol{\eta} + \mathbf{c} = \mathbf{0}.$$

We call this solution the *equilibrium solution.*

In general, it is not enough to know that a system has an equilibrium solution. One also needs to know something about the *stability* of the system. Ideally one would like *all* non-constant solutions $\mathbf{y}$ of the system to converge on the equilibrium solution – i.e. $\mathbf{y} \to \boldsymbol{\eta}$ as $x \to \infty$. One can then be sure that, if the system is disturbed by some external shock, then the system will act to restore the lost equilibrium. In this section we study the conditions under which stability is guaranteed.

We begin by introducing the change of variable

$$\mathbf{z} = \mathbf{y} - \boldsymbol{\eta}.$$

Then

$$D\mathbf{z} = M\mathbf{z}$$

i.e.

$$Dz_1 = m_{11}z_1 + m_{12}z_2$$

$$Dz_2 = m_{21}z_1 + m_{22}z_2$$

which we rewrite in the form

(1)
$$(m_{11} - D)z_1 + m_{12}z_2 = 0$$
$$m_{21}z_1 + (m_{22} - D)z_2 = 0.$$

From these equations we obtain

$$m_{21}(m_{11} - D)z_1 + m_{21}m_{12}z_2 = 0$$

$$m_{21}(m_{11} - D)z_1 + (m_{11} - D)(m_{22} - D)z_2 = 0.$$

Subtracting these equations, we have that

$$P(D)z_2 = ((m_{11} - D)(m_{22} - D) - m_{21}m_{12})z_2 = 0.$$

But

$$P(\lambda) = (m_{11} - \lambda)(m_{22} - \lambda) - m_{21}m_{12}$$

$$= \begin{vmatrix} m_{11} - \lambda & m_{12} \\ m_{21} & m_{22} - \lambda \end{vmatrix} = \det(M - \lambda I).$$

It follows (§5.4) that the roots of $P(\lambda)$ are the *eigenvalues* λ_1 and λ_2 of M. If these are distinct, we may conclude that

$$z_2 = a_2 e^{\lambda_1 x} + b_2 e^{\lambda_2 x}.$$

Similarly

$$z_1 = a_1 e^{\lambda_1 x} + b_1 e^{\lambda_2 x}.$$

(If $\lambda_1 = \lambda_2$, $e^{\lambda_2 x}$ must be replaced by $xe^{\lambda_1 x}$.)

It is now evident that the requisite condition for stability is

$$\mathscr{R} \lambda_1 < 0 \quad \text{and} \quad \mathscr{R} \lambda_2 < 0.$$

Under this condition it is guaranteed that $z \to 0$ as $x \to +\infty$ and hence $y \to \eta$ as $x \to +\infty$.

If $\mathscr{R}\lambda_1 \geq 0$ or $\mathscr{R}\lambda_2 \geq 0$, the solution y may or may not evolve toward the equilibrium solution depending on the values of a_1, b_1, a_2 and b_2. In general, however, one must expect that the solution will not converge to η and it is frequently the case that y behaves very wildly indeed.

Assuming that λ_1 and λ_2 are distinct and substituting $z_1 = a_1 e^{\lambda_1 x} + b_1 e^{\lambda_2 x}, z_2 = a_2 e^{\lambda_1 x} + b_2 e^{\lambda_2 x}$ in equations (1), we obtain that

$$e^{\lambda_1 x}\{(m_{11} - \lambda_1)a_1 + m_{12}a_2\} + e^{\lambda_2 x}\{(m_{11} - \lambda_2)b_1 + m_{12}b_2\} = 0$$

$$e^{\lambda_1 x}\{m_{12}a_1 + (m_{22} - \lambda_1)a_2\} + e^{\lambda_2 x}\{m_{12}b_1 + (m_{22} - \lambda_2)b_2\} = 0$$

and so

$$m_{11}a_1 + m_{12}a_2 = \lambda_1 a_1 \qquad m_{11}b_1 + m_{12}b_2 = \lambda_2 b_1$$

$$m_{12}a_1 + m_{22}a_2 = \lambda_1 a_2 \qquad m_{12}b_1 + m_{22}b_2 = \lambda_2 b_2$$

i.e. $M\mathbf{a} = \lambda_1 \mathbf{a}$ \qquad\qquad i.e. $M\mathbf{b} = \lambda_2 \mathbf{b}$.

Thus **a** is either **0** or else an *eigenvector* of M corresponding to λ_1. In either case we may write $\mathbf{a} = \alpha\mathbf{u}$ where $\mathbf{u}$ is a fixed eigenvector corresponding to λ_1. Similarly, $\mathbf{b} = \beta\mathbf{v}$ where $\mathbf{v}$ is a fixed eigenvector corresponding to λ_2. Thus

$$z_1 = \alpha u_1 e^{\lambda_1 x} + \beta v_1 e^{\lambda_2 x}$$

$$z_2 = \alpha u_2 e^{\lambda_1 x} + \beta v_2 e^{\lambda_2 x}$$

where α and β are arbitrary constants and $\mathbf{u}$ and $\mathbf{v}$ are eigenvectors of M corresponding to the distinct eigenvalues λ_1 and λ_2 respectively. Note that this general solution involves only *two* arbitrary constants α and β.

The same results hold in the n-dimensional case but these are best proved using matrix analysis. Systems of difference equations can be treated in precisely the same way. For stability in the latter case, however, we require that $|\lambda_1| < 1, |\lambda_2| < 1, \ldots, |\lambda_n| < 1$ instead of $\mathcal{R}\lambda_1 < 0, \mathcal{R}\lambda_2 < 0, \ldots, \mathcal{R}\lambda_n < 0$.

Example 13.21.[†]
Consider the system

$$Dy_1 = -y_1 - y_2 + 1$$

$$Dy_2 = -y_1 + y_2 + 1$$

which we write as

$$D\begin{pmatrix} y_1 \\ y_2 \end{pmatrix} = \begin{pmatrix} -1 & -1 \\ 1 & -1 \end{pmatrix}\begin{pmatrix} y_1 \\ y_2 \end{pmatrix} + \begin{pmatrix} 1 \\ 1 \end{pmatrix}.$$

The equilibrium solution $\mathbf{y} = \boldsymbol{\eta}$ is found by solving

$$\begin{pmatrix} -1 & -1 \\ 1 & -1 \end{pmatrix}\begin{pmatrix} \eta_1 \\ \eta_2 \end{pmatrix} + \begin{pmatrix} 1 \\ 1 \end{pmatrix} = \begin{pmatrix} 0 \\ 0 \end{pmatrix}$$

which gives $\eta_1 = 0$ and $\eta_2 = 1$.

To discuss stability we require the eigenvalues of the matrix. These are found by solving

$$\begin{vmatrix} -1-\lambda & -1 \\ 1 & -1-\lambda \end{vmatrix} = (1+\lambda)^2 + 1 = 0$$

which gives $\lambda_1 = -1 + i$ and $\lambda_2 = -1 - i$. Since $\mathcal{R}\lambda_1 < 0$ and $\mathcal{R}\lambda_2 < 0$, the system is *stable* – i.e. for *any* solution $\mathbf{y}$, $\mathbf{y} \to \boldsymbol{\eta}$ as $x \to +\infty$.

We can obtain the general solution explicitly by computing eigenvectors $\mathbf{u}$ and $\mathbf{v}$ corresponding to λ_1 and λ_2. To find $\mathbf{u}$, we consider

$$\begin{pmatrix} -1 & -1 \\ 1 & -1 \end{pmatrix} \begin{pmatrix} u_1 \\ u_2 \end{pmatrix} = (-1+i) \begin{pmatrix} u_1 \\ u_2 \end{pmatrix}$$

i.e.

$$-u_1 - u_2 = -u_1 + iu_1$$

$$u_1 - u_2 = -u_2 + iu_2.$$

Both equations yield $u_1 = iu_2$ and hence the eigenvectors are of the form

$$\begin{pmatrix} iu_2 \\ u_2 \end{pmatrix}.$$

We choose the eigenvector **u** with $u_2 = 1$.

Similarly,

$$\begin{pmatrix} -1 & -1 \\ 1 & -1 \end{pmatrix} \begin{pmatrix} v_1 \\ v_2 \end{pmatrix} = (-1-i) \begin{pmatrix} v_1 \\ v_2 \end{pmatrix}$$

i.e.

$$-v_1 - v_2 = -v_1 - iv_1$$

$$v_1 - v_2 = -v_2 - iv_2.$$

Both equations yield $v_1 = -iv_2$ and so the eigenvectors are of the form

$$\begin{pmatrix} -iv_2 \\ v_2 \end{pmatrix}.$$

We choose the eigenvector **v** with $v_2 = 1$.

It follows that

$$y_1 = 0 + \alpha i e^{(-1+i)x} - \beta i e^{(-1-i)x}$$

$$y_2 = 1 + \alpha e^{(-1+i)x} + \beta e^{(-1-i)x}.$$

If we are only interested in real solutions, we may rewrite these expressions in the form

$$y_1 = 0 + e^{-x}\{i(\alpha - \beta)\cos x - (\alpha + \beta)\sin x\}$$

$$y_2 = 1 + e^{-x}\{(\alpha + \beta)\cos x + i(\alpha - \beta)\sin x\}.$$

Taking $A = i(\alpha - \beta)$ and $B = (\alpha + \beta)$, we obtain

$$y_1 = (A\cos x - B\sin x)e^{-x}$$

$$y_2 = 1 + (B\cos x + A\sin x)e^{-x}$$

where A and B are arbitrary constants.

Example 13.22.†

Consider the system

$$Ey_1 = y_2 - 1$$
$$Ey_2 = -y_1 - 1$$

which we write as

$$E \begin{pmatrix} y_1 \\ y_2 \end{pmatrix} = \begin{pmatrix} 0 & 1 \\ -1 & 0 \end{pmatrix} \begin{pmatrix} y_1 \\ y_2 \end{pmatrix} + \begin{pmatrix} -1 \\ -1 \end{pmatrix}.$$

The equilibrium solution $\mathbf{y} = \boldsymbol{\eta}$ is found by solving

$$\begin{pmatrix} 0 & 1 \\ -1 & 0 \end{pmatrix} \begin{pmatrix} \eta_1 \\ \eta_2 \end{pmatrix} + \begin{pmatrix} -1 \\ -1 \end{pmatrix} = \begin{pmatrix} 0 \\ 0 \end{pmatrix}.$$

This gives $\eta_1 = -1$ and $\eta_2 = 1$.

To discuss stability we require the eigenvalues of the matrix. These are found by solving

$$\begin{vmatrix} -\lambda & 1 \\ -1 & -\lambda \end{vmatrix} = \lambda^2 + 1 = 0$$

which gives $\lambda_1 = i$ and $\lambda_2 = -i$. Since $|\lambda_1| = 1$ and $|\lambda_2| = 1$, the system is *not* stable.

We can obtain the general solution explicitly by computing eigenvectors corresponding to λ_1 and λ_2. We have that

$$\begin{pmatrix} 0 & 1 \\ -1 & 0 \end{pmatrix} \begin{pmatrix} u_1 \\ u_2 \end{pmatrix} = i \begin{pmatrix} u_1 \\ u_2 \end{pmatrix}$$

and

$$\begin{pmatrix} 0 & 1 \\ -1 & 0 \end{pmatrix} \begin{pmatrix} v_1 \\ v_2 \end{pmatrix} = -i \begin{pmatrix} v_1 \\ v_2 \end{pmatrix}.$$

Suitable eigenvectors are therefore

$$\mathbf{u} = \begin{pmatrix} 1 \\ i \end{pmatrix}; \quad \mathbf{v} = \begin{pmatrix} 1 \\ -i \end{pmatrix}$$

and the general solution is

$$y_1 = -1 + \alpha(i)^x + \beta(-i)^x$$
$$y_2 = 1 + \alpha i(i)^x - \beta i(-i)^x$$

which we may rewrite as

$$y_1 = -1 + \alpha e^{i\pi x/2} + \beta e^{-i\pi x/2}$$

$$y_2 = 1 + \alpha e^{i\pi(x+1)/2} + \beta e^{-i\pi(x+1)/2}.$$

Or, in real terms,

$$y_1 = -1 + A \cos\frac{\pi x}{2} + B \sin\frac{\pi x}{2}$$

$$y_2 = 1 + A \cos\frac{\pi(x+1)}{2} + B \sin\frac{\pi(x+1)}{2}$$

Observe that the solutions oscillate finitely unless $A = B = 0$.

Exercises 13.23

1. Find the general solutions of the following differential equations:

(i) $3\dfrac{d^3y}{dx^3} + 5\dfrac{d^2y}{dx^2} - 2\dfrac{dy}{dx} = 0$

(ii) $\dfrac{d^3y}{dx^3} - 4\dfrac{d^2y}{dx^2} + \dfrac{dy}{dx} + 6y = 0$

(iii) $\dfrac{d^4y}{dx^4} - 2\dfrac{d^3y}{dx^3} - 3\dfrac{d^2y}{dx^2} + 4\dfrac{dy}{dx} + 4y = 0$

(iv) $\dfrac{d^5y}{dx^5} + 3\dfrac{d^4y}{dx^4} + 7\dfrac{d^3y}{dx^3} + 13\dfrac{d^2y}{dx^2} + 12\dfrac{dy}{dx} + 4y = 0.$

2.* Find the general solutions of the following differential equations:

(i) $\dfrac{d^2y}{dx^2} + 2\dfrac{dy}{dx} - 3y = 0$

(ii) $\dfrac{d^4y}{dx^4} + 2\dfrac{d^3y}{dx^2} + \dfrac{d^2y}{dx^2} = 0$

(iii) $\dfrac{d^2y}{dx^2} + \dfrac{dy}{dx} + y = 0$

(iv) $\dfrac{d^3y}{dx^3} - 3\dfrac{d^2y}{dx^2} + 9\dfrac{dy}{dx} + 13y = 0.$

3. Find the general solutions of the following difference equations:

(i) $y_{x+2} + 2y_{x+1} - 3y_x = 0$

(ii) $y_{x+4} + 2y_{x+3} + y_{x+2} = 0$

(iii) $y_{x+2} + y_{x+1} + y_x = 0$

(iv) $y_{x+3} - 3y_{x+2} + 9y_{x+1} + 13y_x = 0.$

4.* Find the general solutions of the following difference equations:

(i) $3y_{x+3} + 5y_{x+2} - 2y_{x+1} = 0$

(ii) $y_{x+3} - 4y_{x+2} + y_{x+1} + 6y_x = 0$

(iii) $y_{x+4} - 2y_{x+3} - 3y_{x+2} + 4y_{x+1} + 4y_x = 0$

(iv) $y_{x+5} + 3y_{x+4} + 7y_{x+3} + 13y_{x+2} + 12y_{x+1} + 4y_x = 0.$

5. Find the general solution of the differential equation

$$3\frac{d^3y}{dx^3} + 5\frac{d^2y}{dx^2} - 2\frac{dy}{dx} = q(x)$$

when

(i) $q(x) = e^x$

(ii) $q(x) = \cos x$

(iii) $q(x) = x^2$

(iv) $q(x) = e^{-2x}.$

6.* Find the general solution of the differential equation

$$3\frac{d^3y}{dx^3} + 5\frac{d^2y}{dx^2} - 2\frac{dy}{dx} = q(x)$$

when

(i) $q(x) = e^{-x}$

(ii) $q(x) = \sin x$

(iii) $q(x) = x$

(iv) $q(x) = e^{-x/3}.$

7. Find the general solution of the difference equation

$$y_{x+3} - 5y_{x+2} + 8y_{x+1} - 4y_x = 2^x + x.$$

8.* Find the general solution of the difference equation

$$y_{x+3} - 3y_{x+2} + 9y_{x+1} + 13y_x = q_x$$

when

(i) $q_x = 1$

(ii) $q_x = x$

(iii) $q_x = (-1)^x$.

9.[†] Find a solution of the differential equation of question 1 (i) which satisfies the boundary conditions

(i) $y(0) = 0$

(ii) $y(x) \to 1$ as $x \to +\infty$.

10.*[†] If y_x is a solution of

$$y_{x+2} = \tfrac{1}{2}(y_{x+1} + y_x) \quad (x = 0, 1, 2, \dots)$$

which satisfies the boundary conditions

(i) $y_0 = a$

(ii) $y_1 = b$,

prove that $y_x \to \tfrac{1}{3}(a + 2b)$ as $x \to +\infty$.
11.[†] Show that the system

$$Dy_1 = y_1 - y_2 - 1$$
$$Dy_2 = y_1 + y_2 - 1$$

is not stable. Obtain formulae for the solutions which satisfy the boundary conditions

(i) $y_1(0) = 1$

(ii) $y_2(0) = 1$.

12.*[†] Show that the system

$$Ey_1 = -y_1 - 3y_2 + 1$$
$$Ey_2 = -\tfrac{1}{4}y_1 - y_2 + 1$$

is stable. Obtain formulae for the solutions which satisfy the boundary conditions

(i) $y_1(0) = 4$.

(ii) $y_2(0) = 1$.

SOME APPLICATIONS (OPTIONAL)

13.24   Cobweb model

In §11.31 we considered two versions of the cobweb model. The first version was a discrete model in which the price for the current period is predicted solely on the basis of the price in the previous period. In this version, stability occurs only for a restricted range of the parameters of the model. The second version was a continuous model in which the current price is predicted as a weighted average of all previous prices. In this more sophisticated version, stability is obtained for all values of the parameters.

An intermediate version was also mentioned in §11.31. The equations for this third version are

(1) $D_x = a - bP_x$

(2) $S_x = D_x$

(3) $S_{x+2} = c + d\left(\dfrac{P_x + P_{x+1}}{2}\right).$

Here the current price is predicted as the average of the prices in the two previous periods.

These equations lead to the second order difference equation

$$2bQ_{x+2} + dQ_{x+1} + dQ_x = 0$$

in which $Q_x = P_x - \bar{P}$ where $\bar{P}$ is the equilibrium price (i.e. $a - b\bar{P} = c + d\bar{P}$). Writing the difference equation in the form

$$(2bE^2 + dE + d)Q = 0,$$

we see that we require the roots of the auxiliary equation $2bz^2 + dz + d = 0$. These are given by

$$z = \frac{-d \pm \sqrt{(d^2 - 8bd)}}{4b}.$$

We leave the case $d^2 - 8bd \geqslant 0$ to the reader and study the case $d^2 - 8bd < 0$ (i.e. $d < 8b$) since this leads to complex roots of the auxiliary equation. If these complex roots are $\alpha = \lambda e^{i\mu}$ and $\beta = \lambda e^{-i\mu}$, then the general solution of the difference equation is

$$Q = A\lambda^x \cos \mu x + B\lambda^x \sin \mu x.$$

For stability, we require that $\lambda < 1$. But

$$\lambda^2 = \frac{1}{16b^2}\{d^2 + (8bd - d^2)\} = \frac{d}{2b}.$$

Stability is therefore obtained when $d < 2b$. This result should be compared with that obtained in the original cobweb model in which the condition for stability is $d < b$.

What sort of predictor for the current price will always lead to stability in the discrete case? In view of our experience with the continuous model we shall consider a model in which the current price is predicted as a weighted average of the prices over the previous n periods. The appropriate equations are

(1) $D_x = a - bP_x$

(2) $S_x = D_x$

(3) $S_{x+n} = c + d\left(\dfrac{P_{x+n-1} + \theta P_{x+n-2} + \ldots + \theta^{n-1}P_x}{n}\right).$

Here the 'weights' are based on a discounting factor θ which we assume to satisfy $0 < \theta < 1$. These equations lead to the difference equation

$$bQ_{x+n} + \frac{d}{n}(Q_{x+n-1} + \theta Q_{x+n-2} + \ldots + \theta^{n-1}Q_x) = 0$$

in which $Q_x = P_x - \bar{P}$ where $\bar{P}$ is the equilibrium price given by

$$\alpha - b\bar{P} = c + d\bar{P}\left(\frac{1-\theta^n}{1-\theta}\right).$$

Writing the difference equation in the form

$$\left\{bE^n + \frac{d}{n}(E^{n-1} + \theta E^{n-2} + \ldots + \theta^{n-1})\right\}Q = 0,$$

we see that it is necessary to examine the roots of the auxiliary equation

$$bz^n + \frac{d}{n}(z^{n-1} + \theta z^{n-2} + \ldots + \theta^{n-1}) = 0.$$

For stability, these roots should all have modulus less than one.

To discover anything about the roots of a polynomial of large degree is not often easy. However, Rouché's theorem is sometimes a help with such problems. This asserts that if f and g are functions of a complex variable which are analytic inside and on a closed curve C in the complex plane and the inequality

$$|f(z) - g(z)| < |g(z)|$$

is satisfied for each z on the curve C, then $f(z) = 0$ and $g(z) = 0$ have the same number of roots inside the curve C. We apply Rouché's theorem with $f(z) = bz^n + dn^{-1}(z^{n-1} + \theta z^{n-2} + \ldots + \theta^{n-1})$ and $g(z) = bz^n$. We take the closed curve C to be the circle $|z| = r$ where r is chosen so that $\theta < r < 1$. For values of z on the curve C (i.e. with $|z| = r$), we have that

$$
\begin{aligned}
|f(z) - g(z)| &= |dn^{-1}(z^{n-1} + \theta z^{n-2} + \ldots + \theta^{n-1})| \\
&\leqslant dn^{-1}(r^{n-1} + \theta r^{n-2} + \ldots + \theta^{n-1}) \\
&= \frac{d}{n} \frac{r^n - \theta^n}{r - \theta} \\
&< \frac{d}{n} \frac{r^n}{(r - \theta)} = \frac{d}{n(r - \theta)} |z^n| \\
&= \frac{d}{bn(r - \theta)} |g(z)|.
\end{aligned}
$$

It follows that $|f(z) - g(z)| < |g(z)|$ for each z on the circle C, provided that n is chosen sufficiently large to ensure that

$$
\frac{d}{bn(r - \theta)} < 1.
$$

From Rouché's theorem, we deduce that $f(z) = 0$ has the same number of roots inside the circle C as $g(z) = 0$. But $g(z) = 0$ has n roots inside C (all at $z = 0$). Hence the n roots of $f(z) = 0$ are all inside C as well.

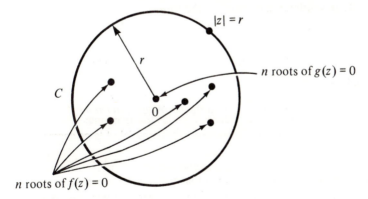

In particular we obtain that the roots of $f(z) = 0$ all have modulus less than one (i.e. the model is stable) provided that n is sufficiently large.

13.25   Gambler's ruin

A gambler enters a casino with $r in his pocket. If this is lost, he is ruined. He decides to continue gambling until he has won $s or else is ruined. He plays the same game repeatedly. At each play he wins $1 with probability p $(0 < p < 1)$ and loses $1 with probability q $(p + q = 1)$. What is the probability that he will be ruined?

Let the probability that the gambler will eventually be ruined, given that his winnings so far amount to $x, be y_x. We want to calculate y_0. The diagram on the left below illustrates the definition of y_x. The diagram on the right examines the situation in the case $-r < x < s$ more closely. Either the gambler will win the next play and his total winnings will become $x + 1$ or else he will lose and his total winnings will become $x - 1$. The probability of ruin via the first route is py_{x+1} and the probability of ruin via the second route is qy_{x-1}. The total probability of ruin is therefore

$$y_x = py_{x+1} + qy_{x-1}.$$

We rewrite this difference equation in the form

$$y_{x+1} = py_{x+2} + qy_x$$

i.e.

$$py_{x+2} - y_{x+1} + qy_x = 0$$

$$(pE^2 - E + q)y = 0.$$

This has to be solved in accordance with the *boundary conditions*

$$y_{-r} = 1$$

$$y_s = 0$$

which are obtained from the observation that the gambler is certain of ruin when his winnings are $-r$ and certain of success when his winnings have hit the prespecified target of $s.

Since $p + q = 1$, the auxiliary equation is

$$pz^2 - z + 1 - p = 0$$

$$p(z^2 - 1) - (z - 1) = 0$$

$$p(z - 1)(z + 1) - (z - 1) = 0$$

$$(z - 1)(pz + p - 1) = 0$$

$$p(z - 1)(z - qp^{-1}) = 0.$$

The roots are $\alpha = 1$ and $\beta = qp^{-1}$. Two cases need to be distinguished. The first is when the roots are unequal ($p \neq q$) and the second is when the roots are equal ($p = q = \frac{1}{2}$).

(i) $p \neq q$. In this case the general solution is

$$y_x = A + B\beta^x.$$

We use the boundary conditions to evaluate the constants A and B. We have that

$$1 = A + B\beta^{-r}$$

and so

$$0 = A + B\beta^s$$

$$A = -\beta^{r+s}(1 - \beta^{r+s})^{-1}; \quad B = \beta^r(1 - \beta^{r+s})^{-1}.$$

It follows that

$$y_x = \frac{\beta^{r+x} - \beta^{r+s}}{1 - \beta^{r+s}}$$

and so the probability of ruin on entering the casino is

$$y_0 = \frac{\beta^r - \beta^{r+s}}{1 - \beta^{r+s}}$$

provided that $\beta \neq 1$ (i.e. $p \neq q$).

Suppose, for example, that the game chosen by the gambler is favourable to him in the sense that $p > q$ (i.e. $\beta < 1$). Then $1 - \beta^{r+s} > 1 - \beta$ and $1 - \beta^s < 1$. Thus

$$y_0 = \beta^r \frac{(1 - \beta^s)}{(1 - \beta^{r+s})} < \frac{\beta^r}{1 - \beta}.$$

This means that, if r is reasonably large, the probability of ruin will be negligible no matter how large the target sum of $\$s$ is made. In particular, if $p = \frac{2}{3}, q = \frac{1}{3}$ and $r = 101$, we have that

$$y_0 < (\tfrac{1}{2})^{100}.$$

(ii) $p = q = \frac{1}{2}$. In this case the general solution is

$$y_x = A + Bx.$$

Using the boundary conditions to evaluate A and B we obtain that

$$A = \frac{s}{r+s}; \quad B = -\frac{1}{r+s}$$

and so

$$y_x = \frac{s-x}{s+r}$$

In particular

$$y_0 = \frac{s}{r+s}.$$

This is perhaps the more interesting case since when $p = q = \frac{1}{2}$ the game is 'fair'. Suppose that the gambler risks \$9000 in such a game in an attempt to win \$1000. His probability of success is then

$$1 - \frac{s}{r+s} = \frac{r}{r+s} = \frac{9000}{9000+1000} = \frac{9}{10}.$$

This large probability explains why the idle rich are able to boast of paying for their holidays in the casinos at Las Vegas or Monte Carlo. But they are not to be congratulated on their good business sense. All that they have done is to balance a high probability of a small win against a low probability of a large loss. Note that their expected gain is

$$1000 \times \frac{9}{10} + (-9000) \times \frac{1}{10} = 0.$$

It is also instructive to observe that the probability of being successful on each occasion if they attempt the coup on ten successive occasions is only

$$\left(\frac{9}{10}\right)^{10} = \left(1 - \frac{1}{10}\right)^{10} \doteqdot \frac{1}{e} = .37.$$

One must also, of course, take into account that casino games are seldom fair.

Returning to the general case, we observe that the duration of the game is also a variable of interest. If $\mathscr{E}_x$ denotes the expected duration of the play to come, given that \$x have been won so far, then we obtain the difference equation

$$\mathscr{E}_x = p\,\mathscr{E}_{x+1} + q\,\mathscr{E}_{x-1} + 1$$

with boundary conditions $\mathscr{E}_s = 0$, $\mathscr{E}_{-r} = 0$. In the 'fair' case with $p = q = \frac{1}{2}$, this leads to

$$\mathscr{E}_0 = rs.$$

Observe that a gambler starting with $1 hoping to win $1000 will last for 1000 plays on average!

13.26 Samuelson multiplier–accelerator model

In this model, national income Y_x at time x depends on consumption C_x, investment Y_x and government expenditure G (assumed constant). The appropriate equations are

(1) $Y_x = C_x + I_x + G$

(2) $C_{x+1} = \gamma Y_x$

(3) $I_{x+1} = \alpha(C_{x+1} - C_x)$.

Here γ $(0 < \gamma < 1)$ is a 'multiplier' and α $(\alpha > 0)$ is an 'accelerator'. The idea is that consumption is proportional to income in the previous period and that investment is proportional to the *increase* in current consumption as compared with consumption in the previous period.

We eliminate national income from these equations and obtain

$$C_{x+1} = \gamma C_x + \gamma I_x + \gamma G$$

$$I_{x+1} = \alpha(\gamma - 1)C_x + \alpha\gamma I_x + \alpha\gamma G$$

$$E\begin{pmatrix} C \\ I \end{pmatrix} = \begin{pmatrix} \gamma & \gamma \\ \alpha(\gamma - 1) & \alpha\gamma \end{pmatrix}\begin{pmatrix} C \\ I \end{pmatrix} + \begin{pmatrix} \gamma G \\ \alpha\gamma G \end{pmatrix}$$

The condition for stability is that the eigenvalues of the matrix have modulus less than one. We consider the case in which the eigenvalues are complex. (This happens when $\gamma(\alpha + 1)^2 < 4\alpha$.) The eigenvalues then take the form $\lambda_1 = re^{i\theta}$, $\lambda_2 = re^{-i\theta}$ and so

$$r^2 = \lambda_1\lambda_2 = \begin{vmatrix} \gamma & \gamma \\ \alpha(\gamma - 1) & \alpha\gamma \end{vmatrix} = \alpha\gamma.$$

Thus, if $\gamma(\alpha + 1)^2 < 4\alpha$, stability occurs when $\alpha\gamma < 1$. Stability can also occur when $\gamma(\alpha + 1)^2 \geqslant 4\alpha$ but we leave this case as an exercise.

Solutions to unstarred problems

Chapter 1. Vectors and matrices

Exercise 1.15

1.

(i) $2D = 2 \begin{pmatrix} 1 & 2 & 1 \\ 2 & 1 & 2 \\ 3 & 3 & 3 \end{pmatrix} = \begin{pmatrix} 2 & 4 & 2 \\ 4 & 2 & 4 \\ 6 & 6 & 6 \end{pmatrix}$.

(ii) $A + B$ is nonsense because A is 2×3 and B is 3×2 (see §1.3).

(iii) $B - C$ is nonsense because B is 3×2 and C is 3×3 (see §1.3).

(iv) $C + D = \begin{pmatrix} 1 & 2 & 3 \\ 4 & 1 & 2 \\ 3 & 5 & 1 \end{pmatrix} + \begin{pmatrix} 1 & 2 & 1 \\ 2 & 1 & 2 \\ 3 & 3 & 3 \end{pmatrix} = \begin{pmatrix} 2 & 4 & 4 \\ 6 & 2 & 4 \\ 6 & 8 & 4 \end{pmatrix}$.

(v) $2C - 3D = 2 \begin{pmatrix} 1 & 2 & 3 \\ 4 & 1 & 2 \\ 3 & 5 & 1 \end{pmatrix} - 3 \begin{pmatrix} 1 & 2 & 1 \\ 2 & 1 & 2 \\ 3 & 3 & 3 \end{pmatrix}$

$= \begin{pmatrix} 2 & 4 & 6 \\ 8 & 2 & 4 \\ 6 & 10 & 2 \end{pmatrix} - \begin{pmatrix} 3 & 6 & 3 \\ 6 & 3 & 6 \\ 9 & 9 & 9 \end{pmatrix} = \begin{pmatrix} -1 & -2 & 3 \\ 2 & -1 & -2 \\ -3 & 1 & -7 \end{pmatrix}$.

(vi) $2A - 3D$ is nonsense because $2A$ is 2×3 and $3D$ is 3×3 (see §1.3).

(vii) $AB = \begin{pmatrix} 6 & 1 & 0 \\ 1 & 0 & 3 \end{pmatrix} \begin{pmatrix} 3 & 1 \\ 1 & 0 \\ 0 & 2 \end{pmatrix} = \begin{pmatrix} 19 & 6 \\ 3 & 7 \end{pmatrix}$.

(viii) $BA = \begin{pmatrix} 3 & 1 \\ 1 & 0 \\ 0 & 2 \end{pmatrix} \begin{pmatrix} 6 & 1 & 0 \\ 1 & 0 & 3 \end{pmatrix} = \begin{pmatrix} 19 & 3 & 3 \\ 6 & 1 & 0 \\ 2 & 0 & 6 \end{pmatrix}.$

(ix) $AC = \begin{pmatrix} 6 & 1 & 0 \\ 1 & 0 & 3 \end{pmatrix} \begin{pmatrix} 1 & 2 & 3 \\ 4 & 1 & 2 \\ 3 & 5 & 1 \end{pmatrix} = \begin{pmatrix} 10 & 13 & 20 \\ 10 & 17 & 6 \end{pmatrix}.$

(x) CA is nonsense because C is 3 x 3 and A is 2 x 3 (see §1.4).

(xi) BC is nonsense because B is 3 x 2 and C is 3 x 3 (see §1.4).

(xii) $CB = \begin{pmatrix} 1 & 2 & 3 \\ 4 & 1 & 2 \\ 3 & 5 & 1 \end{pmatrix} \begin{pmatrix} 3 & 1 \\ 1 & 0 \\ 0 & 2 \end{pmatrix} = \begin{pmatrix} 5 & 7 \\ 13 & 8 \\ 14 & 5 \end{pmatrix}.$

(xiii) $CD = \begin{pmatrix} 1 & 2 & 3 \\ 4 & 1 & 2 \\ 3 & 5 & 1 \end{pmatrix} \begin{pmatrix} 1 & 2 & 1 \\ 2 & 1 & 2 \\ 3 & 3 & 3 \end{pmatrix} = \begin{pmatrix} 14 & 13 & 14 \\ 12 & 15 & 12 \\ 16 & 14 & 16 \end{pmatrix}.$

(xiv) $DC = \begin{pmatrix} 1 & 2 & 1 \\ 2 & 1 & 2 \\ 3 & 3 & 3 \end{pmatrix} \begin{pmatrix} 1 & 2 & 3 \\ 4 & 1 & 2 \\ 3 & 5 & 1 \end{pmatrix} = \begin{pmatrix} 12 & 9 & 8 \\ 12 & 15 & 10 \\ 24 & 24 & 21 \end{pmatrix}.$

(xv) det (A) is nonsense because A is not square (see §1.8).

(xvi) det (B) is nonsense because B is not square (see §1.8).

(xvii) det $(C) = 46$.

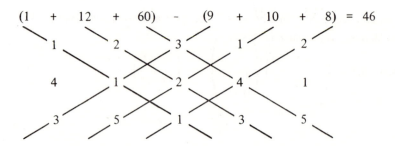

(xviii) det $(D) = 0$.

(3 + 12 + 6) − (3 + 6 + 12) = 0

(xix) det $(AB) = \begin{vmatrix} 19 & 6 \\ 13 & 7 \end{vmatrix} = 19 \times 7 - 13 \times 6 = 55.$

(xx) det $(CD) = 0$ (calculations suppressed!).

(xxi) $A^T = \begin{pmatrix} 6 & 1 & 0 \\ 1 & 0 & 3 \end{pmatrix}^T = \begin{pmatrix} 6 & 1 \\ 1 & 0 \\ 0 & 3 \end{pmatrix}.$

(xxii) $B^T = \begin{pmatrix} 3 & 1 \\ 1 & 0 \\ 0 & 2 \end{pmatrix}^T = \begin{pmatrix} 3 & 1 & 0 \\ 1 & 0 & 2 \end{pmatrix}.$

(xxiii) $C^T = \begin{pmatrix} 1 & 2 & 3 \\ 4 & 1 & 2 \\ 3 & 5 & 1 \end{pmatrix}^T = \begin{pmatrix} 1 & 4 & 3 \\ 2 & 1 & 5 \\ 3 & 2 & 1 \end{pmatrix}.$

(xxiv) $D^T = \begin{pmatrix} 1 & 2 & 1 \\ 2 & 1 & 2 \\ 3 & 3 & 3 \end{pmatrix}^T = \begin{pmatrix} 1 & 2 & 3 \\ 2 & 1 & 3 \\ 1 & 2 & 3 \end{pmatrix}.$

2. (i) Matrix A is non-singular because

$$\det (A) = \begin{vmatrix} 2 & 1 \\ 4 & 3 \end{vmatrix} = 6 - 4 = 2 \neq 0$$

$$A^{-1} = \frac{1}{2}\begin{pmatrix} 3 & -1 \\ -4 & 2 \end{pmatrix} = \begin{pmatrix} \frac{3}{2} & -\frac{1}{2} \\ -2 & 1 \end{pmatrix}.$$

To check that A^{-1} is the inverse of A, observe that

$$AA^{-1} = \begin{pmatrix} 2 & 1 \\ 4 & 3 \end{pmatrix}\begin{pmatrix} \frac{3}{2} & -\frac{1}{2} \\ -2 & 1 \end{pmatrix} = \begin{pmatrix} 1 & 0 \\ 0 & 1 \end{pmatrix} = I$$

$$A^{-1}A = \begin{pmatrix} \frac{3}{2} & -\frac{1}{2} \\ -2 & 1 \end{pmatrix}\begin{pmatrix} 2 & 1 \\ 4 & 3 \end{pmatrix} = \begin{pmatrix} 1 & 0 \\ 0 & 1 \end{pmatrix} = I.$$

(ii) Matrix B is singular because

$$\det (B) = \begin{vmatrix} 1 & 2 \\ 2 & 4 \end{vmatrix} = 4 - 4 = 0$$

and hence B^{-1} makes no sense (see §1.10).

(iii) Matrix C is non-singular because

$$\det (C) = \begin{vmatrix} 1 & 2 & 3 \\ 4 & 1 & 2 \\ 3 & 5 & 1 \end{vmatrix} = 46 \neq 0$$

(see question 1 (xvii)). Using Cramer's rule,

$$C^{-1} = \frac{1}{46}\begin{pmatrix} -9 & 13 & 1 \\ 2 & -8 & 10 \\ 17 & 1 & -7 \end{pmatrix}.$$

To check that C^{-1} is the inverse of C, observe that

$$C^{-1}C = \frac{1}{46}\begin{pmatrix} -9 & 13 & 1 \\ 2 & -8 & 10 \\ 17 & 1 & -7 \end{pmatrix}\begin{pmatrix} 1 & 2 & 3 \\ 4 & 1 & 2 \\ 3 & 5 & 1 \end{pmatrix} = \frac{1}{46}\begin{pmatrix} 46 & 0 & 0 \\ 0 & 46 & 0 \\ 0 & 0 & 46 \end{pmatrix} = I.$$

Similarly, $CC^{-1} = I$.

(iv) Matrix D is singular because $\det (D) = 0$ (see question 1 (xviii)) and hence D^{-1} makes no sense (see §1.10).

3.
$$AB = \begin{pmatrix} 2 & 1 \\ 4 & 3 \end{pmatrix}\begin{pmatrix} 1 & 2 \\ 2 & 4 \end{pmatrix} = \begin{pmatrix} 4 & 8 \\ 10 & 20 \end{pmatrix}$$

$$BA = \begin{pmatrix} 1 & 2 \\ 2 & 4 \end{pmatrix}\begin{pmatrix} 2 & 1 \\ 4 & 3 \end{pmatrix} = \begin{pmatrix} 10 & 7 \\ 20 & 14 \end{pmatrix}.$$

Thus $AB \neq BA$.

4.

$$BC = \begin{pmatrix} 1 & 2 \\ 3 & 4 \end{pmatrix}\begin{pmatrix} 3 \\ 5 \end{pmatrix} = \begin{pmatrix} 13 \\ 29 \end{pmatrix}.$$

Hence

$$A(BC) = (2,4)\begin{pmatrix} 13 \\ 29 \end{pmatrix} = 26 + 116 = 142$$

$$AB = (2,4)\begin{pmatrix} 1 & 2 \\ 3 & 4 \end{pmatrix} = (14, 20).$$

Hence

$$(AB)C = (14, 20)\begin{pmatrix} 3 \\ 5 \end{pmatrix} = 42 + 100 = 142.$$

Thus $A(BC) = (AB)C$.

5.

(i) $A^{-1} = \dfrac{1}{2}\begin{pmatrix} 3 & -1 \\ -4 & 2 \end{pmatrix}$; $B^{-1} = -\dfrac{1}{2}\begin{pmatrix} 4 & -2 \\ -3 & 1 \end{pmatrix}$.

Hence

$$B^{-1}A^{-1} = -\frac{1}{4}\begin{pmatrix} 4 & -2 \\ -3 & 1 \end{pmatrix}\begin{pmatrix} 3 & -1 \\ -4 & 2 \end{pmatrix} = -\frac{1}{4}\begin{pmatrix} 20 & -8 \\ -13 & 5 \end{pmatrix}.$$

On the other hand,

$$AB = \begin{pmatrix} 2 & 1 \\ 4 & 3 \end{pmatrix}\begin{pmatrix} 1 & 2 \\ 3 & 4 \end{pmatrix} = \begin{pmatrix} 5 & 8 \\ 13 & 20 \end{pmatrix}.$$

It follows that $\det(AB) = 100 - 104 = -4$ and so

$$(AB)^{-1} = -\frac{1}{4}\begin{pmatrix} 20 & -8 \\ -13 & 5 \end{pmatrix}.$$

Thus $(AB)^{-1} = B^{-1}A^{-1}$.

(ii) $A^T = \begin{pmatrix} 2 & 4 \\ 1 & 3 \end{pmatrix}$; $B^T = \begin{pmatrix} 1 & 3 \\ 2 & 4 \end{pmatrix}$.

Hence

$$B^T A^T = \begin{pmatrix} 1 & 3 \\ 2 & 4 \end{pmatrix}\begin{pmatrix} 2 & 4 \\ 1 & 3 \end{pmatrix} = \begin{pmatrix} 5 & 13 \\ 8 & 20 \end{pmatrix}.$$

On the other hand, by part (i),

$$(AB)^T = \begin{pmatrix} 5 & 8 \\ 13 & 20 \end{pmatrix}^T = \begin{pmatrix} 5 & 13 \\ 8 & 20 \end{pmatrix}.$$

Thus $(AB)^T = B^T A^T$.

6. We have that

$$AB = I.$$

Hence

$$C(AB) = CI = C.$$

Thus

$$(CA)B = C.$$

But

$$CA = I$$

and therefore

$$B = IB = C.$$

Exercise 1.32

1. Let $\mathbf{x} = (\frac{1}{6}, \frac{1}{3}, \frac{1}{2})^T$, $\mathbf{y} = (\frac{1}{3}, \frac{1}{3}, -\frac{1}{3})^T$, $\mathbf{z} = (\frac{1}{3}, \frac{2}{3}, -\frac{2}{3})^T$.

Then

$$\|\mathbf{x}\| = \left\{ \frac{1}{36} + \frac{1}{9} + \frac{1}{4} \right\}^{1/2} = \left\{ \frac{1+4+9}{36} \right\}^{1/2} = \frac{\sqrt{(14)}}{6}$$

$$\|\mathbf{y}\| = \left\{ \frac{1}{9} + \frac{1}{9} + \frac{1}{9} \right\}^{1/2} = \frac{1}{\sqrt{3}}.$$

$$\|\mathbf{z}\| = \left\{ \frac{1}{9} + \frac{4}{9} + \frac{4}{9} \right\}^{1/2} = 1.$$

Thus only $\mathbf{z}$ is a unit vector.

Also

$$\langle \mathbf{x}, \mathbf{y} \rangle = \frac{1}{6} \cdot \frac{1}{3} + \frac{1}{3} \cdot \frac{1}{3} + \frac{1}{2} \left(-\frac{1}{3} \right) = \frac{1+2-3}{18} = 0$$

$$\langle \mathbf{y}, \mathbf{z} \rangle = \frac{1}{3} \cdot \frac{1}{3} + \frac{1}{3} \cdot \frac{2}{3} + \left(-\frac{1}{3} \right)\left(-\frac{2}{3} \right) = \frac{5}{9}$$

$$\langle \mathbf{z}, \mathbf{x} \rangle = \frac{1}{3} \cdot \frac{1}{6} + \frac{2}{3} \cdot \frac{1}{3} - \frac{2}{3} \cdot \frac{1}{2} = \frac{1+4-6}{18} = -\frac{1}{18}.$$

Thus only $\mathbf{x}$ and $\mathbf{y}$ are orthogonal.

2. The given line passes through $(3, 1, 2)^T$ in the direction of $\mathbf{v} = (1, 2, 1)^T$ (see §1.23). A unit vector which points in the same direction as $\mathbf{v}$ (and hence is parallel to the line) is

$$\mathbf{u} = \|\mathbf{v}\|^{-1}\mathbf{v} = \left(\frac{1}{\sqrt{6}}, \frac{2}{\sqrt{6}}, \frac{1}{\sqrt{6}}\right)^T$$

since $\|\mathbf{v}\| = \{1 + 4 + 1\}^{1/2} = \sqrt{6}$.

3. A line through the point $\boldsymbol{\xi} = (1, 2, 1)^T$ and $\boldsymbol{\eta} = (2, 1, 2)^T$ has the equations

$$\frac{x_1 - \xi_1}{v_1} = \frac{x_2 - \xi_2}{v_2} = \frac{x_3 - \xi_3}{v_3}$$

where $\mathbf{v} = \boldsymbol{\eta} - \boldsymbol{\xi} = (2 - 1, 1 - 2, 2 - 1)^T = (1, -1, 1)^T$ (see §1.23). The required equations are therefore

$$\frac{x_1 - 1}{1} = \frac{x_2 - 2}{-1} = \frac{x_3 - 1}{1}.$$

4. (i) This is the equation of the 'half-line' illustrated below:

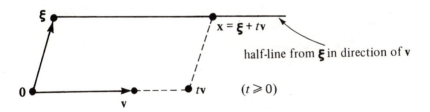

(ii) $\mathbf{x} = (1 - t)\boldsymbol{\xi} + t\boldsymbol{\eta}$ is the equation of a line through $\boldsymbol{\xi}$ and $\boldsymbol{\eta}$. When $t = 0$, $\mathbf{x} = \boldsymbol{\xi}$. When $t = 1$, $\mathbf{x} = \boldsymbol{\eta}$. Thus $\mathbf{x} = (1 - t)\boldsymbol{\xi} + t\boldsymbol{\eta}$ $(0 \leqslant t \leqslant 1)$ represents the line segment joining $\boldsymbol{\xi}$ and $\boldsymbol{\eta}$.

$\boldsymbol{\xi} \bullet\!\!-\!\!\!-\!\!\!-\!\!\!-\!\!\bullet\!\!-\!\!\!-\!\!\!-\!\!\!-\!\!\!-\!\!\!-\!\!\!-\!\!\!-\!\!\bullet \boldsymbol{\eta}$

$\mathbf{x} = (1 - t)\boldsymbol{\xi} + t\boldsymbol{\eta} \ (0 \leqslant t \leqslant 1)$

5. From §1.27 we have that the equation of a plane through ξ with normal v is $\langle x - \xi, v \rangle = 0$ – i.e.

$$(x_1 - \xi_1)v_1 + (x_2 - \xi_2)v_2 + (x_3 - \xi_3)v_3 = 0.$$

The equation required is therefore

$$(x_1 - 1)2 + (x_2 - 2)1 + (x_3 - 1)2 = 0$$

or

$$2x_1 + x_2 + 2x_3 = 6.$$

Observe that a unit normal to the plane is $(\frac{2}{3}, \frac{1}{3}, \frac{2}{3})^T$. We therefore rewrite the equation of the plane in the form

$$\tfrac{2}{3}x_1 + \tfrac{1}{3}x_2 + \tfrac{2}{3}x_3 = 2$$

and deduce that the distance of the plane from the origin is 2.

To find the distance of the plane from the point $(1, 2, 3)^T$, we shift the origin to $(1, 2, 3)^T$ by making the change of variable $x_1 = X_1 + 1$, $x_2 = X_2 + 2, x_3 = X_3 + 3$. The equation of the plane then becomes

$$\tfrac{2}{3}X_1 + \tfrac{1}{3}X_2 + \tfrac{2}{3}X_3 = 2 - \tfrac{10}{3} = -\tfrac{4}{3}.$$

Thus the distance of the plane from the point $(1, 2, 3)^T$ is $\frac{4}{3}$.

6. We first need a vector v orthogonal to $(1, 2)^T$ – i.e.

$$1v_1 + 2v_2 = 0.$$

The choice $v_1 = -2, v_2 = 1$ suffices. The equation of a line orthogonal to $(1, 2)^T$ is then

$$\frac{x_1 - \xi_1}{-2} = \frac{x_2 - \xi_2}{1}.$$

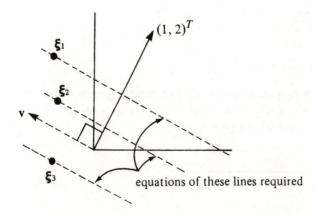

equations of these lines required

7. As in example 1.30, we let the plane be $ux + vy + wz = c$. Then

$$\begin{pmatrix} u \\ v \\ w \end{pmatrix} = \begin{pmatrix} 1 & 1 & 2 \\ 1 & 2 & 1 \\ 2 & 1 & 1 \end{pmatrix}^{-1} c \begin{pmatrix} 1 \\ 1 \\ 1 \end{pmatrix} = \frac{c}{-4} \begin{pmatrix} 1 & 1 & -3 \\ 1 & -3 & 1 \\ -3 & 1 & 1 \end{pmatrix} \begin{pmatrix} 1 \\ 1 \\ 1 \end{pmatrix}.$$

Hence

$$\begin{pmatrix} u \\ v \\ w \end{pmatrix} = \frac{c}{4} \begin{pmatrix} 1 \\ 1 \\ 1 \end{pmatrix}.$$

If $(u, v, w)^T$ is to be a unit vector, then

$$1 = u^2 + v^2 + w^2 = \tfrac{3}{16} c^2.$$

Hence $c = 4/\sqrt{3}$ is the required distance.

8. We have that $x_1 + 2x_2 + 3x_3 = \langle \mathbf{x}, \mathbf{v} \rangle$ where $\mathbf{v} = (1, 2, 3)^T$. Hence $(1, 2, 3)^T$ is normal to the plane (§1.27). A point $\boldsymbol{\xi}$ in the plane must satisfy

$$\xi_1 + 2\xi_2 + 3\xi_3 = 6.$$

The point $\boldsymbol{\xi} = (1, 1, 1)^T$ suffices.

9. The second lies at distance $6/\sqrt{(14)}$ and the first at distance $5/\sqrt{(14)}$ from the origin.

10. Let $\mathbf{p} = (p_1, p_2, p_3)^T$, $\mathbf{q} = (q_1, q_2, q_3)^T$ and $\mathbf{r} = (r_1, r_2, r_3)^T$. The two equations given then reduce to

$$\left. \begin{aligned} \langle \mathbf{p}, \mathbf{r} \rangle &= 0 \\ \langle \mathbf{q}, \mathbf{r} \rangle &= 0 \end{aligned} \right\}$$

which mean, respectively, that $\mathbf{p}$ and $\mathbf{r}$ are orthogonal and that $\mathbf{q}$ and $\mathbf{r}$ are orthogonal. Taking $r_1 = p_2 q_3 - p_3 q_2$, $r_2 = -p_1 q_3 + p_3 q_1$, $r_3 = p_1 q_2 - p_2 q_1$, we obtain

$$\left. \begin{aligned} p_1(p_2 q_3 - p_3 q_2) + p_2(-p_1 q_3 + p_3 q_1) + p_3(p_1 q_2 - p_2 q_1) &= 0 \\ q_1(p_2 q_3 - p_3 q_2) + q_2(-p_1 q_3 + p_3 q_1) + q_3(p_1 q_2 - p_2 q_1) &= 0 \end{aligned} \right\}$$

and the result follows.

11. (i) We first require a vector $\mathbf{r}$ orthogonal to the vectors $\mathbf{p} = (1, 1, 1)^T$ and $\mathbf{q} = (1, -1, 1)^T$. Using question 10, we have that $\mathbf{r}$ may be taken to be

$$\begin{aligned} \mathbf{r} &= (1 \times 1 - 1 \times (-1), -1 \times 1 + 1 \times 1, 1 \times (-1) - 1 \times 1)^T \\ &= (2, 0, -2)^T. \end{aligned}$$

The required plane then has equation

$$2(x_1 - 3) + 0(x_2 - 1) - 2(x_3 - 2) = 0$$

i.e.

$$2x_1 - 2x_3 = 2$$

$$x_1 - x_3 = 1.$$

(ii) The required plane will pass through $(3, 1, 2)^T$ and be parallel to $(1, 1, 1)^T$ and $(3, 1, 2)^T - (1, 2, 3)^T = (2, -1, -1)^T$. As in (i), we therefore take **r** to be

$$\mathbf{r} = (1 \times (-1) - 1 \times (-1), -1 \times (-1) + 1 \times 2,$$

$$1 \times (-1) - 1 \times 2)^T = (0, 3, -3)^T.$$

The required plane is therefore

$$0(x_1 - 3) + 3(x_2 - 1) - 3(x_3 - 2) = 0$$

i.e.

$$x_2 - x_3 = -1.$$

12. The required line is common to the planes

$$\left. \begin{array}{l} x_1 + 2x_2 + 3x_3 = 1 \\ 2x_1 + x_2 = 1. \end{array} \right\}$$

Its direction is therefore orthogonal to the normals $\mathbf{p} = (1, 2, 3)^T$ and $\mathbf{q} = (2, 1, 0)^T$. It therefore points in the direction of the vector **r** given by

$$\mathbf{r} = (2 \times 0 - 3 \times 1, -1 \times 0 + 3 \times 2, 1 \times 1 - 2 \times 2)^T$$

$$= (-3, 6, -3)^T.$$

We also require a point $\boldsymbol{\xi}$ on the line. Such a point is $\boldsymbol{\xi} = (\frac{1}{3}, \frac{1}{3}, 0)^T$. The parametric equation of the line is then

$$\mathbf{x} = \boldsymbol{\xi} + t\mathbf{r}$$

or, written in full,

$$\left. \begin{array}{l} x_1 = \frac{1}{3} - 3t \\ x_2 = \frac{1}{3} + 6t \\ x_3 = 0 - 3t. \end{array} \right\}$$

(It may be instructive to recover the original pair of equations by eliminating t from these final three equations.)

13.
$$\begin{pmatrix} x_1 \\ x_2 \\ x_3 \end{pmatrix} = \begin{pmatrix} 1 & 2 & 3 \\ 3 & 1 & 2 \\ 2 & 3 & 1 \end{pmatrix}^{-1} \begin{pmatrix} 1 \\ 2 \\ 3 \end{pmatrix} = \frac{1}{18} \begin{pmatrix} -5 & 7 & 1 \\ 1 & -5 & 7 \\ 7 & 1 & -5 \end{pmatrix} \begin{pmatrix} 1 \\ 2 \\ 3 \end{pmatrix}$$

$$\begin{pmatrix} x_1 \\ x_2 \\ x_3 \end{pmatrix} = \frac{1}{18} \begin{pmatrix} 12 \\ 12 \\ -6 \end{pmatrix} = \begin{pmatrix} \frac{2}{3} \\ \frac{2}{3} \\ -\frac{1}{3} \end{pmatrix}.$$

14. The vectors $(1, 1, 1)^T$ and $(-1, -1, -1)^T$ point in opposite directions because $(-1, -1, -1)^T = (-1)(1, 1, 1)^T$ (see §1.16).

(i) Writing the matrix equation in full, we obtain that

$$x_1 + x_2 + x_3 = 2$$

$$-x_1 - x_2 - x_3 = -2.$$

The second equation is essentially the same as the first and is therefore redundant. The flat is therefore the plane $x_1 + x_2 + x_3 = 2$.

(ii) Writing the matrix equation in full, we obtain that

$$x_1 + x_2 + x_3 = 2$$

$$-x_1 - x_2 - x_3 = 2.$$

Adding the equations yields that $0 = 4$ and so the equations are inconsistent.

15.
$$\begin{vmatrix} 1 & 2 & 3 \\ 2 & 1 & 0 \\ 3 & 3 & 3 \end{vmatrix} = (3 + 0 + 18) - (9 + 0 + 12) = 0.$$

Writing the matrix equation out in full, we obtain that

$$x_1 + 2x_2 + 3x_3 = 1$$

$$2x_1 + x_2 + 0x_3 = 1$$

$$3x_1 + 3x_2 + 3x_3 = 2.$$

The final equation is the sum of the other two and hence is redundant. The flat is therefore the intersection of two planes and hence is a line.

Chapter 2. Differentiation

Exercise 2.7

1.
$$f'(1) = \lim_{h \to 0} \frac{f(1 + h) - f(1)}{h}$$

$$= \lim_{h \to 0} \frac{\{(1 + h)^3 + 2(1 + h) + 3\} - \{(1)^3 + 2(1) + 3\}}{h}$$

$$= \lim_{h \to 0} \frac{1}{h} \{1 + 3h + 3h^2 + h^3 + 2 + 2h + 3 - 1 - 2 - 3\}$$

$$= \lim_{h \to 0} \{3 + 3h + h^2 + 2\} = 5.$$

2.

(i) $\dfrac{d}{dx}\{x^4 - 3x^2 + 1\} = 4x^3 - 6x.$

(ii) $\dfrac{d}{dx}(\sqrt{x}) = \dfrac{d}{dx}(x^{1/2}) = \dfrac{1}{2}x^{1/2-1} = \dfrac{1}{2}x^{-1/2} = \dfrac{1}{2\sqrt{x}}.$

(iii) Put $z = \sqrt{y}$ and $y = x^2 + 1$. Then

$$\frac{d}{dx}\{\sqrt{(x^2 + 1)}\} = \frac{dz}{dy}\frac{dy}{dx} = \frac{1}{2\sqrt{y}}2x = \frac{x}{\sqrt{(x^2 + 1)}}.$$

(iv) Put $z = y^{3/2}$ and $y = x + 1$. Then

$$\frac{d}{dx}\{(x + 1)^{3/2}\} = \frac{dz}{dy}\frac{dy}{dx} = \frac{3}{2}y^{1/2} = \frac{3}{2}\sqrt{(x + 1)}.$$

(v) Put $y = x$ and $z = \sin x$. Then

$$\frac{d}{dx}\{x \sin x\} = \frac{d}{dx}\{yz\} = z\frac{dy}{dx} + y\frac{dz}{dx} = \sin x + x \cos x.$$

(vi) Put $y = \tan x$ and $z = x$. Then

$$\frac{d}{dx}\left(\frac{\tan x}{x}\right) = \frac{1}{z^2}\left(z\frac{dy}{dx} - y\frac{dz}{dx}\right) = \frac{1}{x^2}\{x \sec^2 x - \tan x\}.$$

3. Each of these results is obtained by using the formula

$$\frac{dz}{dx} = \frac{dz}{dy}\frac{dy}{dx}$$

with $y = ax + b$. (In case (ii), $a = -1$ and $b = 0$.)

4. The tangent line to $y = f(x)$ at $(\xi, f(\xi))^T$ has equation

$$y = f(\xi) + f'(\xi)(x - \xi).$$

In each case to be considered, $\xi = 1$ and $f(\xi) = 1$. The problem therefore reduces to calculating $f'(1)$.

(i) $f'(x) = 2x$. Hence $f'(1) = 2$ and the tangent line has equation $y = 1 + 2(x - 1)$.

(ii) $f'(x) = 3x^2$. Hence $f'(1) = 3$ and the tangent line has equation $y = 1 + 3(x - 1)$.

(iii) $f'(x) = \frac{1}{2}\{(x - 1)^2 + 1\}^{-1/2}2(x - 1)$. Hence $f'(1) = 0$ and the tangent line has equation $y = 1$.

5. The required price is $f'(2) = \frac{1}{2}\{(2-1)^2 + 1\}^{-1/2} 2(2-1) = 1/\sqrt{2}$.

6. The speed v and acceleration a at time t are

$$v = \frac{dx}{dt} = 3t^2$$

and

$$a = \frac{dv}{dt} = 6t.$$

When $t = 3$, we obtain that $v = 27$ m/sec and $a = 18$ m/sec^2.

Exercise 2.22

1. (i) $f(x, y) = x^2 + 3y^2$.

$$\frac{\partial f}{\partial x} = 2x; \qquad \frac{\partial f}{\partial y} = 6y.$$

(ii) $f(x, y) = \dfrac{x-y}{x+y}$.

$$\frac{\partial f}{\partial x} = \frac{(x+y) \cdot 1 - (x-y) \cdot 1}{(x+y)^2} = \frac{2y}{(x+y)^2}$$

$$\frac{\partial f}{\partial y} = \frac{(x+y)(-1) - (x-y) \cdot 1}{(x+y)^2} = \frac{-2x}{(x+y)^2}.$$

(iii) $f(x, y) = e^{xy^2}$.

$$\frac{\partial f}{\partial x} = y^2 e^{xy^2}; \qquad \frac{\partial f}{\partial y} = 2xy e^{xy^2}.$$

(iv) $f(x, y) = \cos(x^2 y)$.

$$\frac{\partial f}{\partial x} = -2xy \sin(x^2 y); \qquad \frac{\partial f}{\partial y} = -x^2 \sin(x^2 y).$$

3. The equation of the tangent plane to the surface $z = f(x, y)$ where $x = 1$ and $y = 1$ is

$$z - Z = f_x(1, 1)(x - 1) + f_y(1, 1)(y - 1)$$

where $Z = f(1, 1)$. The required equations are therefore

(i) $z - 4 = 2(x - 1) + 6(y - 1)$.

(ii) $z = \frac{1}{2}(x - 1) - \frac{1}{2}(y - 1)$.

(iii) $z - e = e(x - 1) - 2e(y - 1)$.

(iv) $z - \cos(1) = -2 \sin(1)(x - 1) - \sin(1)(y - 1)$.

5. $\dfrac{dz}{dt} = \dfrac{\partial z}{\partial x}\dfrac{dx}{dt} + \dfrac{\partial z}{\partial y}\dfrac{dy}{dt}$

$= (-2xy \sin (x^2 y))3t^2 + (-x^2 \sin (x^2 y))2t$

$= -6t^7 \sin (t^8) - 2t^7 \sin (t^8) = -8t^7 \sin (t^8).$

7. The vector $\mathbf{v} = (-3, 4)^T$ is not a unit vector because $(-3)^2 + 4^2 = 25$. We therefore replace $\mathbf{v}$ by the unit vector $\mathbf{u} = (-\frac{3}{5}, \frac{4}{5})^T$ which points in the same direction. The rate of increase of f at $(1, 1)^T$ in the direction $(-3, 4)^T$ is then

$$\langle \nabla f, \mathbf{u} \rangle = -\tfrac{3}{5} f_x(1, 1) + \tfrac{4}{5} f_y(1, 1).$$

The direction of maximum rate of increase at $(1, 1)^T$ is $\nabla f = (f_x(1, 1), f_y(1, 1))^T$ and the rate of increase in this direction is

$$\|\nabla f\| = \{f_x(1, 1)^2 + f_y(1, 1)^2\}^{1/2}.$$

(i) When $(x, y)^T = (1, 1)^T$,

$$\nabla f = (2x, 6y)^T = (2, 6)^T.$$

Hence

(a) $\langle \nabla f, \mathbf{u} \rangle = -\tfrac{3}{5} \cdot 2 + \tfrac{4}{5} \cdot 6 = \tfrac{18}{5}$

(b) The direction of maximum rate of increase is $\nabla f = (2, 6)^T$ and the rate is $\|\nabla f\| = \{4 + 36\}^{1/2} = \sqrt{(40)}$.

(ii) When $(x, y)^T = (1, 1)^T$,

$$\nabla f = \left(\dfrac{2y}{(x+y)^2}, -\dfrac{2x}{(x+y)^2} \right)^T = \left(\dfrac{1}{2}, -\dfrac{1}{2} \right)^T.$$

Hence

(a) $\langle \nabla f, \mathbf{u} \rangle = (\tfrac{1}{2})(-\tfrac{3}{5}) + (-\tfrac{1}{2})(\tfrac{4}{5}) = -\tfrac{7}{10}$

(b) The direction of maximum rate of increase is $\nabla f = (\tfrac{1}{2}, -\tfrac{1}{2})^T$ and the rate is $\|\nabla f\| = \{\tfrac{1}{4} + \tfrac{1}{4}\}^{1/2} = 1/\sqrt{2}$.

(iii) When $(x, y)^T = (1, 1)^T$,

$$\nabla f = (y^2 e^{xy^2}, 2xye^{xy^2})^T = (e, 2e)^T.$$

Hence

(a) $\langle \nabla f, \mathbf{u} \rangle = -\tfrac{3}{5} \cdot e + \tfrac{4}{5} \cdot 2e = e$

(b) The direction of maximum rate of increase is $\nabla f = (e, 2e)^T$ – which is the same direction as $(1, 2)^T$ – and the rate is $\|\nabla f\| = \{e^2 + 4e^2\}^{1/2} = e\sqrt{5}$.

(iv) When $(x, y)^T = (1, 1)^T$,

$$\nabla f = (-2xy \sin (x^2 y), -x^2 \sin (x^2 y))^T = (-2 \sin (1), -\sin (1))^T.$$

Hence

(a) $\langle \nabla f, \mathbf{u} \rangle = -\tfrac{3}{5}(-2 \sin (1)) + \tfrac{4}{5}(-\sin (1)) = \tfrac{2}{5} \sin (1)$.

(b) The direction of maximum rate of increase is

$\nabla f = (-2 \sin (1), -\sin (1))^T$ – which is the same direction as $(-2, -1)^T$ – and the rate is $\|\nabla f\| = \{4(\sin (1))^2 + (\sin (1))^2 \}^{1/2} = \sqrt{5} \sin (1)$.

9.
$$\langle \nabla f, \mathbf{i} \rangle = \frac{\partial f}{\partial x} \cdot 1 + \frac{\partial f}{\partial y} \cdot 0 = \frac{\partial f}{\partial x}$$

$$\langle \nabla f, \mathbf{j} \rangle = \frac{\partial f}{\partial x} \cdot 0 + \frac{\partial f}{\partial y} \cdot 1 = \frac{\partial f}{\partial y}.$$

These results confirm that $\partial f/\partial x$ is the rate of increase of f in the direction of $\mathbf{i}$ (which is the direction in which the x axis points) and $\partial f/\partial y$ is the rate of increase of f in the direction of $\mathbf{j}$ (which is the direction in which the y axis points).

11. $f(x_1, x_2, x_3) = x_1^2 + x_2 x_3$.

$$\frac{\partial f}{\partial x_1} = 2x_1; \qquad \frac{\partial f}{\partial x_2} = x_3; \qquad \frac{\partial f}{\partial x_3} = x_2.$$

We have that $f(1, 0, 2) = 1$ and so the required tangent hyperplane is

$$z - 1 = 2(x_1 - 1) + 2(x_2 - 0) + 0(x_3 - 2)$$
i.e.
$$z = 2x_1 + 2x_2 - 1.$$

The normal to the contour $1 = x_1^2 + x_2 x_3$ at $(1, 0, 2)^T$ is just the value of the gradient

$$\nabla f = (2x_1, x_3, x_2)^T$$

and hence is equal to $(2, 2, 0)^T$. The tangent plane to the contour $1 = x_1^2 + x_2 x_3$ therefore has equation

$$0 = 2(x_1 - 1) + 2(x_2 - 0) + 0(x_3 - 2)$$
i.e.
$$0 = 2x_1 + 2x_2 - 2$$

$$x_1 + x_2 = 1.$$

The vector $\mathbf{v} = (2, -1, 2)^T$ is not a unit vector because $\|\mathbf{v}\| = \{4 + 1 + 4\}^{1/2} = 3$. A unit vector $\mathbf{u}$ which points in the same direction is $\mathbf{u} = (\frac{2}{3}, -\frac{1}{3}, \frac{2}{3})^T$. The rate of increase of f in the direction $(2, -1, 2)^T$ is therefore

$$\langle \nabla f, \mathbf{u} \rangle = 2 \cdot \tfrac{2}{3} + 2(-\tfrac{1}{3}) + 0(\tfrac{2}{3}) = \tfrac{2}{3}$$

at the point $(1, 0, 2)^T$.

13. If $z = f(x, y)$ and $x = tX, y = tY$, we have from the chain rule that

$$\frac{dz}{dt} = \frac{\partial z}{\partial x} \frac{dx}{dt} + \frac{\partial z}{\partial y} \frac{dy}{dt} = Xf_x(tX, tY) + Yf_y(tX, tY).$$

But

$$\frac{dz}{dt} = \frac{d}{dt} f(tX, tY) = \frac{d}{dt} t^\alpha f(X, Y) = \alpha t^{\alpha-1} f(X, Y).$$

Thus

$$\alpha t^{\alpha-1} f(X, Y) = X f_x(tX, tY) + Y f_y(tX, tY).$$

Now put $t = 1$. We obtain that

$$\alpha f(X, Y) = X f_x(X, Y) + Y f_y(X, Y).$$

Replacing $(X, Y)^T$ by $(x, y)^T$, we obtain the (less precise) equation

$$\alpha f = x \frac{\partial f}{\partial x} + y \frac{\partial f}{\partial y}.$$

Chapter 3. Stationary points

Exercise 3.12
1. We have that $f(x) = e^{2x} + e^{-3x}$ and so

$$f'(x) = 2e^{2x} - 3e^{-3x}.$$

To find the stationary points, we must therefore solve the equation

$$2e^{2x} - 3e^{-3x} = 0$$
$$2e^{2x} = 3e^{-3x}$$
$$e^{2x} e^{3x} = \tfrac{3}{2} e^{-3x} e^{3x}$$
$$e^{5x} = \tfrac{3}{2} e^0 = \tfrac{3}{2}$$
$$5x = \log\left(\tfrac{3}{2}\right)$$
$$x = \tfrac{1}{5} \log\left(\tfrac{3}{2}\right) = \log\left(\tfrac{3}{2}\right)^{1/5}.$$

(Recall that we use log to denote the *natural* logarithm.)

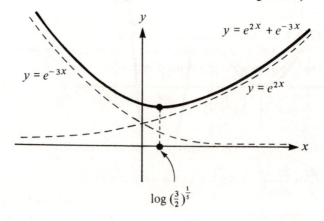

$$\log\left(\tfrac{3}{2}\right)^{\frac{1}{5}}$$

The single stationary point is evidently a global minimum for the function.

3. We have that $f(x) = x \sin x$ and so

$$f'(x) = \sin x + x \cos x.$$

To find the stationary points, we must therefore solve the equation

$$\sin x + x \cos x = 0$$

$$\tan x = -x.$$

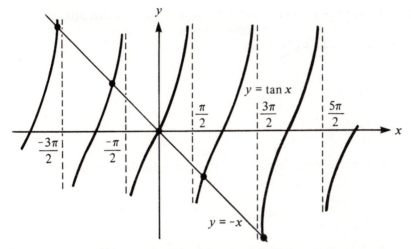

The diagram indicates that the curves $y = -x$ and $y = \tan x$ cross in an infinite number of places. Hence there are an infinite number of stationary points.

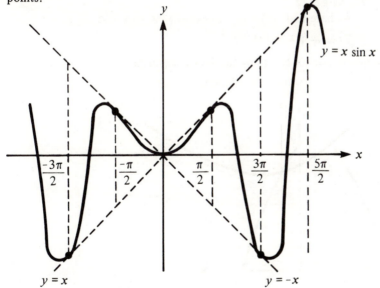

As the diagram shows, the stationary point at $x = 0$ is a *local* minimum. The function has *no* global maximum and *no* global minimum.

5. The substitution $x = X + 1, y = Y + 2$ in $f(x, y) = 2x + 4y - x^2 - y^2 - 3$ yields that

$$\phi(X, Y) = f(X + 1, Y + 2) =$$
$$2(X + 1) + 4(Y + 2) - (X + 1)^2 - (Y + 2)^2 - 3 = 2 - X^2 - Y^2.$$

Since $X^2 \geq 0$ and $Y^2 \geq 0$, it follows that $\phi(X, Y) \leq 2$ and that $\phi(0, 0) = 2$. We conclude that ϕ achieves a global maximum of 2 when $(X, Y)^T = (0, 0)^T$. Hence f achieves a global maximum of 2 when $(x, y)^T = (1, 2)^T$.

Observe that

$$\frac{\partial f}{\partial x} = 2 - 2x$$

$$\frac{\partial f}{\partial y} = 4 - 2y.$$

To find the stationary points we must therefore solve the simultaneous equations

$$2 - 2x = 0$$
$$4 - 2y = 0.$$

These have the unique solution $(x, y)^T = (1, 2)^T$.

Exercise 3.22

1. We have that $f(x) = 6 - x - x^2 = (3 + x)(2 - x)$ and so

$$f'(x) = -1 - 2x.$$

To find the stationary points, we must solve the equation $-1 - 2x = 0$. It follows that $x = -\frac{1}{2}$ is the single stationary point.

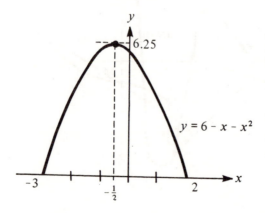

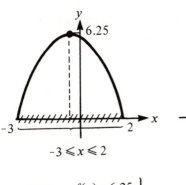

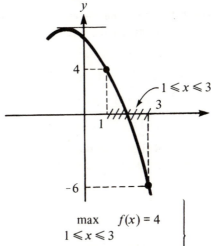

$$\begin{array}{c} \max\limits_{-3\,\leqslant\,x\,\leqslant\,2}\; f(x) = 6.25 \\[4pt] \min\limits_{-3\,\leqslant\,x\,\leqslant\,2}\; f(x) = 0 \end{array}$$

$$\begin{array}{c} \max\limits_{1\,\leqslant\,x\,\leqslant\,3}\; f(x) = 4 \\[4pt] \min\limits_{1\,\leqslant\,x\,\leqslant\,3}\; f(x) = -6 \end{array}$$

3. We considered a similar problem in example 3.6. In this case

$$\pi(x) = -x^3 + 3x^2 + (p-3)x$$

and so

$$\pi'(x) = -3x^2 + 6x + (p-3).$$

The stationary points are therefore found by solving

$$-3x^2 + 6x + (p-3) = 0$$

i.e.

$$x^2 - 2x + \left(1 - \frac{p}{3}\right) = 0$$

$$x = 1 \pm \sqrt{\left(1 - \left(1 - \frac{p}{3}\right)\right)} = 1 \pm \sqrt{\left(\frac{p}{3}\right)}.$$

The two cases to be distinguished are illustrated below.

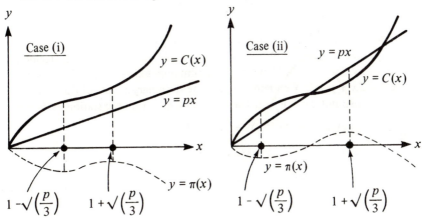

The difference in the two cases is that, in the second, there are points $x > 0$ such that $\pi(x) = 0$ – i.e.

$$\pi(x) = -x^3 + 3x^2 + (p - 3)x = 0$$

has solutions $x > 0$. But then

$$x^2 - 3x + (3 - p) = 0$$

has solutions and so '$b^2 - 4ac \geqslant 0$' – i.e.

$$9 - 4(3 - p) \geqslant 0$$

$$4p \geqslant 3$$

$$p \geqslant \tfrac{3}{4}.$$

Thus, if $0 \leqslant p < \tfrac{3}{4}$, then case (i) applies and

$$\max_{x \geqslant 0} \pi(x) = \pi(0) = 0$$

and, if $p \geqslant \tfrac{3}{4}$, then case (ii) applies and

$$\max_{x \geqslant 0} \pi(x) = \pi(\xi)$$

where $\xi = 1 + \sqrt{(p/3)}$.

5. We have that $f(x, y) = (x - 2)^2 + (y - 3)^2$ and so

$$\left. \begin{aligned} \frac{\partial f}{\partial x} &= 2(x - 2) \\[2mm] \frac{\partial f}{\partial y} &= 2(y - 3). \end{aligned} \right\}$$

The stationary points are found by solving the simultaneous equations

$$\left. \begin{aligned} x - 2 &= 0 \\ y - 3 &= 0 \end{aligned} \right\}$$

and so there is a single stationary point, namely $(2, 3)^T$.

(i) Since $2 \times 2 + 3 = 7 > 2$, the point $(2, 3)^T$ does not lie in the region defined by $2x + y \leqslant 2$. If the required minimum exists, it therefore lies on the line $2x + y = 2$. We therefore seek the stationary points of $f(x, y)$ subject to the constraint $2x + y = 2$. These are obtained from the stationary points of

$$\phi(x) = f(x, 2 - 2x) = (x - 2)^2 + (2x + 1)^2.$$

But

$$\phi'(x) = 2(x - 2) + 4(2x + 1)$$

$$= 10x.$$

and hence ϕ has a single stationary point at $x = 0$. The corresponding value of y is $y = 2 - 2 \times 0 = 2$ and so the required minimum is attained at the point $(0, 2)^T$. Thus

$$\min_{2x+y \leqslant 2} f(x, y) = (0 - 2)^2 + (2 - 3)^2 = 4 + 1 = 5.$$

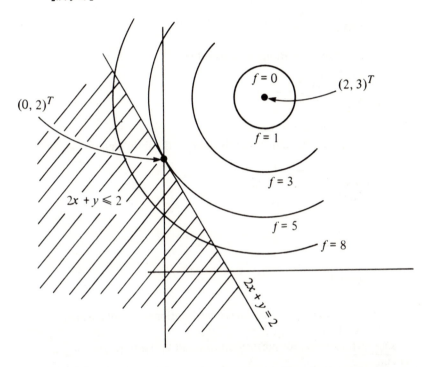

(ii) Since $3 \times 2 + 2 \times 3 = 12 > 6$, the point $(2, 3)^T$ lies in the region defined by $3x + 2y \geqslant 6$. Since $(2, 3)^T$ is a global minimum, it also minimises $f(x, y)$ subject to the constraint $3x + 2y \geqslant 6$. Thus

$$\min_{3x+2y \geqslant 6} f(x, y) = (2 - 2)^2 + (3 - 3)^2 = 0.$$

A *local* maximum will be found by evaluating the stationary points of $f(x, y)$ subject to the constraint $3x + 2y = 6$.

7. The Lagrangian is

$$L = x^2 + y^2 + \lambda(5x^2 + 6xy + 5y^2 - 8).$$

We therefore have to solve the simultaneous equations

$$\begin{cases} \dfrac{\partial L}{\partial x} = 2x + \lambda(10x + 6y) = (10\lambda + 2)x + 6\lambda y = 0 \\[2mm] \dfrac{\partial L}{\partial y} = 2y + \lambda(6x + 10y) = 6\lambda x + (10\lambda + 2)y = 0 \\[2mm] \dfrac{\partial L}{\partial \lambda} = 5x^2 + 6xy + 5y^2 - 8 = 0. \end{cases}$$

From the first two equations we obtain that

$$\frac{y}{x} = -\frac{(10\lambda + 2)}{6\lambda} = \frac{x}{y}.$$

Thus $y^2 = x^2$ and so $y = \pm x$.

$\underline{\text{Case (i)}}$. $y = x$. We substitute this result in the third equation and obtain

$$5x^2 + 6x^2 + 5x^2 = 8$$

$$16x^2 = 8$$

$$x^2 = \frac{1}{2}$$

$$x = \pm \frac{1}{\sqrt{2}}.$$

This case therefore yields two constrained stationary points, namely $(1/\sqrt{2}, 1/\sqrt{2})^T$ and $(-1/\sqrt{2}, -1/\sqrt{2})^T$.

$\underline{\text{Case (ii)}}$. $y = -x$. We substitute this result in the third equation and obtain

$$5x^2 - 6x^2 + 5x^2 = 8$$

$$4x^2 = 8$$

$$x = \pm \sqrt{2}.$$

This case therefore yields two constrained stationary points, namely $(\sqrt{2}, -\sqrt{2})^T$ and $(-\sqrt{2}, \sqrt{2})^T$.

Now

$$\left. \begin{array}{l} f\left(\dfrac{1}{\sqrt{2}}, \dfrac{1}{\sqrt{2}}\right) = f\left(-\dfrac{1}{\sqrt{2}}, -\dfrac{1}{\sqrt{2}}\right) = \dfrac{1}{2} + \dfrac{1}{2} = 1 \\[3mm] f(\sqrt{2}, -\sqrt{2}) = f(-\sqrt{2}, \sqrt{2}) = 2 + 2 = 4. \end{array} \right\}$$

The required maximum is therefore 4 and the minimum is 1.

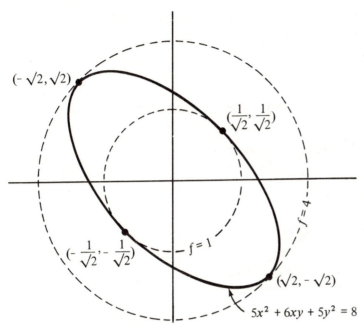

$$(-\sqrt{2}, \sqrt{2})$$

$$\left(\frac{1}{\sqrt{2}}, \frac{1}{\sqrt{2}}\right)$$

$$f = 4$$

$$\left(-\frac{1}{\sqrt{2}}, -\frac{1}{\sqrt{2}}\right)$$

$$f = 1$$

$$(\sqrt{2}, -\sqrt{2})$$

$$5x^2 + 6xy + 5y^2 = 8$$

9. We have to maximize

$$A = xy$$

subject to the constraint

$$2x + 2y = p.$$

The Lagrangian is

$$L = xy + \lambda(2x + 2y - p)$$

and so we solve the simultaneous equations

$$\frac{\partial L}{\partial x} = y + 2\lambda = 0$$

$$\frac{\partial L}{\partial y} = x + 2\lambda = 0$$

$$\frac{\partial L}{\partial \lambda} = 2x + 2y - p = 0.$$

From the first two equations we obtain that $x = y$. Substituting in the third equation yields that

$$4x = p$$

$$x = \tfrac{1}{4}p.$$

The rectangle of largest area with perimeter p is therefore the square of side $\tfrac{1}{4}p$.

11. The Lagrangian is

$$L = (x - X)^2 + (y - Y)^2 + \lambda(xy - 1) + \lambda(X + 2Y - 1)$$

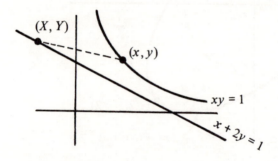

and so we solve the simultaneous equations

(1) $\quad \dfrac{\partial L}{\partial x} = 2(x - X) + \lambda y = 0$

(2) $\quad \dfrac{\partial L}{\partial y} = 2(y - Y) + \lambda x = 0$

(3) $\quad \dfrac{\partial L}{\partial X} = -2(x - X) + \mu = 0$

(4) $\quad \dfrac{\partial L}{\partial Y} = -2(y - Y) + 2\mu = 0$

(5) $\quad \dfrac{\partial L}{\partial \lambda} = xy - 1 = 0$

(6) $\quad \dfrac{\partial L}{\partial \mu} = X + 2Y - 1 = 0.$

From equations (1) and (2),

$$\frac{x - X}{y - Y} = \frac{y}{x}.$$

and from equations (3) and (4),

(7) $\dfrac{x - X}{y - Y} = \dfrac{1}{2}.$

Thus

$$\dfrac{1}{2} = \dfrac{y}{x}$$

(8) $x = 2y.$

Substitute this result in equation (5). Then

$$2y^2 = 1$$

$$y = \pm \dfrac{1}{\sqrt{2}}$$

Using equation (8) we obtain

case (i) $x = \sqrt{2}; \quad y = \dfrac{1}{\sqrt{2}}$

OR

case (ii) $x = -\sqrt{2}; \quad y = -\dfrac{1}{\sqrt{2}}.$

In case (i), we substitute in equation (7) and obtain that

$$2(\sqrt{2} - X) = \dfrac{1}{\sqrt{2}} - Y$$

$$2X - Y = \dfrac{3}{\sqrt{2}}$$

$$Y = 2X - \dfrac{3}{\sqrt{2}}.$$

Substituting in equation (6)

$$X + 4X - 6/\sqrt{2} = 1$$

$$5X = 1 + 3\sqrt{2}$$

$$X = \tfrac{1}{5}(1 + 3\sqrt{2}).$$

Hence

$$Y = \tfrac{1}{2}(1 - X) = \tfrac{1}{2}(1 - \tfrac{1}{5} - \tfrac{3}{5}\sqrt{2}) = \tfrac{1}{10}(4 - 3\sqrt{2}).$$

Case (i) therefore gives rise to the constrained stationary point

$$(x, y, X, Y)^T = \left(\sqrt{2}, \dfrac{1}{\sqrt{2}}, \dfrac{1}{5}(1 + 3\sqrt{2}), \dfrac{1}{10}(4 - 3\sqrt{2}) \right)^T.$$

Case (ii) gives rise to a second constrained stationary point but we need not calculate this since Case (i) is clearly the result we are seeking.

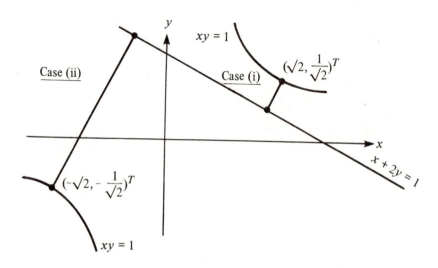

The required minimum is therefore

$$\left\{\sqrt{2} - \frac{1}{5}(1 + 3\sqrt{2})\right\}^2 + \left\{\frac{1}{\sqrt{2}} - \frac{1}{10}(4 - 3\sqrt{2})\right\}^2$$

(The minimum distance is the square root of this.)

13. The Lagrangian is

$$L = xyz + \lambda(x + 2y + 3z - 1) + \mu(3x + 2y + z - 1)$$

and so we have to solve the simultaneous equations

(1) $\dfrac{\partial L}{\partial x} = yz + \lambda + 3\mu = 0$

(2) $\dfrac{\partial L}{\partial y} = xz + 2\lambda + 2\mu = 0$

(3) $\dfrac{\partial L}{\partial z} = xy + 3\lambda + \mu = 0$

(4) $\dfrac{\partial L}{\partial \lambda} = x + 2y + 3z - 1 = 0$

(5) $\dfrac{\partial L}{\partial \mu} = 3x + 2y + z - 1 = 0.$

These equations can be solved without much difficulty but Lagrange's method is actually not the simplest way to proceed for this problem. It is easier to use the method of § 3.15.

We begin by solving the constraint equations for x and z in terms of y. This yields that

(6) $4x = 4z = 1 - 2y.$

Substituting for x and z in the expression xyz, we then seek the *unconstrained* stationary points for

$$\phi(y) = (1 - 2y)^2 y / 16$$

These unconstrained stationary points occur where $\phi'(y) = 0$. We therefore require the values of y for which

$$2(1 - 2y)(-2)y + (1 - 2y)^2 = 0.$$

Thus $y = \frac{1}{2}$ or $-4y + 1 - 2y = 0$, *i.e.* $y = \frac{1}{2}$ or $y = \frac{1}{6}$.

Case (i). $y = \frac{1}{2}$. From (6), we obtain that $(0, \frac{1}{2}, 0)^T$ is a constrained stationary point for the original problem. Note that

$$f(0, \tfrac{1}{2}, 0) = 0.$$

Case (ii). $y = \frac{1}{6}$. From (6), we obtain that $(\frac{1}{6}, \frac{1}{6}, \frac{1}{6})^T$ is a constrained stationary point for the original problem. Note that

$$f(\tfrac{1}{6}, \tfrac{1}{6}, \tfrac{1}{6}) = \tfrac{1}{216}.$$

Observe that xyz is larger in case (*ii*) but that xyz is *not* maximised at $(\frac{1}{6}, \frac{1}{6}, \frac{1}{6})^T$ subject to the constraints. To see this, not that $\phi(y)$ can be made as large as we choose by making y sufficiently large and positive.

15. Since $p = 1/\sqrt{x}$, we have that

$$\pi(x) = px - C(x)$$
$$= \sqrt{x} - 2x^2.$$

Thus
$$\pi'(x) = \tfrac{1}{2}x^{-1/2} - 4x.$$

The stationary points are found by solving

$$\tfrac{1}{2}x^{-1/2} - 4x = 0$$

$$1 - 8x^{3/2} = 0$$

$$x^{3/2} = \tfrac{1}{8}$$

$$x^{1/2} = (\tfrac{1}{8})^{1/3} = \tfrac{1}{2}$$

$$x = \tfrac{1}{4}.$$

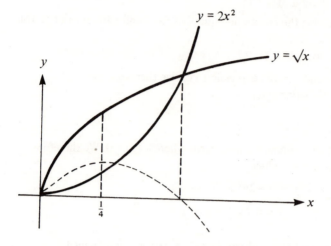

Profit is therefore maximized when $x = \tfrac{1}{4}$. The corresponding price is $p = 2$.

Chapter 4. Vector functions

Exercise 4.25

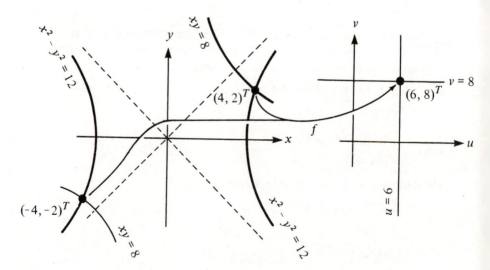

3.
$$\begin{pmatrix} \dfrac{\partial u}{\partial x} & \dfrac{\partial u}{\partial y} & \dfrac{\partial u}{\partial z} \\[2mm] \dfrac{\partial v}{\partial x} & \dfrac{\partial v}{\partial y} & \dfrac{\partial v}{\partial z} \end{pmatrix} = \begin{pmatrix} y^{-1} & -xy^{-2} & 0 \\ 0 & z^{-1} & -yz^{-2} \end{pmatrix}.$$

It follows that

$$f'(1, 2, 3) = \begin{pmatrix} \frac{1}{2} & -\frac{1}{4} & 0 \\ 0 & \frac{1}{3} & -\frac{2}{9} \end{pmatrix}.$$

The tangent flat where $(x, y, z)^T = (1, 2, 3)^T$ is given by

$$\begin{pmatrix} u - \frac{1}{2} \\ v - \frac{2}{3} \end{pmatrix} = \begin{pmatrix} \frac{1}{2} & -\frac{1}{4} & 0 \\ 0 & \frac{1}{3} & -\frac{2}{9} \end{pmatrix} \begin{pmatrix} x - 1 \\ y - 2 \\ z - 3 \end{pmatrix}$$

i.e.
$$\left. \begin{array}{l} u - \frac{1}{2} = \frac{1}{2}(x - 1) - \frac{1}{4}(y - 2) \\ v - \frac{2}{3} = \frac{1}{3}(y - 2) - \frac{2}{9}(z - 3). \end{array} \right\}$$

$$\left. \begin{array}{l} 4u = 2x - y + 2 \\ 9v = 3y - 2z + 6. \end{array} \right\}$$

5. Put $t = x^2 + y^2$. Then $z = f(t)$. If $\mathbf{u} = (x, y)^T$, we have that

$$\left(\dfrac{\partial z}{\partial x}, \dfrac{\partial z}{\partial y} \right) = \dfrac{dz}{d\mathbf{u}} = \dfrac{dz}{dt}\dfrac{dt}{d\mathbf{u}} = f'(t)\left(\dfrac{\partial t}{\partial x}, \dfrac{\partial t}{\partial y} \right)$$

and so

$$\left. \begin{array}{l} \dfrac{\partial z}{\partial x} = f'(t)\dfrac{\partial t}{\partial x} = f'(x^2 + y^2)2x \\[3mm] \dfrac{\partial z}{\partial y} = f'(t)\dfrac{\partial t}{\partial y} = f'(x^2 + y^2)2y. \end{array} \right\}$$

(These formulae can also, of course, be obtained directly.) Eliminating $f'(x^2 + y^2)$, we obtain that

$$\frac{1}{x} \cdot \frac{\partial z}{\partial x} = \frac{1}{y} \cdot \frac{\partial z}{\partial y}$$

as required.

7. The stationary points are found by solving the equation

$$\frac{dy}{dx} = 0.$$

Observe that, using rules (II) and (VI),

$$\frac{dy}{dx} = \frac{d}{dx}(A\mathbf{x} + \mathbf{a})^T(B\mathbf{x} + \mathbf{b})$$

$$= (B\mathbf{x} + \mathbf{b})^T A + (A\mathbf{x} + \mathbf{a})^T B$$

$$= (\mathbf{x}^T B^T + \mathbf{b}^T)A + (\mathbf{x}^T A^T + \mathbf{a}^T)B$$

$$= \mathbf{x}^T(B^T A + A^T B) + \mathbf{b}^T A + \mathbf{a}^T B.$$

The required result is therefore

$$\mathbf{x}^T(B^T A + A^T B) = -(\mathbf{b}^T A + \mathbf{a}^T B).$$

or, transposing,

$$(A^T B + B^T A)\mathbf{x} = -(A^T \mathbf{b} + B^T \mathbf{a}).$$

9. The Lagrangian is

$$L = \mathbf{x}^T \mathbf{x} + \boldsymbol{\lambda}^T(B\mathbf{x} - \mathbf{c}).$$

The constrained stationary points are found by solving

$$\left. \begin{array}{l} \dfrac{\partial L}{\partial \mathbf{x}} = 0 \\[2em] \dfrac{\partial L}{\partial \boldsymbol{\lambda}} = 0. \end{array} \right\}$$

Now

$$\frac{\partial L}{\partial \mathbf{x}} = 2\mathbf{x}^T + \boldsymbol{\lambda}^T B.$$

Since L is a scalar, $L = L^T = \mathbf{x}^T \mathbf{x} + (B\mathbf{x} - \mathbf{c})^T \boldsymbol{\lambda}$ and thus

$$\frac{\partial L}{\partial \boldsymbol{\lambda}} = (B\mathbf{x} - \mathbf{c})^T$$

After transposing these results, we obtain

(1) $2\mathbf{x} + B^T \boldsymbol{\lambda} = 0$

(2) $B\mathbf{x} - \mathbf{c} = 0$.

(As always, the second equation is just the constraint equation.)
Multiplying through (1) by B, we obtain that

$$2B\mathbf{x} + BB^T \boldsymbol{\lambda} = 0.$$

Substituting from (2) yields that

$$2\mathbf{c} + BB^T\boldsymbol{\lambda} = 0$$

$$\boldsymbol{\lambda} = -2(BB^T)^{-1}\mathbf{c}.$$

This result may be substituted back in (1)

$$2\mathbf{x} - 2B^T(BB^T)^{-1}\mathbf{c} = 0$$

and so the problem has the unique solution

$$\mathbf{x} = B^T(BB^T)^{-1}\mathbf{c}.$$

11. $\dfrac{\partial z}{\partial x} = e^{-y}; \quad \dfrac{\partial z}{\partial y} = -xe^{-y}.$

Hence the second derivative is

$$\begin{pmatrix} \dfrac{\partial^2 z}{\partial x^2} & \dfrac{\partial^2 z}{\partial x \partial y} \\[2mm] \dfrac{\partial^2 z}{\partial y \partial x} & \dfrac{\partial^2 z}{\partial y^2} \end{pmatrix} = \begin{pmatrix} 0 & -e^{-y} \\ -e^{-y} & xe^{-y} \end{pmatrix}.$$

We have that

$$f(1,0) = 1; \quad f'(1,0) = (1,-1)$$

$$f''(1,0) = \begin{pmatrix} 0 & -1 \\ -1 & 1 \end{pmatrix}.$$

The first three terms of the Taylor expression are therefore

$$f(1,0) + \frac{1}{1!}f'(1,0)\begin{pmatrix} x-1 \\ y-0 \end{pmatrix} + \frac{1}{2!}(x-1, y-0)f''(1,0)\begin{pmatrix} x-1 \\ y-0 \end{pmatrix}$$

$$= 1 + (1,-1)\begin{pmatrix} x-1 \\ y \end{pmatrix} + \frac{1}{2}(x-1, y)\begin{pmatrix} 0 & -1 \\ -1 & 1 \end{pmatrix}\begin{pmatrix} x-1 \\ y \end{pmatrix}$$

$$= 1 + \{x-1-y\} + \frac{1}{2}\{-2(x-1)y + y^2\}$$

$$= x - xy + \frac{1}{2}y^2.$$

This result can also be obtained by observing that

$$xe^{-y} = \{1 + (x-1)\}\left\{1 - y + \frac{y^2}{2!} - \frac{y^3}{3!} + \ldots\right\}$$

$$= \left\{1 - y + \frac{y^2}{2!} - \frac{y^3}{3!} + \ldots\right\} + \{(x-1) - (x-1)y + \ldots\}$$

$$= \left(1 - y + \frac{y^2}{2} + (x-1) - (x-1)y\right) + \ldots$$

$$= \left(x - xy + \frac{1}{2}y^2\right) + \ldots.$$

Chapter 5. Maxima and minima

Exercise 5.14

1. We know from the solution to exercise 4.12 (1) that
$f(x) = e^{2x} + e^{-3x}$ has a unique stationary point at

$$x = \log \left(\tfrac{3}{2}\right)^{1/5}.$$

Since

$$f'(x) = 2e^{2x} - 3e^{-3x},$$

$$f''(x) = 4e^{2x} + 9e^{-3x}.$$

This is positive for *all* values of x. Thus the stationary point is a local minimum. (Note that the stationary point is actually a global minimum but this fact cannot be deduced from the inequality $f''(\log \left(\tfrac{3}{2}\right)^{1/5}) > 0$ alone.)

3. We know from example 3.8 that the stationary points are

$$(0, 0)^T, (1, 0)^T, (0, 1)^T, (0, -1)^T, \left(\frac{2}{5}, \frac{1}{\sqrt{5}}\right)^T, \left(\frac{2}{5}, -\frac{1}{\sqrt{5}}\right)^T.$$

The second derivative is

$$\begin{pmatrix} f_{xx} & f_{xy} \\ f_{yx} & f_{yy} \end{pmatrix} = \begin{pmatrix} 2y & 2x + 3y^2 - 1 \\ 2x + 3y^2 - 1 & 6xy \end{pmatrix}.$$

The principal minors are

$$\Delta = \begin{vmatrix} 2y & 2x + 3y^2 - 1 \\ 2x + 3y^2 - 1 & 6xy \end{vmatrix} \quad \text{and} \quad \delta = 2y$$

(i) At $(0, 0)^T$,

$$\Delta = \begin{vmatrix} 0 & -1 \\ -1 & 0 \end{vmatrix} = -1 < 0.$$

and hence $(0, 0)^T$ is a saddle point.

(ii) At $(1, 0)^T$,

$$\Delta = \begin{vmatrix} 0 & 1 \\ 1 & 0 \end{vmatrix} = -1 < 0$$

and hence $(1, 0)^T$ is a saddle point.

(iii) At $(0, 1)^T$,

$$\Delta = \begin{vmatrix} 2 & 2 \\ 2 & 0 \end{vmatrix} = -4 < 0$$

and hence $(0, 1)^T$ is a saddle point.

(iv) At $(0, -1)^T$,

$$\Delta = \begin{vmatrix} -2 & 2 \\ 2 & 0 \end{vmatrix} = -4 < 0$$

and hence $(0, -1)^T$ is a saddle point.

(v) At $(\frac{2}{3}, 1/\sqrt{5})^T$,

$$\Delta = \begin{vmatrix} \dfrac{2}{\sqrt{5}} & 0 \\ 0 & \dfrac{12}{5\sqrt{5}} \end{vmatrix} = \frac{24}{25} > 0.$$

Also $\delta = 2/\sqrt{5} > 0$. Hence $(\frac{2}{3}, 1/\sqrt{5})^T$ is a local minimum.

(vi) At $(\frac{2}{3}, -1/\sqrt{5})^T$,

$$\Delta = \begin{vmatrix} \dfrac{-2}{\sqrt{5}} & 0 \\ 0 & \dfrac{-12}{5\sqrt{5}} \end{vmatrix} = \frac{24}{25} > 0.$$

Also $\delta = -2/\sqrt{5} < 0$. Hence $(\frac{2}{3}, -1/\sqrt{5})^T$ is a local maximum.

5. The stationary points are found by solving the simultaneous equations

$$(1) \begin{cases} \dfrac{\partial f}{\partial x} = 10x - 6z = 0 & \text{— i.e. } 3z = 5x \\[2mm] \dfrac{\partial f}{\partial y} = 10y - 12z = 0 & \text{— i.e. } 5y = 6z \\[2mm] \dfrac{\partial f}{\partial z} = 18z - 6x - 12y = 0 & \text{— i.e. } 3z - x - 2y = 0. \end{cases}$$

Substituting for x and y in the third equation, we obtain that

$$`3z - \tfrac{3}{5}z - \tfrac{12}{5}z = 0$$

which is satisfied for *all* values of z. Thus, if we write $z = 5\alpha$, we obtain that the point

$$\alpha(3, 6, 5)^T$$

is a stationary point for all values of α.

The second derivative is

$$\begin{pmatrix} f_{xx} & f_{xy} & f_{xz} \\ f_{yz} & f_{yy} & f_{yz} \\ f_{zx} & f_{zy} & f_{zz} \end{pmatrix} = \begin{pmatrix} 10 & 0 & -6 \\ 0 & 10 & -12 \\ -6 & -12 & 18 \end{pmatrix}.$$

Since the determinant of this matrix is zero (which can be deduced, incidentally, from the fact that equations (1) have multiple solutions), the standard method does not work.

However, in this case, the given function is a quadratic form. In fact

$$f(x, y, z) = (x, y, z) \begin{pmatrix} 5 & 0 & -3 \\ 0 & 5 & -6 \\ -3 & -6 & 9 \end{pmatrix} \begin{pmatrix} x \\ y \\ z \end{pmatrix}.$$

The eigenvalues of the matrix are obtained by solving

$$\begin{vmatrix} 5-\lambda & 0 & -3 \\ 0 & 5-\lambda & -6 \\ -3 & -6 & 9-\lambda \end{vmatrix} = 0$$

i.e.

$$(5-\lambda)^2(9-\lambda) - (5-\lambda)(9+36) = 0.$$

Thus $\lambda = 5$ <u>OR</u>

$$(5-\lambda)(9-\lambda) - 45 = 0$$

$$45 - 14\lambda + \lambda^2 - 45 = 0$$

$$\lambda(\lambda - 14) = 0$$

and so

$$\lambda = 0 \quad \underline{OR} \quad \lambda = 14.$$

By an appropriate change of variable, we may therefore write

$$f(x, y, z) = 0X^2 + 5Y^2 + 14Z^2$$

from which it follows that f has local minima where $Y = Z = 0$ – i.e. along the line where $(x, y, z)^T = \alpha(3, 6, 5)^T$.

7. We have that

$$z = xy = (x, y)\begin{pmatrix} 0 & \frac{1}{2} \\ \frac{1}{2} & 0 \end{pmatrix}\begin{pmatrix} x \\ y \end{pmatrix}.$$

The eigenvalues of the matrix are obtained by solving

$$\begin{vmatrix} -\lambda & \frac{1}{2} \\ \frac{1}{2} & -\lambda \end{vmatrix} = \lambda^2 - \frac{1}{4} = 0.$$

We obtain that $\lambda_1 = \frac{1}{2}$ and $\lambda_2 = -\frac{1}{2}$. Thus

$$z = \frac{1}{2}X^2 - \frac{1}{2}Y^2.$$

To find the directions in which the X and Y axes point, we require eigenvectors e_1 and e_2 corresponding to λ_1 and λ_2. The equation for e_1 is

$$\begin{pmatrix} 0 & \frac{1}{2} \\ \frac{1}{2} & 0 \end{pmatrix}\begin{pmatrix} e \\ f \end{pmatrix} = \frac{1}{2}\begin{pmatrix} e \\ f \end{pmatrix}$$

which yields the result $e = f$. Thus the X axis passes through $(1, 1)^T$. The equation for e_2 is

$$\begin{pmatrix} 0 & \frac{1}{2} \\ \frac{1}{2} & 0 \end{pmatrix}\begin{pmatrix} e \\ f \end{pmatrix} = -\frac{1}{2}\begin{pmatrix} e \\ f \end{pmatrix}$$

which yields the result $e = -f$. Thus the Y axis passes through $(-1, 1)^T$.

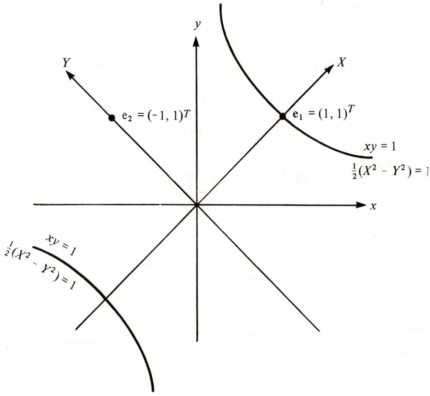

We have that

$$f(x, y) = g(X, Y) = \tfrac{1}{2}X^2 - \tfrac{1}{2}Y^2.$$

Observe that $g(X, 0)$ has a maximum at $X = 0$ while $g(0, Y)$ has a minimum at $Y = 0$. Thus $(0, 0)^T$ is a *saddle point*.

Chapter 6. Inverse functions

Exercise 6.9

1. (i) Write $u = \log x$. Then $y = \log (\log x) = \log u$. Hence

$$D(\log \log x) = \frac{dy}{dx} = \frac{dy}{du}\frac{du}{dx} = \frac{1}{u}\cdot\frac{1}{x} = \frac{1}{x \log x} \quad (x > 1).$$

(ii) Write $u = \log (x + 1)$. Then $y = (\log (x + 1))^2 = u^2$. Hence

$$D(\log (x + 1))^2 = \frac{dy}{dx} = \frac{dy}{du}\frac{du}{dx} = 2u \cdot \frac{1}{x + 1} = \frac{2 \log (x + 1)}{(x + 1)} \quad (x > -1).$$

(iii) Write $u = \log x$ and $v = u^{1/2}$. Then

$$y = e^{(\log x)^{1/2}} = e^{u^{1/2}} = e^v$$

and so

$$D(e^{(\log x)^{1/2}}) = \frac{dy}{dx} = \frac{dy}{du}\frac{du}{dx} = \frac{dy}{dv}\frac{dv}{du}\frac{du}{dx}$$

$$= e^v \frac{1}{2}u^{-1/2}\frac{1}{x} = \frac{1}{2x(\log x)^{1/2}}e^{(\log x)^{1/2}} \quad (x > 1).$$

(iv) Write $2^x = e^{x \log 2}$. Then

$$D(2^x) = D(e^{x \log 2}) = (\log 2)e^{x \log 2} = (\log 2)2^x.$$

3.

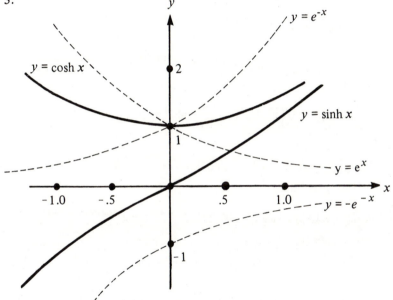

(i) $\dfrac{d}{dx}(\sinh x) = \dfrac{d}{dx}\left\{\dfrac{1}{2}(e^x - e^{-x})\right\} = \dfrac{1}{2}(e^x - (-1)e^{-x}) = \cosh x.$

(ii) $\dfrac{d}{dx}(\cosh x) = \dfrac{d}{dx}\left\{\dfrac{1}{2}(e^x + e^{-x})\right\} = \dfrac{1}{2}(e^x - e^{-x}) = \sinh x.$

The equation $y = \sinh x$ always has a unique solution for x in terms of y because $D(\sinh x)$ is always positive. (Note that $y = \cosh x$ does *not* always have a unique solution for x in terms of y.)

$$\cosh^2 x - \sinh^2 x = \tfrac{1}{4}(e^{2x} + 2 + e^{-2x}) - \tfrac{1}{4}(e^{2x} - 2 + e^{-2x})$$
$$= \tfrac{2}{4} + \tfrac{2}{4} = 1.$$

Write $y = \sinh x$. Then $x = \sinh^{-1} y$. Hence

$$\frac{d}{dy}(\sinh^{-1} y) = \frac{dx}{dy} = \left(\frac{dy}{dx}\right)^{-1} = \frac{1}{\cosh x} = \frac{1}{\sqrt{(1 + \sinh^2 x)}} = \frac{1}{\sqrt{(1 + y^2)}}.$$

(We take the positive square root because we know that $\sinh^{-1} y$ has a positive derivative.)

We have seen that the function $f: \mathbb{R} \to \mathbb{R}$ defined by $f(x) = \cosh x$ does *not* have an inverse function $g: \mathbb{R} \to \mathbb{R}$. To proceed in a similar way with $\cosh x$ it would be necessary to restrict attention to values of x and y satisfying $x > 0$ and $y > 1$.

5.
$$\frac{d}{dy} \log \{y + \sqrt{(y^2 + 1)}\}$$

$$= \frac{1}{\{y + \sqrt{(y^2 + 1)}\}} \left\{1 + \frac{1}{2}(y^2 + 1)^{-1/2} 2y\right\} = \frac{1}{\sqrt{(y^2 + 1)}}.$$

Observe that

$$\frac{d}{dy} \log \{y + \sqrt{(y^2 + 1)}\} = \frac{d}{dy} \sinh^{-1} y.$$

In fact,

$$\log \{y + \sqrt{(y^2 + 1)}\} = \sinh^{-1} y \quad (y > -1).$$

To verify this we consider

$$\sinh (\log \{y + \sqrt{(y^2 + 1)}\}) = \frac{1}{2}(e^{\log (y + \sqrt{(y^2+1)})} - e^{-\log (y + \sqrt{(y^2+1)})})$$

$$= \frac{1}{2}\left(\{y + \sqrt{(y^2 + 1)}\} - \frac{1}{\{y + \sqrt{(y^2 + 1)}\}}\right)$$

$$= \frac{y^2 + 2y\sqrt{(y^2 + 1)} + (y^2 + 1) - 1}{2(y + \sqrt{(y^2 + 1)})}$$

$$= \frac{2y(y + \sqrt{(y^2 + 1)})}{2(y + \sqrt{(y^2 + 1)})} = y.$$

7. The Taylor series of a function $f: \mathbb{R} \to \mathbb{R}$ about the point 0 is

$$f(0) + \frac{x}{1!}f'(0) + \frac{x^2}{2!}f''(0) + \frac{x^3}{3!}f'''(0) + \dots.$$

In the case $f(x) = e^x$, $1 = f(0) = f'(0) = f''(0) = \dots$ and so

$$e^x = 1 + x + \frac{x^2}{2!} + \frac{x^3}{3!} + \dots.$$

If $x > 0$, it follows that, for any n,

$$e^x > \frac{x^{n+1}}{(n+1)!}.$$

Thus

$$0 < x^n e^{-x} < \frac{(n+1)!}{x} \to 0 \quad \text{as} \quad x \to \infty.$$

Exercise 6.23

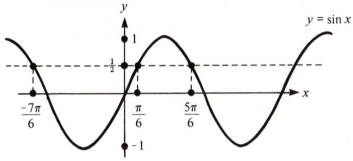

Sketched below are three appropriate local inverses
   (i) $g_1 : [-1, 1] \to [-\pi/2, \pi/2]$
  (ii) $g_2 : [-1, 1] \to [\pi/2, 3\pi/2]$
 (iii) $g_3 : [-1, 1] \to [-\pi/2, -3\pi/2].$

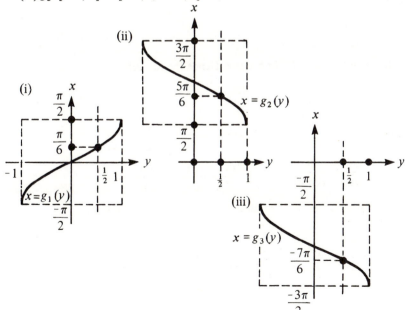

The required formulae are

(i) $x = g_1(y) = \arcsin y$

(ii) $x = g_2(y) = \pi - \arcsin y$

(iii) $x = g_3(y) = -\pi - \arcsin y.$

We have that

(i) $g_1'(y) = \dfrac{1}{\sqrt{(1-y^2)}}(-1 < y < 1).$

Thus

$$g_1'\left(\frac{1}{2}\right) = \frac{2}{\sqrt{3}}.$$

(ii) $g_2'(y) = \dfrac{d}{dy}(\pi - \arcsin y) = -\dfrac{1}{\sqrt{(1-y^2)}}(-1 < y < 0).$

Thus

$$g_2'\left(\frac{1}{2}\right) = -\frac{2}{\sqrt{3}}.$$

(iii) $g_3'(y) = \dfrac{d}{dy}(-\pi - \arcsin y) = -\dfrac{1}{\sqrt{(1-y^2)}}(-1 < y < 1).$

Thus

$$g_3'\left(\frac{1}{2}\right) = -\frac{2}{\sqrt{3}}.$$

If $y = \sin x$, then $dy/dx = \cos x$. Hence the formula

$$\frac{dx}{dy} = \left(\frac{dy}{dx}\right)^{-1}$$

reduces to

$$\frac{dx}{dy} = \frac{1}{\cos x}.$$

Observe that

(i) $\left(\cos\dfrac{\pi}{6}\right)^{-1} = \left(\dfrac{\sqrt{3}}{2}\right)^{-1} = g_1'\left(\dfrac{1}{2}\right).$

(ii) $\left(\cos\dfrac{5\pi}{6}\right)^{-1} = \left(\cos\left(\pi - \dfrac{\pi}{6}\right)\right)^{-1} = \left(\dfrac{-\sqrt{3}}{2}\right)^{-1} = g_2'\left(\dfrac{1}{2}\right).$

(iii) $\left(\cos\left(\dfrac{-\pi}{6}\right)\right)^{-1} = \left(\cos\left(-\pi - \dfrac{\pi}{6}\right)\right)^{-1} = \left(\dfrac{-\sqrt{3}}{2}\right)^{-1} = g_3'\left(\dfrac{1}{2}\right).$

No local inverse exists at $x = \pi/2$ because this is a critical point – i.e. $f'(\pi/2) = \cos(\pi/2) = 0$.

3. The derivative of the function $f : \mathbb{R}^2 \to \mathbb{R}^2$ defined by

$$
\left.\begin{array}{l}
u = xe^y \\
v = xe^{-y}
\end{array}\right\}
$$

is the matrix

$$
(1) \quad \begin{pmatrix} \dfrac{\partial u}{\partial x} & \dfrac{\partial u}{\partial y} \\[2mm] \dfrac{\partial v}{\partial x} & \dfrac{\partial v}{\partial y} \end{pmatrix} = \begin{pmatrix} e^y & xe^y \\ e^{-y} & -xe^{-y} \end{pmatrix}.
$$

Local inverses exist except at critical points – i.e. where

$$
0 = \begin{vmatrix} \dfrac{\partial u}{\partial x} & \dfrac{\partial u}{\partial y} \\[2mm] \dfrac{\partial v}{\partial x} & \dfrac{\partial v}{\partial y} \end{vmatrix} = \begin{vmatrix} e^y & xe^y \\ e^{-y} & -xe^{-y} \end{vmatrix} = -2x.
$$

Thus local inverses exist at all points $(x, y)^T$ with $x \neq 0$. The derivative of a local inverse is given by

$$
(2) \quad \begin{pmatrix} \dfrac{\partial x}{\partial u} & \dfrac{\partial x}{\partial v} \\[2mm] \dfrac{\partial y}{\partial u} & \dfrac{\partial y}{\partial v} \end{pmatrix} = \begin{pmatrix} \dfrac{\partial u}{\partial x} & \dfrac{\partial u}{\partial y} \\[2mm] \dfrac{\partial v}{\partial x} & \dfrac{\partial v}{\partial y} \end{pmatrix}^{-1}
$$

$$
= \begin{pmatrix} e^y & xe^y \\ e^{-y} & -xe^{-y} \end{pmatrix}^{-1} = \dfrac{1}{-2x}\begin{pmatrix} -xe^{-y} & -xe^y \\ -e^{-y} & e^y \end{pmatrix}.
$$

From (1) we observe that

$$\left(\frac{\partial u}{\partial x}\right)_y = e^y$$

and from (2) that

$$\left(\frac{\partial x}{\partial u}\right)_v = \frac{1}{2}e^{-y}.$$

5.
$$\left.\begin{array}{l} u = x + y \\ v = x - y \end{array}\right\}$$

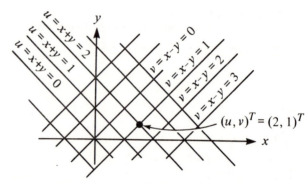

$(u, v)^T = (2, 1)^T$

$$\frac{\partial}{\partial x} = \frac{\partial u}{\partial x}\frac{\partial}{\partial u} + \frac{\partial v}{\partial x}\frac{\partial}{\partial v}$$

$$= \frac{\partial}{\partial u} + \frac{\partial}{\partial v}.$$

Hence

$$\frac{\partial^2 z}{\partial x^2} = \frac{\partial}{\partial x}\left(\frac{\partial z}{\partial u} + \frac{\partial z}{\partial v}\right) = \frac{\partial}{\partial u}\left(\frac{\partial z}{\partial u} + \frac{\partial z}{\partial v}\right) + \frac{\partial}{\partial v}\left(\frac{\partial z}{\partial u} + \frac{\partial z}{\partial v}\right)$$

$$= \frac{\partial^2 z}{\partial u^2} + 2\frac{\partial^2 z}{\partial u \partial v} + \frac{\partial^2 z}{\partial v^2}.$$

7.
$$\frac{\partial}{\partial x} = \frac{\partial u}{\partial x}\frac{\partial}{\partial u} + \frac{\partial v}{\partial x}\frac{\partial}{\partial v}$$

$$\frac{\partial}{\partial y} = \frac{\partial u}{\partial y}\frac{\partial}{\partial u} + \frac{\partial v}{\partial y}\frac{\partial}{\partial v}$$

$$\begin{pmatrix} \dfrac{\partial u}{\partial x} & \dfrac{\partial u}{\partial y} \\ \dfrac{\partial v}{\partial x} & \dfrac{\partial v}{\partial y} \end{pmatrix} = \begin{pmatrix} \dfrac{\partial x}{\partial u} & \dfrac{\partial x}{\partial v} \\ \dfrac{\partial y}{\partial u} & \dfrac{\partial y}{\partial v} \end{pmatrix}^{-1} = \begin{pmatrix} 1 & 1 \\ 1 & -1 \end{pmatrix}^{-1} = -\frac{1}{2}\begin{pmatrix} -1 & -1 \\ -1 & 1 \end{pmatrix}.$$

Hence

$$\frac{\partial}{\partial x} = \frac{1}{2}\frac{\partial}{\partial u} + \frac{1}{2}\frac{\partial}{\partial v}$$

$$\frac{\partial}{\partial y} = \frac{1}{2}\frac{\partial}{\partial u} - \frac{1}{2}\frac{\partial}{\partial v}.$$

These equations can also, of course, be obtained by observing that

$$\left. \begin{array}{l} u = \frac{1}{2}x + \frac{1}{2}y \\ v = \frac{1}{2}x - \frac{1}{2}y \end{array} \right\}.$$

$$\frac{\partial^2 z}{\partial x^2} = \frac{\partial}{\partial x}\left(\frac{\partial z}{\partial x}\right) = \frac{\partial}{\partial x}\left(\frac{1}{2}\frac{\partial z}{\partial u} + \frac{1}{2}\frac{\partial z}{\partial v}\right)$$

$$= \frac{1}{4}\frac{\partial^2 z}{\partial u^2} + \frac{1}{2}\frac{\partial^2 z}{\partial u \partial v} + \frac{1}{4}\frac{\partial^2 z}{\partial v^2}$$

$$\frac{\partial^2 z}{\partial y^2} = \frac{\partial}{\partial y}\left(\frac{\partial z}{\partial y}\right) = \frac{\partial}{\partial y}\left(\frac{1}{2}\frac{\partial z}{\partial u} - \frac{1}{2}\frac{\partial z}{\partial v}\right)$$

$$= \frac{1}{4}\frac{\partial^2 z}{\partial u^2} - \frac{1}{2}\frac{\partial^2 z}{\partial u \partial v} + \frac{1}{4}\frac{\partial^2 z}{\partial v^2}.$$

Thus

$$\frac{\partial^2 z}{\partial x^2} - \frac{\partial^2 z}{\partial y^2} = \frac{\partial^2 z}{\partial u \partial v}.$$

9.

$$\frac{\partial}{\partial r} = \frac{\partial x}{\partial r}\frac{\partial}{\partial x} + \frac{\partial y}{\partial r}\frac{\partial}{\partial y}$$

$$= \cos\theta \frac{\partial}{\partial x} + \sin\theta \frac{\partial}{\partial y}$$

$$\frac{\partial}{\partial \theta} = \frac{\partial x}{\partial \theta}\frac{\partial}{\partial x} + \frac{\partial y}{\partial \theta}\frac{\partial}{\partial y}$$

$$= -r\sin\theta \frac{\partial}{\partial x} + r\cos\theta \frac{\partial}{\partial y}.$$

It follows that

$$\frac{\partial f}{\partial r} = \cos\theta \frac{\partial f}{\partial x} + \sin\theta \frac{\partial f}{\partial y} = \langle \nabla f, \mathbf{e}_1 \rangle$$

$$\frac{1}{r}\frac{\partial f}{\partial \theta} = -\sin\theta \frac{\partial f}{\partial x} + \cos\theta \frac{\partial f}{\partial y} = \langle \nabla f, \mathbf{e}_2 \rangle.$$

where $e_1 = (\cos\theta, \sin\theta)^T$ and $e_2 = (-\sin\theta, \cos\theta)^T$ are the unit vectors indicated below:

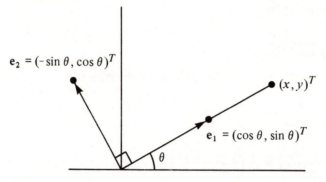

$e_2 = (-\sin\theta, \cos\theta)^T$

$(x,y)^T$

$e_1 = (\cos\theta, \sin\theta)^T$

θ

Observe that $\langle \nabla f, e_1 \rangle$ is the directional derivative of f in the e_1 direction. Thus

$$\frac{\partial f}{\partial r}$$

is the rate at which f increases at $(x,y)^T$ in the *radial* direction. Similarly

$$\frac{1}{r}\frac{\partial f}{\partial\theta}$$

is the rate at which f increases at $(x,y)^T$ in the *tangential* direction.

Chapter 7. Implicit functions

Exercise 7.7

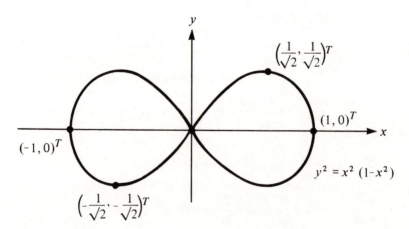

y

$\left(\frac{1}{\sqrt{2}}, \frac{1}{\sqrt{2}}\right)^T$

$(1,0)^T$

x

$(-1,0)^T$

$y^2 = x^2(1-x^2)$

$\left(-\frac{1}{\sqrt{2}}, -\frac{1}{\sqrt{2}}\right)^T$

Write $\phi(x, y) = y^2 - x^2(1 - x^2)$. Then

$$\frac{\partial \phi}{\partial y} = 2y.$$

This is non-zero for $(x, y)^T = (\xi, \eta)^T$ if and only if $\eta \neq 0$. Thus a unique local differentiable solution $y = g(x)$ exists at each $(\xi, \eta)^T$ satisfying $\eta^2 = \xi^2(1 - \xi^2)$ when $\eta = 0$. Also

$$g'(x) = \frac{dy}{dx} = -\left(\frac{\partial \phi}{\partial y}\right)^{-1}\left(\frac{\partial \phi}{\partial x}\right)$$

$$= -\frac{1}{2y}(-2x(1 - x^2) - x^2(-2x))$$

$$= \frac{x}{y}(1 - 2x^2).$$

The local solution $y = g_1(x)$ at $(1/\sqrt{2}, \frac{1}{2})^T$ is given by

$$y = g_1(x) = \sqrt{(x^2(1 - x^2))} \quad (0 \leqslant x \leqslant 1)$$

and the local solution $y = g_2(x)$ at $(-1/\sqrt{2}, -1/\sqrt{2})^T$ is given by

$$y = g_2(x) = -\sqrt{(x^2(1 - x^2))} \quad (-1 \leqslant x \leqslant 0).$$

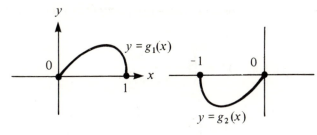

3. The point $(x, y)^T = (-1, 1)^T$ lies on the curve $\sin(\pi x^2 y) = 1 + xy^2$ because

$$\sin(\pi(-1)^2 1) = \sin \pi = 0$$

$$1 + (-1)(1)^2 = 1 - 1 = 0.$$

Write $\phi(x, y) = \sin(\pi x^2 y) - 1 - xy^2$. Then

$$\frac{\partial \phi}{\partial y} = \pi x^2 \cos(\pi x^2 y) - 2xy$$

$$= -\pi + 2 \neq 0$$

when $(x, y)^T = (-1, 1)^T$.

Thus a unique local differentiable solution $y = g(x)$ exists at $(x, y)^T = (-1, 1)^T$ and

$$g'(x) = \frac{dy}{dx} = -\left(\frac{\partial \phi}{\partial y}\right)^{-1} \left(\frac{\partial \phi}{\partial x}\right)$$

$$= \frac{-1}{\pi x^2 \cos (\pi x^2 y) - 2xy} (2\pi xy \cos (\pi x^2 y) - y^2).$$

Thus

$$g'(-1) = \frac{-1}{2 - \pi}(2\pi - 1) = \frac{2\pi - 1}{\pi - 2}.$$

5. Write

$$\phi(x, y, z) = \sin (yz) + \sin (zx) + \sin (xy).$$

Then

$$\frac{\partial \phi}{\partial z} = y \cos (yz) + x \cos (zx).$$

We need that $\phi(\xi, \eta, \zeta) = 0$ but $\phi_z(\xi, \eta, \zeta) \neq 0$. An appropriate choice is $(\xi, \eta, \zeta)^T = (0, \pi, 1)^T$. At this point a unique local differentiable solution $z = g(x, y)$ exists. We have that

$$g'(x, y) = \left(\frac{\partial z}{\partial x}, \frac{\partial z}{\partial y}\right) = -\left(\frac{\partial \phi}{\partial z}\right)^{-1} \left(\frac{\partial \phi}{\partial x}, \frac{\partial \phi}{\partial y}\right)$$

$$= \frac{-1}{y \cos (yz) + x \cos (zx)} (z \cos (zx) + y \cos (xy), z \cos (yz) + x \cos (xy)).$$

Thus

$$g'(0, \pi) = \frac{-1}{-\pi}(1 + \pi, -1) = \left(1 + \frac{1}{\pi}, -\frac{1}{\pi}\right).$$

7. Write

$$\phi(x, y, z) = \begin{pmatrix} \phi_1(x, y, z) \\ \phi_2(x, y, z) \end{pmatrix} = \begin{pmatrix} xy^2 + y^2 z^3 + z^4 x^5 - 1 \\ zx^2 + x^2 y^3 + y^4 z^5 + 1 \end{pmatrix}.$$

Then

$$\det \begin{pmatrix} \dfrac{\partial \phi_1}{\partial x} & \dfrac{\partial \phi_1}{\partial y} \\[2mm] \dfrac{\partial \phi_2}{\partial x} & \dfrac{\partial \phi_2}{\partial y} \end{pmatrix} = \begin{vmatrix} y^2 + 5x^4 z^4 & 2xy + 2yz^3 \\ 2zx + 2xy^3 & 3x^2 y^2 + 4y^3 z^5 \end{vmatrix}$$

$$= \begin{vmatrix} 6 & 0 \\ 0 & -1 \end{vmatrix} = -6.$$

when $(x, y, z)^T = (1, 1, -1)^T$.

It follows that a unique local differentiable solution

$$\begin{pmatrix} x \\ y \end{pmatrix} = g(z) = \begin{pmatrix} g_1(z) \\ g_2(z) \end{pmatrix}$$

exists at $(x, y, z)^T = (1, 1, -1)^T$ and

$$g'(z) = \begin{pmatrix} \dfrac{dx}{dz} \\ \dfrac{dy}{dz} \end{pmatrix} = -\begin{pmatrix} \dfrac{\partial \phi_1}{\partial x} & \dfrac{\partial \phi_1}{\partial y} \\ \dfrac{\partial \phi_2}{\partial x} & \dfrac{\partial \phi_2}{\partial y} \end{pmatrix}^{-1} \begin{pmatrix} \dfrac{\partial \phi_1}{\partial z} \\ \dfrac{\partial \phi_2}{\partial z} \end{pmatrix}$$

$$= -\begin{pmatrix} y^2 + 5x^4 z^4 & 2xy + 2yz^3 \\ 2zx + 2xy^3 & 3x^2 y^2 + 4y^3 z^5 \end{pmatrix}^{-1} \begin{pmatrix} 3y^2 z^2 + 4z^3 x^5 \\ x^2 + 5y^4 z^4 \end{pmatrix}.$$

Thus

$$g'(-1) = -\begin{pmatrix} 6 & 0 \\ 0 & -1 \end{pmatrix}^{-1} \begin{pmatrix} -1 \\ 6 \end{pmatrix}$$

$$= \frac{1}{6}\begin{pmatrix} -1 & 0 \\ 0 & 6 \end{pmatrix}\begin{pmatrix} -1 \\ 6 \end{pmatrix} = \frac{1}{6}\begin{pmatrix} 1 \\ 36 \end{pmatrix} = \begin{pmatrix} \frac{1}{6} \\ 6 \end{pmatrix}.$$

9. Write

$$\phi_1 = x^2 + u + e^v$$
$$\phi_2 = y^2 + v + e^w$$
$$\phi_3 = z^2 + w + e^u.$$

We need to calculate

$$\det \begin{pmatrix} \dfrac{\partial \phi_1}{\partial u} & \dfrac{\partial \phi_1}{\partial v} & \dfrac{\partial \phi_1}{\partial w} \\[2mm] \dfrac{\partial \phi_2}{\partial u} & \dfrac{\partial \phi_2}{\partial v} & \dfrac{\partial \phi_2}{\partial w} \\[2mm] \dfrac{\partial \phi_3}{\partial u} & \dfrac{\partial \phi_3}{\partial v} & \dfrac{\partial \phi_3}{\partial w} \end{pmatrix} = \begin{vmatrix} 1 & e^v & 0 \\ 0 & 1 & e^w \\ e^u & 0 & 1 \end{vmatrix} = 1 + e^{u+v+w}.$$

Since this determinant is never zero, a unique local differentiable solution

$$\begin{pmatrix} u \\ v \\ w \end{pmatrix} = g(x, y, z) = \begin{pmatrix} g_1(x, y, z) \\ g_2(x, y, z) \\ g_3(x, y, z) \end{pmatrix}$$

exists at every point $(x, y, z, u, v, w)^T$ which satisfies the equations. Also

$$g'(x, y, z) = \begin{pmatrix} \dfrac{\partial u}{\partial x} & \dfrac{\partial u}{\partial y} & \dfrac{\partial u}{\partial z} \\[2mm] \dfrac{\partial v}{\partial x} & \dfrac{\partial v}{\partial y} & \dfrac{\partial v}{\partial z} \\[2mm] \dfrac{\partial w}{\partial x} & \dfrac{\partial w}{\partial y} & \dfrac{\partial w}{\partial z} \end{pmatrix}$$

$$= - \begin{pmatrix} \dfrac{\partial \phi_1}{\partial u} & \dfrac{\partial \phi_1}{\partial v} & \dfrac{\partial \phi_1}{\partial w} \\[2mm] \dfrac{\partial \phi_2}{\partial u} & \dfrac{\partial \phi_2}{\partial v} & \dfrac{\partial \phi_2}{\partial w} \\[2mm] \dfrac{\partial \phi_3}{\partial u} & \dfrac{\partial \phi_3}{\partial v} & \dfrac{\partial \phi_3}{\partial w} \end{pmatrix}^{-1} \begin{pmatrix} \dfrac{\partial \phi_1}{\partial x} & \dfrac{\partial \phi_1}{\partial y} & \dfrac{\partial \phi_1}{\partial z} \\[2mm] \dfrac{\partial \phi_2}{\partial x} & \dfrac{\partial \phi_2}{\partial y} & \dfrac{\partial \phi_2}{\partial z} \\[2mm] \dfrac{\partial \phi_3}{\partial x} & \dfrac{\partial \phi_3}{\partial y} & \dfrac{\partial \phi_3}{\partial z} \end{pmatrix}$$

$$= - \begin{pmatrix} 1 & e^v & 0 \\ 0 & 1 & e^w \\ e^u & 0 & 1 \end{pmatrix}^{-1} \begin{pmatrix} 2x & 0 & 0 \\ 0 & 2y & 0 \\ 0 & 0 & 2z \end{pmatrix}$$

$$= - \frac{1}{1 + e^{u+v+w}} \begin{pmatrix} 1 & -e^v & e^{v+w} \\ e^{u+w} & 1 & -e^w \\ -e^u & e^{u+v} & 1 \end{pmatrix} \begin{pmatrix} 2x & 0 & 0 \\ 0 & 2y & 0 \\ 0 & 0 & 2z \end{pmatrix}$$

$$= - \frac{1}{1 + e^{u+v+w}} \begin{pmatrix} 2x & -2ye^v & 2ze^{v+w} \\ 2xe^{u+w} & 2y & -2ze^w \\ -2xe^u & 2ye^{u+v} & 2z \end{pmatrix}.$$

Chapter 8. Differentials

Exercise 8.8
1. We have that

$$f_x dx + f_y dy + f_z dz = 0$$
$$g_x dx + g_y dy + g_z dz = 0.$$

Also

$$dy = h'(x)dx.$$

One can eliminate dz from the first two equations and compare the result with the third equation. Alternatively, one can observe that the condition for the three linear equations to have a non-trivial solution for dx, dy and dz is

$$\begin{vmatrix} f_x & f_y & f_z \\ g_x & g_y & g_z \\ h'(x) & -1 & 0 \end{vmatrix} = 0.$$

Hence

$$h'(x)\begin{vmatrix} f_y & f_z \\ g_y & g_z \end{vmatrix} - (-1)\begin{vmatrix} f_x & f_z \\ g_x & g_z \end{vmatrix} + 0\begin{vmatrix} f_x & f_y \\ g_x & g_y \end{vmatrix} = 0$$

and so

$$h'(x) = -\begin{vmatrix} f_x & f_z \\ g_x & g_z \end{vmatrix} \Bigg/ \begin{vmatrix} f_y & f_z \\ g_y & g_z \end{vmatrix} = \frac{f_z g_x - f_x g_z}{f_y g_z - f_z g_y}.$$

3. We have that

$$dx = udu - vdv$$

$$dy = vdu + udv$$

and hence

$$du = \left(\frac{u}{u^2 + v^2}\right)dx + \left(\frac{v}{u^2 + v^2}\right)dy.$$

But

$$du = \left(\frac{\partial u}{\partial x}\right)_y dx + \left(\frac{\partial u}{\partial y}\right)_x dy.$$

Thus

$$\left(\frac{\partial u}{\partial x}\right)_y = \frac{u}{u^2 + v^2}; \quad \left(\frac{\partial u}{\partial y}\right)_x = \frac{v}{u^2 + v^2}.$$

5. There are functions $k: \mathbb{R}^2 \to \mathbb{R}, l: \mathbb{R}^2 \to \mathbb{R}, m: \mathbb{R}^2 \to \mathbb{R}$ and $n: \mathbb{R}^2 \to \mathbb{R}$ such that

$$T = k(u, p) = l(p, v)$$

and

$$p = m(T, v) = n(T, u).$$

Hence

$$dT = \left(\frac{\partial T}{\partial u}\right)_p du + \left(\frac{\partial T}{\partial p}\right)_u dp$$

$$dT = \left(\frac{\partial T}{\partial p}\right)_v dp + \left(\frac{\partial T}{\partial v}\right)_p dv$$

$$dp = \left(\frac{\partial p}{\partial T}\right)_v dT + \left(\frac{\partial p}{\partial v}\right)_T dv$$

$$dp = \left(\frac{\partial p}{\partial T}\right)_u dT + \left(\frac{\partial p}{\partial u}\right)_T du.$$

The condition that this system of linear equations have a non-trivial solution for du, dT, dp and dv is that the determinant given in the question is zero.

7. We have that

$$dx = udu - vdv$$

$$dv = \frac{u}{v}du - \frac{1}{v}dx.$$

Hence

$$\delta v \doteq \frac{u}{v}\delta u - \frac{1}{v}\delta x.$$

Chapter 9. Sums and integrals

Exercise 9.23

1. Pascal's triangle as far as the relevant row is given below.

```
                 1
              1     1
           1     2     1
        1     3     3     1
     1     4     6     4     1
   1     5    10    10     5     1
 1    6    15    20    15     6     1
1   7   21   35   35   21    7    1
1  8  28   56   70   56   28   8   1
```

The required expansion is therefore

$$(1 + x)^8 = 1 + 8x + 28x^2 + 56x^3 + 70x^4 + 56x^5 + 28x^6 + 8x^7 + x^8.$$

3. The sum

$$\sum_{k=0}^{n-1} f\left(\frac{k}{n}\right)\frac{1}{n}$$

represents the shaded area in the diagram and hence is an approximation to the area

$$\int_0^1 f(x)dx$$

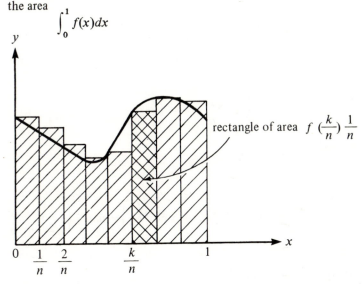

rectangle of area $f\left(\dfrac{k}{n}\right)\dfrac{1}{n}$

This approximation improves as n increases (provided that f is continuous). In the limit, we have that

$$\int_0^1 f(x)dx = \lim_{n \to \infty} \sum_{k=0}^{n-1} f\left(\frac{k}{n}\right)\frac{1}{n}.$$

Applying this result in the case when $f(x) = e^x$, we obtain that

$$\int_0^1 e^x dx = \lim_{n \to \infty} \sum_{k=0}^{n-1} e^{k/n}\frac{1}{n}.$$

Using the formula for the sum of a geometric progression,

$$\sum_{k=0}^{n-1} e^{k/n}\frac{1}{n} = \frac{1}{n} \sum_{k=0}^{n-1} (e^{1/n})^k$$

$$= \frac{1}{n}\left\{\frac{(e^{1/n})^n - 1}{e^{1/n} - 1}\right\} = \frac{1}{n}\left(\frac{e - 1}{e^{1/n} - 1}\right).$$

We therefore have to evaluate,

$$\lim_{n \to \infty}\left\{\frac{e^{1/n} - 1}{1/n}\right\}.$$

But this is the same as

$$\lim_{h \to 0} \left\{ \frac{e^h - 1}{h} \right\} = \lim_{h \to 0} \left\{ \frac{e^h - e^0}{h} \right\}$$

which is the value of the derivative of the exponential function at the point 0. Since this is equal to one, it follows that

$$\int_0^1 e^x dx = \lim_{n \to \infty} \sum_{k=0}^{n-1} e^{k/n} \frac{1}{n} = e - 1.$$

5. (i) By exercise 6.9 (3),

$$\frac{d}{dy}(\sinh^{-1} y) = \frac{1}{\sqrt{(1 + y^2)}}.$$

Hence

$$\int_{-1/2}^{1/2} \frac{dy}{\sqrt{(1 + y^2)}} = [\sinh^{-1} y]_{-1/2}^{1/2} = \sinh^{-1}\left(\frac{1}{2}\right) - \sinh^{-1}\left(-\frac{1}{2}\right).$$

(ii) By exercise 6.9 (4)

$$\frac{d}{dy}(\tanh^{-1} y) = \frac{1}{1 - y^2} \quad (-1 < y < 1).$$

Hence

$$\int_{-1/2}^{1/2} \frac{dy}{1 - y^2} = [\tanh^{-1} y]_{-1/2}^{1/2} = \tanh^{-1}\left(\frac{1}{2}\right) - \tanh^{-1}\left(-\frac{1}{2}\right).$$

(An alternative method is to use partial fractions which will yield the answer in terms of logarithms.)

7. The graph of $y = \sec^2 x$ has an unpleasant *discontinuity* at $x = \pi/2$ which lies in the middle of the range of integration. The fundamental theorem of calculus therefore does *not* apply in this case.

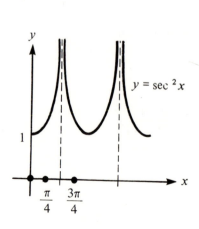

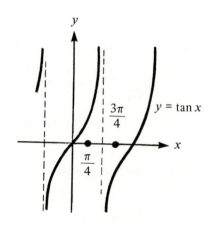

9. We have that

$$x^3 = x(x^2 + x - 2) - x^2 + 2x$$

$$= x(x^2 + x - 2) - (x^2 + x - 2) + 3x - 2.$$

Hence

$$\frac{x^3}{x^2 + x - 2} = (x - 1) + \frac{3x - 2}{x^2 + x - 2}.$$

Alternatively, one can use 'long division' as below:

$$
\begin{array}{r}
x - 1 \\
x^2 + x - 2 \overline{\smash{)}\, x^3} \\
x^3 + x^2 - 2x \\
\overline{-x^2 + 2x} \\
-x^2 - x + 2 \\
\overline{3x - 2.}
\end{array}
$$

Thus

$$\int_2^3 \frac{x^3}{x^2 + x - 2} dx = \int_2^3 (x - 1)dx + \int_2^3 \frac{3x - 2}{x^2 + x - 2} dx.$$

But

$$\frac{3x - 2}{x^2 + x - 2} = \frac{3x - 2}{(x + 2)(x - 1)} = \frac{A}{x + 2} + \frac{B}{x - 1}$$

where

$$A = \lim_{x \to -2} (x + 2)\frac{(3x - 2)}{(x + 2)(x - 1)} = \frac{-8}{-3} = \frac{8}{3}$$

$$B = \lim_{x \to 1} (x - 1)\frac{(3x - 2)}{(x + 2)(x - 1)} = \frac{1}{3}.$$

Hence

$$\int_2^3 \frac{x^3}{x^2 + x - 2} dx = \left[\frac{1}{2}(x - 1)^2 + \frac{8}{3}\log(x + 2) + \frac{1}{3}\log(x - 1)\right]_2^3$$

$$= \left[\frac{1}{2}(x - 1)^2 + \frac{1}{3}\log(x + 2)^8(x - 1)\right]_2^3$$

$$= \left(\frac{1}{2}4^2 + \frac{1}{3}\log 5^8 \cdot 2\right) - \left(\frac{1}{2}1^2 + \frac{1}{3}\log 4^8 \cdot 1\right)$$

$$= 7\tfrac{1}{2} + \frac{1}{3}\log(5^8 \cdot 2) - \frac{1}{3}\log(4^8).$$

11. We have that

$$7 - 2x - x^2 = 8 - 1 - 2x - x^2 = 8 - (1 + x)^2.$$

Hence

$$\int_0^1 \frac{dx}{\sqrt{(7-2x-x^2)}} = \int_0^1 \frac{dx}{\sqrt{(8-(1+x)^2)}}$$

$$= \left[\arcsin \frac{1+x}{\sqrt{8}} \right]_0^1$$

$$= \arcsin \left(\frac{2}{\sqrt{8}} \right) - \arcsin \left(\frac{1}{\sqrt{8}} \right).$$

13. We have that

$$\int_0^{\pi/3} \cos^3 \theta \, d\theta = \int_0^{\pi/3} \frac{\cos 3\theta + 3 \cos \theta}{4} \, d\theta$$

$$= \frac{1}{4} \left[\frac{\sin 3\theta}{3} + 3 \sin \theta \right]_0^{\pi/3}$$

$$= \frac{1}{4} \left(\frac{\sin \pi}{3} + 3 \sin \frac{\pi}{3} \right) = \frac{3\sqrt{3}}{8}.$$

15. An obvious change of variable to try is $x = \sqrt{y}$. Then

$$dx = \tfrac{1}{2} y^{-1/2} dy$$

and so

$$\int_1^2 \frac{\sqrt{(1+\sqrt{y})}}{\sqrt{y}} dy = \int_1^{\sqrt{2}} \sqrt{(1+x)} 2 dx$$

$$= \left[\frac{4}{3}(1+x)^{3/2} \right]_1^{\sqrt{2}}$$

$$= \frac{4}{3}(1+\sqrt{2})^{3/2} - \frac{4}{3}(2)^{3/2}.$$

17. We have that

$$\sin \theta = 2 \sin \frac{1}{2}\theta \cos \frac{1}{2}\theta = 2 \tan \frac{1}{2}\theta \cos^2 \frac{1}{2}\theta$$

$$= 2 \tan \frac{1}{2}\theta \frac{1}{1 + \tan^2 \frac{1}{2}\theta} = \frac{2t}{1+t^2}$$

(because $1 + \tan^2 \frac{1}{2}\theta = \sec^2 \frac{1}{2}\theta$). Thus

$$\cos \theta = \{1 - \sin^2 \theta\}^{1/2} = \left\{ 1 - \frac{4t^2}{(1+t^2)^2} \right\}^{1/2}$$

$$= \left\{ \frac{1 - 2t^2 + t^4}{(1+t^2)^2} \right\}^{1/2} = \left\{ \left(\frac{1-t^2}{1+t^2} \right)^2 \right\}^{1/2} = \frac{1-t^2}{1+t^2}.$$

Note also that, since $t = \tan \frac{1}{2}\theta$,

$$dt = \frac{1}{2}\sec^2 \frac{1}{2}\theta\, d\theta = \frac{1}{2}(1 + \tan^2 \frac{1}{2}\theta)d\theta$$

and so

$$d\theta = \frac{2dt}{1 + t^2}.$$

Using the above results, we obtain that

(i) $\displaystyle\int_{\pi/4}^{\pi/2} \frac{d\theta}{\sin \theta} = \int_{\tan(\pi/8)}^{1} \frac{1 + t^2}{2t} \frac{2dt}{1 + t^2}$

$$= \left[\log t \right]_{\tan(\pi/8)}^{1} = -\log \left\{ \tan \left(\frac{\pi}{8} \right) \right\}.$$

(Note that this result may also be directly obtained from exercise 9.24 (6 (i)).)

(ii) $\displaystyle\int_{0}^{\pi/4} \frac{d\theta}{\cos \theta} = \int_{0}^{\tan(\pi/8)} \frac{1 + t^2}{1 - t^2} \frac{2dt}{1 + t^2}$

$$= \int_{0}^{\tan(\pi/8)} \left(\frac{1}{1 - t} + \frac{1}{1 + t} \right) dt$$

$$= \left[-\log(1 - t) + \log(1 + t) \right]_{0}^{\tan(\pi/8)}$$

$$= \log \left\{ \left| \frac{1 + \tan\left(\dfrac{\pi}{8}\right)}{1 - \tan\left(\dfrac{\pi}{8}\right)} \right| \right\}.$$

(This result can also be expressed in terms of $\tanh^{-1} y$ as in exercise 9.24 (5).)

(iii) $\displaystyle\int_{\pi/6}^{\pi/3} \frac{d\theta}{\sin \theta + \cos \theta} = \int_{\tan(\pi/12)}^{\tan(\pi/6)} \frac{(1 + t^2)}{(1 + 2t - t^2)} \frac{2dt}{(1 + t^2)}$

$$= \int_{\tan(\pi/12)}^{\tan(\pi/6)} \frac{1}{2\sqrt{2}} \left(\frac{1}{t - \sqrt{2} - 1} - \frac{1}{t + \sqrt{2} - 1} \right) dt$$

$$= \left[\frac{1}{2\sqrt{2}} \log \frac{t - \sqrt{2} - 1}{t + \sqrt{2} - 1} \right]_{\tan(\pi/12)}^{\tan(\pi/6)}$$

(Rather than using partial fractions, it is also possible to complete the square by writing $1 + 2t - t^2 = 2 - (t - 1)^2$.)

19. Write $\phi(x) = x^3 + x^2 + x + 1$. Then

$$\int_0^1 \frac{3x^2 + 2x + 1}{x^3 + x^2 + x + 1} dx = \int_0^1 \frac{\phi'(x)}{\phi(x)} dx$$

$$= [\log \phi(x)]_0^1 = \log 4.$$

21. We have that

(i) $\displaystyle\int_0^\pi x^3 \cos x\, dx = [x^3 \sin x]_0^\pi - \int_0^\pi 3x^2 \sin x\, dx$

$$= -[3x^2(-\cos x)]_0^\pi + \int_0^\pi 6x(-\cos x)dx$$

$$= -3\pi^2 - [6x \sin x]_0^\pi + \int_0^\pi 6 \sin x\, dx$$

$$= -3\pi^2 + 6[-\cos x]_0^\pi$$

$$= -3\pi^2 + 12.$$

(ii) $\displaystyle\int_0^X x^2 e^{-x} dx = [x^2(-e^{-x})]_0^X - \int_0^X 2x(-e^{-x})dx$

$$= -X^2 e^{-X} + [2x(-e^{-x})]_0^X - \int_0^X 2(-e^{-x})dx$$

$$= -X^2 e^{-X} - 2Xe^{-X} + [2(-e^{-x})]_0^X$$

$$= -X^2 e^{-X} - 2Xe^{-X} - 2e^{-X} + 2.$$

Exercise 9.37

1. From exercise 6.9 (7) we have that

$$x^{n-1} e^{-x} < \frac{(n+1)!}{x^2} \quad (x > 0).$$

We may therefore take $g(x) = (n+1)!/x^2$ in the comparison test (§9.30) since the integral

$$\int_1^\infty \frac{dx}{x^2}$$

converges (example 9.27). We deduce that

$$\int_1^\infty x^{n-1} e^{-x} dx$$

also converges. The convergence of the integral over the range $[0, \infty)$ then follows.

If $n \geqslant 1$,

$$\Gamma(n+1) = \int_0^\infty x^n e^{-x} dx = [x^n(-e^{-x})]_0^\infty - \int_0^\infty nx^{n-1}(-e^{-x}) dx$$

$$= n \int_0^\infty x^{n-1} e^{-x} dx = n\Gamma(n).$$

This calculation is based on the use of integration by parts in the case of an integral with finite limits of integration (§9.20). We have that

$$\int_0^X x^n e^{-x} dx = [x^n(-e^{-x})]_0^X - \int_0^X nx^{n-1}(-e^{-x}) dx$$

$$= -X^n e^{-X} + \int_0^X nx^{n-1} e^{-x} dx.$$

Thus

$$\int_0^\infty x^n e^{-x} dx = \lim_{X \to \infty} \int_0^X x^n e^{-x} dx$$

$$= -\lim_{X \to \infty} X^n e^{-X} + n \lim_{X \to \infty} \int_0^X x^{n-1} e^{-x} dx$$

$$= 0 + n \int_0^\infty x^{n-1} e^{-x} dx.$$

Note that $X^n e^{-X} \to 0$ as $X \to \infty$ (exercise 6.9 (7)).

Finally, observe that

$$\Gamma(n+1) = n\Gamma(n) = n(n-1)\Gamma(n-1) = \ldots = n!\Gamma(1)$$

and

$$\Gamma(1) = \int_0^\infty e^{-x} dx = \lim_{X \to \infty} \int_0^X e^{-x} dx = \lim_{X \to \infty} (1 - e^{-X}) = 1.$$

3. We begin by noting that $x^y = e^{y \log x}$ (§6.7). Thus

$$\frac{\partial}{\partial y}(x^y) = (\log x)x^y$$

$$\frac{\partial^2}{\partial y^2}(x^y) = (\log x)^2 x^y$$

$$\frac{\partial^n}{\partial y^n}(x^y) = (\log x)^n x^y.$$

It follows that

$$\frac{d^n}{dy^n} \int_0^1 x^y dx = \int_0^1 \frac{\partial^n}{\partial y^n} x^y dx = \int_0^1 (\log x)^n x^y dx.$$

But

$$\int_0^1 x^y dx = \left[\frac{x^{y+1}}{y+1}\right]_0^1 = \frac{1}{y+1}$$

and so

$$\frac{d^n}{dy^n} \int_0^1 x^y dx = \frac{d^n}{dy^n} \frac{1}{y+1} = \frac{(-1)^n n!}{(y+1)^{n+1}}.$$

The result follows on taking $y = 0$.

5. We have that

$$\lim_{n \to \infty} \int_0^1 (n+1)x^n dx = \lim_{n \to \infty} \left[(n+1)\frac{x^{n+1}}{(n+1)}\right]_0^1 = \lim_{n \to \infty} 1 = 1.$$

But, if $0 \leqslant x < 1$,

$$\lim_{n \to \infty} (n+1)x^n = 0$$

and so

$$\int_0^1 \left(\lim_{n \to \infty} (n+1)x^n\right) dx = \int_0^1 0 dx = 0.$$

7. We have that

$$\int_{a(y)}^{b(y)} f(x,y)dx = \int_\xi^{b(y)} f(x,y)dx + \int_{a(y)}^\xi f(x,y)dx$$

$$= \int_\xi^{b(y)} f(x,y)dx - \int_\xi^{a(y)} f(x,y)dx.$$

Hence

$$\frac{d}{dy} \int_{a(y)}^{b(y)} f(x,y)dx = \left(f(b(y),y)b'(y) + \int_\xi^{b(y)} \frac{\partial f}{\partial y}(x,y)dx\right)$$

$$- \left(f(a(y),y)a'(y) + \int_\xi^{a(y)} \frac{\partial f}{\partial y}(x,y)dx\right)$$

and so

$$\frac{d}{dy} \int_{a(y)}^{b(y)} f(x,y)dx = \{f(b(y),y)b'(y) - f(a(y),y)a'(y)\}$$

$$+ \int_{a(y)}^{b(y)} \frac{\partial f}{\partial y}(x,y)dx.$$

9. We know that

$$\frac{1}{1-X} = 1 + X + X^2 + X^3 + \dots$$

provided $-1 < X < 1$ (example 9.26(i)). By taking $X = -x^2$, it follows that

$$\frac{1}{1+x^2} = 1 - x^2 + x^4 - x^6 + \dots$$

provided $-1 < x < 1$. Thus, if $-1 < y < 1$,

$$\arctan y = \int_0^y \frac{dx}{1+x^2}$$

$$= \int_0^y \{1 - x^2 + x^4 - x^6 + \ldots\} dx$$

$$= y - \frac{y^3}{3} + \frac{y^5}{5} - \frac{y^7}{7} + \ldots.$$

11. We use the formulae for the radius convergence R given in §9.35.

(i) $\dfrac{1}{R} = \lim_{n \to \infty} \left| \dfrac{a_{n+1}}{a_n} \right| = \lim_{n \to \infty} \dfrac{(n+1)^2}{n^2} = \lim_{n \to \infty} \left(1 + \dfrac{1}{n}\right)^2 = 1.$

Thus $R = 1$.

(ii) $\dfrac{1}{R} = \lim_{n \to \infty} \left| \dfrac{a_{n+1}}{a_n} \right| = \lim_{n \to \infty} \dfrac{2^{n+1}}{2^n} = 2.$

Thus $R = \frac{1}{2}$.

(iii) $\dfrac{1}{R} = \lim_{n \to \infty} |a_n|^{1/n} = \lim_{n \to \infty} (n^n)^{1/n} = \lim_{n \to \infty} n = \infty.$

The formula we have obtained here is interpreted to mean that $R = 0$ – i.e. the power series converges only for $x = 0$.

Chapter 10. Multiple integrals

Exercise 10.18

1. When changing the order of integration in a repeated integral it is first necessary to identify the region D of integration. Wherever possible, it is a good idea to sketch this region:

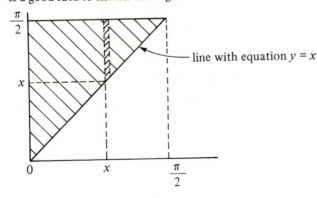

line with equation $y = x$

The region D is found by observing that x varies between 0 and $\pi/2$ while, for any given value of x in this range, y varies between x and $\pi/2$. The diagram below illustrates the situation when the order of integration is reversed.

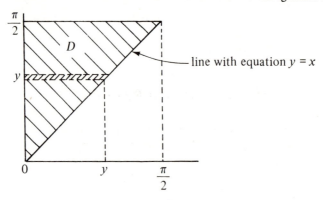

line with equation $y = x$

We obtain that

$$\int_0^{\pi/2} dx \int_x^{\pi/2} \frac{\sin y}{y} dy = \int_0^{\pi/2} dy \int_0^y \frac{\sin y}{y} dx$$

$$= \int_0^{\pi/2} \frac{\sin y}{y} dy \left[x\right]_0^y$$

$$= \int_0^{\pi/2} \sin y\, dy = \left[-\cos y\right]_0^{\pi/2} = 1$$

3. The diagrams show that both repeated integrals are equal to

$$\iint_D f(x, y)dxdy$$

where D is the region illustrated.

curve with equation

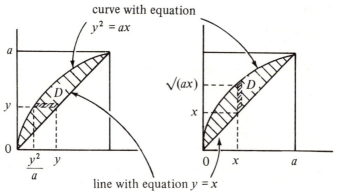

line with equation $y = x$

It follows that

$$\int_0^a y\,dy \int_{y^2/a}^y \frac{dx}{(a-x)\sqrt{(ax-y^2)}} = \int_0^a \frac{dx}{a-x} \int_x^{\sqrt{(ax)}} \frac{y\,dy}{\sqrt{(ax-y^2)}}$$

$$= \int_0^a \frac{dx}{a-x} [-(ax-y^2)^{1/2}]_x^{\sqrt{(ax)}}$$

$$= \int_0^a \frac{dx}{a-x} \{(ax-x^2)^{1/2} - (ax-ax)^{1/2}\}$$

$$= \int_0^a \left\{\frac{x}{a-x}\right\}^{1/2} dx.$$

With some ingenuity this can be evaluated neatly (put $x = a \sin^2 \theta$) but we shall introduce the uninspired and clumsy change of variable

$$u^2 = \frac{x}{a-x}$$

which transforms the integral to

$$2a \int_0^\infty \frac{u^2}{(1+u^2)^2} du.$$

We know that

$$\int_0^\infty \frac{dy}{1+y^2} = [\arctan y]_0^\infty = \frac{\pi}{2}.$$

Hence, writing $y = u\sqrt{t}$, we obtain that

$$\int_0^\infty \frac{du}{1+tu^2} = \frac{\pi}{2} t^{-1/2}.$$

Thus

$$\int_0^\infty -\frac{u^2}{(1+tu^2)^2} du = \int_0^\infty \frac{\partial}{\partial t}\left(\frac{1}{1+tu^2}\right) du$$

$$= \frac{d}{dt} \int_0^\infty \frac{du}{1+tu^2}$$

$$= \frac{d}{dt}\left(\frac{\pi}{2} t^{-1/2}\right) = -\frac{\pi}{4} t^{-3/2}.$$

Taking $t = 1$, we deduce that

$$2a \int_0^\infty \frac{u^2}{(1+u^2)^2} du = \frac{1}{2}\pi a.$$

5. We have that

$$\frac{\partial}{\partial b}\int_a^\infty I(\alpha, b)d\alpha = \frac{\partial}{\partial b}\int_a^\infty d\alpha \int_0^\infty (\sin bx)e^{-\alpha x}dx$$

$$= \frac{\partial}{\partial b}\int_0^\infty (\sin bx)dx \int_a^\infty e^{-\alpha x}d\alpha$$

$$= \frac{\partial}{\partial b}\int_0^\infty (\sin bx)dx \left[-\frac{e^{\alpha x}}{x}\right]_a^\infty$$

$$= \frac{\partial}{\partial b}\int_0^\infty \left(\frac{\sin bx}{x}\right)e^{-ax}dx$$

$$= \int_0^\infty \frac{\partial}{\partial b}\left(\frac{\sin bx}{x}\right)e^{-ax}dx$$

$$= \int_0^\infty \left(\frac{x\cos bx}{x}\right)e^{-ax}dx = J(a, b).$$

Thus

$$J(a, b) = \frac{\partial}{\partial b}\int_a^\infty \frac{b}{\alpha^2 + b^2}d\alpha$$

$$= \frac{\partial}{\partial b}\left[\arctan\frac{\alpha}{b}\right]_a^\infty$$

$$= \frac{\partial}{\partial b}\left\{\frac{\pi}{2} - \arctan\frac{a}{b}\right\}$$

$$= -\frac{1}{1 + \frac{a^2}{b^2}}\left(-\frac{a}{b^2}\right) = \frac{a}{a^2 + b^2}.$$

7. The curves $x^2 + 2y^2 = 1, x^2 + 2y^2 = 4, y = 2x$ and $y = 5x$ are mapped onto $u = 1, u = 4, v = 2$ and $v = 5$ respectively. This allows us to identify Δ as the region indicated below:

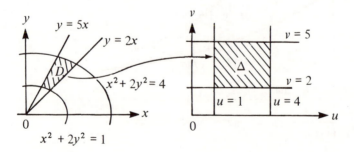

The appropriate Jacobian is

$$\frac{\partial(x,y)}{\partial(u,v)} = \left\{\frac{\partial(u,v)}{\partial(x,y)}\right\}^{-1} = \begin{vmatrix} \dfrac{\partial u}{\partial x} & \dfrac{\partial u}{\partial y} \\ \dfrac{\partial v}{\partial x} & \dfrac{\partial v}{\partial y} \end{vmatrix}^{-1} = \begin{vmatrix} 2x & 4y \\ -\dfrac{y}{x^2} & \dfrac{1}{x} \end{vmatrix}^{-1}$$

$$= \left(2 + \frac{4y^2}{x^2}\right)^{-1} = \frac{1}{2(1 + 2v^2)}.$$

Hence

$$\iint_D \frac{y}{x}\,dx\,dy = \iint_\Delta \frac{v}{2(1 + 2v^2)}\,du\,dv$$

$$= \int_1^4 du \int_2^5 \frac{v}{2(1 + 2v^2)}\,dv$$

$$= \int_1^4 du \frac{1}{2}\left[\frac{1}{4}\log(1 + 2v^2)\right]_2^5$$

$$= \frac{1}{8}\log\frac{51}{9}[u]_1^4 = \frac{3}{8}\log\left(\frac{17}{3}\right).$$

9. The region D in the $(x,y)^T$ plane which is mapped onto the square Δ in the $(u,v)^T$ plane enclosed by the lines $u = 1, u = e, v = 1$ and $v = e$ is sketched below. Note that the curves need to be drawn rather carefully if the correct shape for D is to be obtained. Even more care is necessary in the consideration of the critical points.

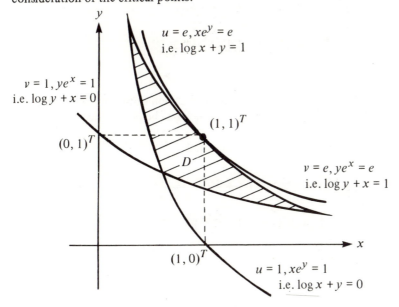

The Jacobian of f is

$$\frac{\partial(u, v)}{\partial(x, y)} = \begin{vmatrix} e^y & xe^y \\ ye^x & e^x \end{vmatrix} = e^{x+y} \begin{vmatrix} 1 & x \\ y & 1 \end{vmatrix} = (1 - xy)e^{x+y}$$

and hence f has critical points on the line $xy = 1$. It is important that none of these critical points lie *inside* the region D. Suppose that a point $(x, y)^T$ on the curve $xy = 1$ lies inside D. Then $x > 0, y > 0, xy = 1, xe^y < e$ and $ye^x < e$. Thus

$$1 = xy < e^{1-y}e^{1-x} = e^{2 - 1/x - x}$$

and so, taking logarithms,

$$0 < 2 - \frac{1}{x} - x$$

$$x + \frac{1}{x} < 2.$$

But it is easy to check that the minimum of the left-hand side for $x > 0$ is equal to 2. Thus the inequality cannot hold. It follows that no critical point of f lies *inside* D (although the critical point $(1, 1)^T$ is on the boundary of D).

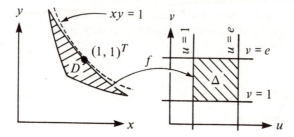

In practice of course the problem of the location of the critical points is usually ignored until it is noticed that the answers being obtained cannot possibly be correct.

We now consider the given double integral

$$\iint_D (x^2y^3 - x^3y^4)e^{4x+3y}\,dxdy$$

$$= \iint_\Delta x^2y^3(1 - xy)e^{4x+3y}\{(1 - xy)e^{x+y}\}^{-1}\,dudv$$

$$= \iint_\Delta x^2y^3 e^{3x}e^{2y}\,dudv$$

$$= \iint_\Delta u^2 v^3 \, du \, dv$$

$$= \int_1^e u^2 \, du \int_1^e v^3 \, dv$$

$$= \left[\frac{1}{3} u^3\right]_1^e \left[\frac{1}{4} v^4\right]_1^e$$

$$= \frac{1}{12}(e^3 - 1)(e^4 - 1).$$

11. The region Δ in this case is the brick shaped region determined by the inequalities

$$\left.\begin{array}{c} 0 \leqslant r \leqslant R \\[4pt] -\pi < \theta \leqslant \pi \\[4pt] 0 \leqslant \phi < \pi. \end{array}\right\}$$

To see this, one must observe that, as $(r, \theta, \phi)^T$ varies subject to these inequalities, the corresponding vector $(x, y, z)^T$ visits each point in the sphere of radius R and centre $(0, 0, 0)^T$ exactly once.

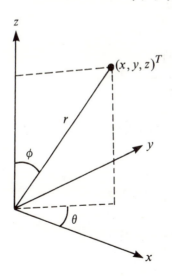

The appropriate Jacobian is

$$\frac{\partial(x, y, z)}{\partial(r, \theta, \phi)} = \begin{vmatrix} \dfrac{\partial x}{\partial r} & \dfrac{\partial x}{\partial \theta} & \dfrac{\partial x}{\partial \phi} \\[2mm] \dfrac{\partial y}{\partial r} & \dfrac{\partial y}{\partial \theta} & \dfrac{\partial y}{\partial \phi} \\[2mm] \dfrac{\partial z}{\partial r} & \dfrac{\partial z}{\partial \theta} & \dfrac{\partial z}{\partial \phi} \end{vmatrix}$$

$$= \begin{vmatrix} \cos \theta \sin \phi & -r \sin \theta \sin \phi & r \cos \theta \cos \phi \\ \sin \theta \sin \phi & r \cos \theta \sin \phi & r \sin \theta \cos \phi \\ \cos \phi & 0 & -r \sin \phi \end{vmatrix}$$

$$= \cos \phi \begin{vmatrix} -r \sin \theta \sin \phi & r \cos \theta \cos \phi \\ r \cos \theta \sin \phi & r \sin \theta \cos \phi \end{vmatrix}$$

$$- r \sin \phi \begin{vmatrix} \cos \theta \sin \phi & -r \sin \theta \sin \phi \\ \sin \theta \sin \phi & r \cos \theta \sin \phi \end{vmatrix}$$

$$= r^2 \cos^2 \phi \sin \phi \begin{vmatrix} -\sin \theta & \cos \theta \\ \cos \theta & \sin \theta \end{vmatrix}$$

$$- r^2 \sin^3 \phi \begin{vmatrix} \cos \theta & -\sin \theta \\ \sin \theta & \cos \theta \end{vmatrix}$$

$$= r^2 \cos^2 \phi \sin \phi (-\sin^2 \theta - \cos^2 \theta)$$

$$- r^2 \sin^3 \phi (\cos^2 \theta + \sin^2 \theta)$$

$$= -r^2 \sin \phi (\cos^2 \phi + \sin^2 \phi) = -r^2 \sin \phi.$$

Thus

$$\iiint_D dx\,dy\,dz = \iiint_\Delta r^2 \sin \phi\, dr\,d\theta\,d\phi.$$

(Note that we take the *modulus* of the Jacobian.)
   Hence

$$\iiint_D dx\,dy\,dz = \int_0^R r^2\,dr \int_{-\pi}^{\pi} d\theta \int_0^{\pi} \sin \phi\,d\phi$$

$$= \int_0^R r^2\,dr \int_{-\pi}^{\pi} d\theta\, [-\cos \phi]_0^{\pi}$$

$$= 2 \int_0^R r^2\,dr\, {}_{-\pi}^{\pi}[\theta]$$

$$= 4\pi \left[\frac{1}{3}r^3\right]_0^R = \frac{4\pi R^3}{3}.$$

13. See §12.13.

Chapter 11. Differential and difference equations (I)

Exercise 11.23
1. (i) order 3; degree 1

(ii) order 2; degree 2.

3.
(i) $(4+x)\dfrac{dy}{dx} = y^3$

$$y^{-3}dy = \frac{dx}{4+x}$$

$$\int y^{-3}dy = \int \frac{dx}{4+x} + c$$

$$-\frac{1}{2}y^{-2} = \log(4+x) + c.$$

(ii) $(xy+y)\dfrac{dy}{dx} = (x-xy)$

$$\frac{ydy}{1-y} = \frac{xdx}{x+1}$$

$$\int \frac{ydy}{1-y} = \int \frac{xdx}{x+1} + c.$$

But

$$\int \frac{ydy}{1-y} = -\int\left(\frac{1-y}{1-y} - \frac{1}{1-y}\right)dy$$

$$= -y - \log(1-y)$$

$$\int \frac{xdx}{x+1} = \int\left(\frac{1+x}{1+x} - \frac{1}{1+x}\right)dx$$

$$= x - \log(1+x).$$

Hence

$$-y - \log(1-y) = x - \log(1+x) + c$$

or

$$\left(\frac{1+x}{1-y}\right) = ae^{x+y}.$$

(iii) $x^2 y \dfrac{dy}{dx} = e^y$

$$ye^{-y} dy = \dfrac{dx}{x^2}$$

$$\int ye^{-y} dy = \int x^{-2} dx + c.$$

But

$$\int ye^{-y} dy = -ye^{-y} - \int -e^{-y} dy$$

$$= -ye^{-y} - e^{-y}$$

and so

$$-ye^{-y} - e^{-y} = -x^{-1} + c$$

or

$$x(y+1)e^{-y} = 1 + ax.$$

(iv) $\dfrac{1}{y} \dfrac{dy}{dx} = (\log x)(\log y)$

$$\dfrac{dy}{y \log y} = (\log x) dx$$

$$\int \dfrac{dy}{y \log y} = \int (\log x) dx + c.$$

But

$$\int \dfrac{dy}{y \log y} = \log (\log y),$$

$$\int (\log x) dx = x \log x - x.$$

(Check these results by differentiating.) Hence

$$\log \log y = x \log x - x + c.$$

5. (i) We have that

$$\dfrac{\partial M}{\partial y} = 3x^2 ; \quad \dfrac{\partial N}{\partial x} = 3x^2$$

and hence the equation is exact.
(ii) We have that

$$\dfrac{\partial M}{\partial y} = 3; \quad \dfrac{\partial N}{\partial x} = -1$$

and so the equation is not exact.

7. Only (ii) is linear. We write this in the form

$$\frac{dy}{dx} - \frac{3y}{x} = x^4.$$

The integrating factor is

$$\mu = e^{\int Pdx} = e^{-\int 3/x\, dx} = e^{-3\log x} = \frac{1}{x^3}.$$

Multiplying through by μ we obtain

$$\frac{1}{x^3}\frac{dy}{dx} - \frac{3y}{x^4} = x$$

$$\frac{d}{dx}\left(\frac{1}{x^3} y\right) = x$$

$$\frac{y}{x^3} = \int x\, dx + c$$

$$y = \frac{1}{2}x^5 + cx^3.$$

9. We first write the equation in standard form as below.

$$\frac{dy}{dx} - \frac{2xy}{1+x^2} = x^2.$$

The integrating factor is

$$\mu = e^{\int Pdx} = e^{-\int 2x/(1+x^2)dx} = e^{-\log(1+x^2)} = \frac{1}{1+x^2}.$$

Multiplying through by μ, we obtain

$$\frac{1}{1+x^2}\frac{dy}{dx} - \frac{2x}{(1+x^2)^2}y = \frac{x^2}{1+x^2}$$

$$\frac{d}{dx}\left(\frac{y}{1+x^2}\right) = 1 - \frac{1}{1+x^2}$$

$$\frac{y}{1+x^2} = \int\left(1 - \frac{1}{1+x^2}\right)dx + c$$

$$y = (1+x^2)(x - \arctan x) + c(1+x^2).$$

11. We have that $du = \sec^2 x dx$ and $dv = -\sin y dy$ and so the given equation reduces to

$$(3u - 2v)du - u dv = 0.$$

This is a linear equation which we write in the standard form

$$\frac{dv}{du} + \frac{2}{u} v = 3.$$

The integrating factor is

$$\mu = e^{\int P du} = e^{\int (2/u) du} = e^{2 \log u} = u^2.$$

Multiplying through by u^2, we obtain

$$u^2 \frac{dv}{du} + 2uv = 3u^2$$

$$\frac{d}{du}(u^2 v) = 3u^2$$

$$u^2 v = u^3 + c$$

i.e.

$$(\tan x)^2 (\cos y) = (\tan x)^3 + c.$$

13. From example 6.20 we know that the given change of variable reduces the partial differential equation to the form

$$\frac{\partial f}{\partial u} = \frac{1}{4}.$$

The general solution is therefore

$$f = \tfrac{1}{4} u + \phi(v)$$

i.e.

$$f = \tfrac{1}{4}(x^2 + y^2) + \phi(x^2 - y^2)$$

where ϕ is an arbitrary function.

15. We have that

$$\frac{\partial M}{\partial y} = 6x^2 y + 2x; \quad \frac{\partial N}{\partial x} = 6x^2 y + 2x$$

and hence the equation is exact. This means that a function f can be found such that

$$\frac{\partial f}{\partial x} = 3x^2 y^2 + 2yx = M$$

$$\frac{\partial f}{\partial y} = 2x^3 y + x^2 = N.$$

The general solutions of these partial differential equations are

$$f = x^3 y^2 + yx^2 + \phi(y)$$
$$f = x^3 y^2 + x^2 y + \psi(x).$$

We need to choose the arbitrary functions ϕ and ψ so that these equations hold simultaneously. The choice $\phi(y) = 0$ and $\psi(x) = 0$ suffices. It follows that the general solution of the original equation is

$$x^3 y^2 + yx^2 = c.$$

Exercise 11.30
1. (i) one (ii) one (iii) one (iv) two (v) two (vi) three.

3. (i) $y_{x+1} - y_x = 2x$

$$\Delta y_x = 2x$$

$$y_x = x(x-1) + c.$$

(ii) $y_{x+1} - y_x = 2^x$

$$\Delta y_x = 2^x$$

$$y_x = 2^x(2-1)^{-1} + c$$

$$y_x = 2^x + c.$$

(iii) $y_{x+1} + 2y_x = x$

Using the method of example 11.28, we write

$$(-2)^{-x-1} y_{x+1} - (-2)^{-x} y_x = (-2)^{-x-1} x$$

$$\Delta((-2)^{-x} y_x) = -\tfrac{1}{2} x (-\tfrac{1}{2})^x.$$

Thus

$$(-\tfrac{1}{2})^x y_x = -\tfrac{1}{2} \{ -\tfrac{4}{9}(-\tfrac{1}{2})^{x+1} - \tfrac{2}{3} x(-\tfrac{1}{2})^x \} + c$$

$$y_x = -\tfrac{1}{9} + \tfrac{1}{3} x + c(-2)^x.$$

Note that the result

$$\Delta \left(\frac{-r^{x+1}}{(r-1)^2} + \frac{xr^x}{(r-1)} \right) = xr^x \quad (r \ne 1)$$

is quoted in exercise 11.30 (2).

(iv) $y_{x+1} - 2y_x = 3^x$

$$2^{-x-1} y_{x+1} - 2^{-x} y_x = 2^{-x-1} 3^x$$

$$\Delta(2^{-x}y_x) = \tfrac{1}{2}(\tfrac{3}{2})^x$$

$$2^{-x}y_x = \tfrac{1}{2}(\tfrac{3}{2}-1)^{-1}(\tfrac{3}{2})^x + c$$

$$y_x = 3^x + c2^x.$$

5. Making the indicated change of variable, we obtain that

$$2^{z_{x+1}}2^{z_x} = 2$$

i.e.

$$z_{x+1} + z_x = 1$$

$$(-1)^{-x-1}z_{x+1} - (-1)^{-x}z_x = (-1)^{-x-1}$$

$$\Delta((-1)^{-x}z_x) = (-1)^{-x-1}$$

$$(-1)^{-x}z_x = -(-2)^{-1}(-1)^{-x} + c$$

$$z_x = \tfrac{1}{2} + c(-1)^x$$

$$y_x = \log_2(\tfrac{1}{2} + c(-1)^x).$$

7. (i) The appropriate difference equation is

$$y_{x+1} = y_x + \frac{10}{100}y_x$$

$$y_{x+1} - (1.1)y_x = 0.$$

This is solved as in example 11.28. We obtain the general solution

$$y_x = c(1.1)^x.$$

To evaluate c the boundary condition

$$y_{100} = 1\,000\,000$$

is used. Then

$$c = \frac{1\,000\,000}{(1.1)^{100}} = 72.57.$$

But $y_0 = c$ and so the amount originally invested is $72.57 (approximately).

(ii) In this problem the boundary condition is $y_0 = 10$ and so $c = 10$. The average held over the first X years is

$$\frac{1}{X}(y_0 + y_1 + \ldots + y_{X-1}) = \frac{10}{X}((1.1)^0 + \ldots + (1.1)^{X-1})$$

$$= \frac{10}{X}\left(\frac{(1.1)^X - 1}{(1.1) - 1}\right) = \frac{100}{X}((1.1)^X - 1)$$

We therefore need to solve the equation

$$15.93 = \frac{100}{X}((1.1)^X - 1)$$

Trial and error shows the solution to be $X = 10$. The amount held at the beginning of the $(X + 1)$th year is therefore

$$y_{11} = 10(1.1)^{11} = 28.53.$$

Chapter 12. Complex numbers

Exercise 12.6

1. (i) $z = \{(1 - 2i) + (-2 + 3i)\}(-3 - 4i) = (-1 + i)(-3 - 4i)$

$$= (3 + 4) + i(-3 + 4) = 7 + i.$$

Thus $\mathcal{R}z = 7$ and $\mathcal{I}z = 1$.

(ii) $z = (1 - 2i)(-2 + 3i) - (-3 - 4i) = (-2 + 6) + i(4 + 3)$

$$+ 3 + 4i = (4 + 3) + i(7 + 4) = 7 + 11i.$$

Thus $\mathcal{R}z = 7$ and $\mathcal{I}z = 11$.

(iii) $z = \dfrac{(1 - 2i) - (-3 - 4i)}{(-2 + 3i)} = \dfrac{4 + 2i}{-2 + 3i} = \dfrac{(4 + 2i)(-2 - 3i)}{(-2 + 3i)(-2 - 3i)}$

$$= \frac{(-8 + 6) + i(-4 - 12)}{(4 + 9) + i(-6 + 6)} = \frac{-2 - 16i}{13}.$$

Thus $\mathcal{R}z = -\frac{2}{13}$ and $\mathcal{I}z = -\frac{16}{13}$.

$$z = \frac{(1 - 2i)(-3 - 4i)}{(-2 + 3i)} = \frac{(-3 - 8) + i(6 - 4)}{(-2 + 3i)}$$

$$= \frac{(-11 + 2i)(-2 - 3i)}{(-2 + 3i)(-2 - 3i)} = \frac{(22 + 6) + i(-4 + 33)}{(4 + 9) + i(-6 + 6)} = \frac{28 + 29i}{13}.$$

Thus $\mathcal{R}z = \frac{28}{13}$ and $\mathcal{I}z = \frac{29}{13}$.

The moduli of z_1, z_2 and z_3 are given by

$$|z_1| = \{1^2 + (-2)^2\}^{1/2} = \{1 + 4\}^{1/2} = \sqrt{5}$$

$$|z_2| = \{(-2)^2 + 3^2\}^{1/2} = \{4 + 9\}^{1/2} = \sqrt{13}$$

$$|z_3| = \{(-3)^2 + (-4)^2\}^{1/2} = \{9 + 16\}^{1/2} = 5.$$

Some diagrams are helpful in calculating the arguments of z_1, z_2 and z_3.

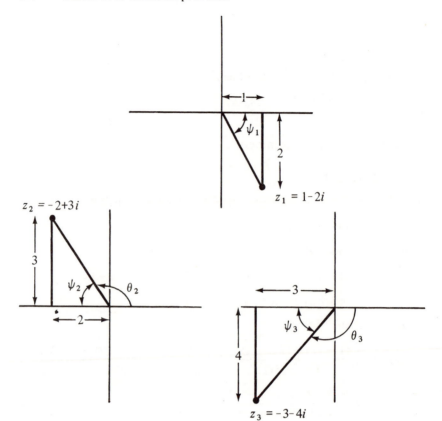

$$z_1 = 1 - 2i$$

$$z_2 = -2 + 3i$$

$$z_3 = -3 - 4i$$

We have that

$$\arg z_1 = \theta_1 = -\psi_1 = -\arctan 2$$

$$\arg z_2 = \theta_2 = \pi - \psi_2 = \pi - \arctan\left(\tfrac{3}{2}\right)$$

$$\arg z_3 = \theta_3 = -\pi + \psi_3 = -\pi + \arctan\left(\tfrac{4}{3}\right).$$

3.
(i) $\dfrac{1}{i} = \dfrac{i}{i^2} = -i.$

(ii) $i^4 = i^2 \cdot i^2 = (-1)(-1) = 1.$

(iii) $i^3 = i^2 \cdot i = (-1)i = -i.$

(iv) $|1 + i| = \{1^2 + 1^2\}^{1/2} = \sqrt{2}; \quad \arg(1 + i) = \arctan\left(\dfrac{1}{1}\right) = \dfrac{\pi}{4}.$

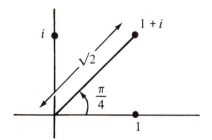

5. DeMoivre's theorem is certainly true for $n = 1$. We assume it to be true for $n = k$ and deduce its truth for $n = k + 1$. This is enough to prove the theorem true for all $n = 1, 2, 3, \ldots$ because of the principle of induction.

We then have that

$$
\begin{aligned}
(\cos \theta + i \sin \theta)^{k+1} &= (\cos \theta + i \sin \theta)^k (\cos \theta + i \sin \theta) \\
&= (\cos k\theta + i \sin k\theta)(\cos \theta + i \sin \theta) \\
&= (\cos k\theta \cos \theta - \sin k\theta \sin \theta) \\
&\quad + i(\sin k\theta \cos \theta + \cos k\theta \sin \theta) \\
&= \cos (k + 1)\theta + i \sin (k + 1)\theta
\end{aligned}
$$

as required

In the case $n = 3$,

$$
\begin{aligned}
\cos 3\theta + i \sin 3\theta &= (\cos \theta + i \sin \theta)^3 \\
&= \cos^3 \theta + 3 \cos^2 \theta (i \sin \theta) \\
&\quad + 3 \cos \theta (i \sin \theta)^2 + (i \sin \theta)^3 \\
&= (\cos^3 \theta - 3 \cos \theta \sin^2 \theta) \\
&\quad + i(3 \cos^2 \theta \sin \theta - \sin^3 \theta).
\end{aligned}
$$

Equating real parts, we obtain that

$$
\begin{aligned}
\cos 3\theta &= \cos^3 \theta - 3 \cos \theta \sin^2 \theta \\
&= \cos^3 \theta - 3 \cos \theta (1 - \cos^2 \theta) \\
&= 4 \cos^3 \theta - 3 \cos \theta.
\end{aligned}
$$

7. (i) $\overline{z_1 + z_2} = \overline{(x_1 + x_2) + i(y_1 + y_2)} = (x_1 + x_2) - i(y_1 + y_2)$
$$= (x_1 - iy_1) + (x_2 - iy_2) = \bar{z}_1 + \bar{z}_2.$$

(ii) $\overline{z_1 z_2} = \overline{(x_1 + iy_1)(x_2 + iy_2)} = \overline{(x_1 x_2 - y_1 y_2) + i(y_1 x_2 + x_1 y_2)}$
$$= (x_1 x_2 - y_1 y_2) - i(y_1 x_2 + x_1 y_2)$$

$$= (x_1 x_2 - (-y_1)(-y_2)) + i((-y_1)x_2 + x_1(-y_2))$$
$$= (x_1 - iy_1)(x_2 - iy_2) = \bar{z}_1 \bar{z}_2.$$

(iii) $\dfrac{z + \bar{z}}{2} = \dfrac{(x + iy) + (x - iy)}{2} = \dfrac{2x}{2} = x.$

(iv) $(\bar{\bar{z}}) = \overline{x - iy} = x + iy = z.$

(v) $\mathcal{R}\left(\dfrac{z}{w}\right) = \mathcal{R}\left(\dfrac{z\bar{w}}{w\bar{w}}\right) = \mathcal{R}\left(\dfrac{z\bar{w}}{|w|^2}\right) = \dfrac{1}{|w|^2}\mathcal{R}(z\bar{w}).$

(For the final calculation, see exercise 12.6 (8 (iv)).)

Exercise 12.15

1. (i) To solve $z^2 = i$, we write $i = e^{i\pi/2}$. The roots may then be expressed as

$$z_0 = e^{i\pi/4} = \left(\frac{1 + i}{\sqrt{2}}\right)$$

$$z_1 = e^{i\pi/4}e^{i\pi} = -e^{i\pi/4} = -\left(\frac{1 + i}{\sqrt{2}}\right).$$

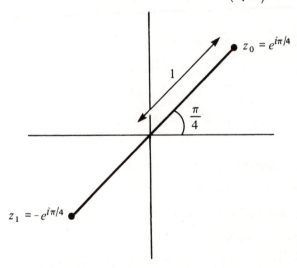

(ii) To solve $z^5 = -1$, we write $-1 = e^{\pi i}$. The roots may then be expressed as

$$z_0 = e^{\pi i/5}$$

$$z_1 \doteq e^{3\pi i/5}$$

$$z_2 = e^{5\pi i/5} = e^{\pi i} = -1$$

$$z_3 = e^{7\pi i/5}$$

$$z_4 = e^{9\pi i/5}$$

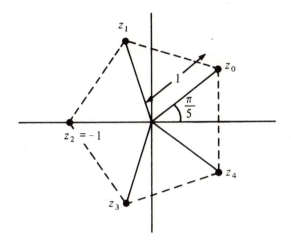

3. (i) $z^3 - 3z^2 - 10z = z(z^2 - 3z - 10)$

$$= z(z-5)(z+2).$$

(ii) $P(z) = z^4 + z^3 - 3z^2 - 5z - 2.$

Observe that $P(0) = -2 \neq 0, P(1) = -15 \neq 0$, but $P(-1) = 0$. Thus $z = -1$ is a root (and $z = 0$ and $z = 1$ are not). It follows that

$$P(z) = (z+1)Q(z).$$

The coefficient of the polynomial $Q(z)$ may be obtained from the following table

n	4	3	2	1	0	-1
p_n	1	1	-3	-5	-2	0
q_n	0	1	0	-3	-2	0

We have that $Q(z) = z^3 - 3z - 2$. Again $z = -1$ is a root since $Q(-1) = 0$.
Thus
$$Q(z) = (z+1)R(z).$$

The coefficients of $R(z)$ may be obtained by continuing our table as indicated below:

q_n	0	1	0	-3	-2	0
r_n	0	0	1	-1	-2	0

This leaves us with $R(z) = z^2 - z - 2 = (z-2)(z+1)$. The required factorization is therefore
$$P(z) = (z+1)^3(z-2).$$

5. Using the results of exercise 12.6 (7),
$$\overline{P(z)} = \bar{a}_n \bar{z}^n + \bar{a}_{n-1}\bar{z}^{n-1} + \ldots + \bar{a}_1 \bar{z} + \bar{a}_0$$
$$= a_n \bar{z}^n + a_{n-1}\bar{z}^{n-1} + \ldots + a_1 \bar{z} + a_0$$
$$= P(\bar{z}).$$

Note that $\bar{a}_k = a_k$ because a_k is real.
   Thus, if $P(\zeta) = 0$, then
$$P(\bar{\zeta}) = \overline{P(\zeta)} = \bar{0} = 0.$$

It follows that the roots which are not real fall into *pairs* of the form ζ and $\bar{\zeta}$. But a polynomial of *odd* degree has an *odd* number of roots. Hence at least one of these must be real.

   How is this argument affected by the consideration that some roots may be repeated?

Chapter 13. Differential and difference equations (II)

Exercise 13.23

1. (i) $(3D^3 + 5D^2 - 2D)y = 0$
$$D(3D^2 + 5D - 2)y = 0$$
$$D(3D - 1)(D + 2)y = 0.$$

The general solution is

$$y = A + Be^{x/3} + Ce^{-2x}.$$

(ii) $P(D)y = (D^3 - 4D + D + 6)y = 0.$

We have that $z = -1$ is a root of $P(z) = 0$. Thus $P(z) = (z+1)Q(z) = (z+1)(z^2 - 5z + 6)$ by the table below. It follows that

$$P(D)y = (D+1)(D-2)(D-3)y = 0$$

and so the general solution is

$$y = Ae^{-x} + Be^{2x} + Ce^{3x}.$$

n	3	2	1	0	-1
p_n	1	-4	1	6	0
q_n	0	1	5	-6	0

(iii) $P(D)y = (D^4 - 2D^3 - 3D^2 + 4D + 4)y = 0.$

Again $z = -1$ is a root of $P(z) = 0$. Thus $P(z) = (z+1)Q(z) = (z+1)(z^3 - 3z^2 + 4)$ by the table below. Observe that $z = -1$ is also a root of $Q(z) = z^3 - 3z^2 + 4$ and so $Q(z) = (z+1)R(z) = (z+1)(z^2 - 4z + 4) = (z+1)(z-2)^2$. Hence

$$P(D)y = (D+1)^2(D-2)^2 y = 0$$

and so the general solution is

$$y = A_0 e^{-x} + A_1 x e^{-x} + B_0 e^{2x} + B_1 x e^{2x}.$$

n	4	3	2	1	0	-1
p_n	1	-2	-3	4	4	0
q_n	0	1	-3	0	4	0
r_n	0	0	1	-4	4	0

(iv) $P(D)y = (D^5 + 3D^4 + 7D^3 + 13D^2 + 12D + 4)y = 0.$

Again $z = -1$ is a root of $P(z)$ and so $P(z) = (z+1)Q(z) = (z+1)(z^4 + 2z^3 + 5z^2 + 8z + 4)$ from the table below. But $z = -1$ is also a root of $Q(z)$. Hence $Q(z) = (z + 1)R(z) = (z + 1)(z^3 + z^2 + 4z + 4)$. Yet again $z = -1$ is a root of $R(z)$ and so $R(z) = (z + 1)S(z) = (z + 1)(z^2 + 4) = (z + 1)(z - 2i)(z + 2i)$. Thus

$$P(D)y = (D + 1)^3(D - 2i)(D + 2i)y = 0.$$

The general solution is therefore

$$y = A_0e^{-x} + A_1xe^{-x} + A_2x_2e^{-x} + Be^{2ix} + Ce^{-2ix}.$$

If only real solutions are of interest, we may rewrite this as

$$y = A_0e^{-x} + A_1xe^{-x} + A_2x^2e^{-x} + b\cos 2x + c\sin 2x.$$

n	5	4	3	2	1	0
p_n	1	3	7	13	12	4	0
q_n	0	1	2	5	8	4	0
r_n	0	0	1	1	4	4	0
s_n	0	0	0	1	0	4	0

3. (i) $P(E)y = (E^2 + 2E - 3)y = 0$

$$(E + 3)(E - 1)y = 0.$$

The general solution is

$$y = A(-3)^x + B.$$

(ii) Write $z_x = y_{x+2}$. The equation then reduces to $z_{x+2} + 2z_{x+1} + z_x = 0$ – i.e.

$$P(E)y = (E^2 + 2E + 1)z = 0$$

$$(E + 1)^2z = 0$$

which has general solution

$$z = A_0(-1)^x + A_1x(-1)^x$$

Thus

$$y_{x+2} = A_0(-1)^x + A_1x(-1)^x \quad (x = 0, 1, 2, \ldots)$$

and we may choose $y_0 = B$ and $y_1 = C$ where B and C are arbitrary.

(iii) $P(E)y = (E^2 + E + 1)y = 0$.

The roots of $P(z) = z^2 + z + 1 = 0$ may be obtained from the formula

$$z = \frac{-1 \pm \sqrt{(1-4)}}{2}.$$

We obtain that

$$P(E)y = \left(E - \left(-\frac{1}{2} + \frac{i\sqrt{3}}{2}\right)\right)\left(E - \left(-\frac{1}{2} - \frac{i\sqrt{3}}{2}\right)\right)y = 0$$

and so the general solution is

$$y = A\left(-\frac{1}{2} + \frac{i\sqrt{3}}{2}\right)^x + B\left(-\frac{1}{2} - \frac{i\sqrt{3}}{2}\right)^x.$$

This may be expressed more neatly by observing that

$$-\frac{1}{2} + \frac{i\sqrt{3}}{2} = e^{2\pi i/3}$$

$$-\frac{1}{2} - \frac{i\sqrt{3}}{2} = e^{-2\pi i/3}$$

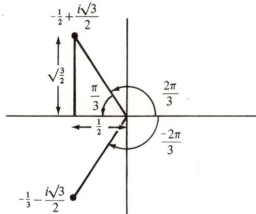

The general solution is therefore

$$y = Ae^{2\pi ix/3} + Be^{-2\pi ix/3}.$$

If only real solutions are of interest, this may be rewritten as

$$y = a\cos\left(\frac{2\pi x}{3}\right) + b\sin\left(\frac{2\pi x}{3}\right).$$

(iv) $P(E)y = (E^3 - 3E^2 + 9E + 13)y = 0.$

Once more $z = -1$ is a root of $P(z)$. Thus $P(z) = (z + 1)Q(z) = (z + 1)(z^2 - 4z + 13)$. The roots of $Q(z)$ may be obtained from the formula

$$z = 2 \pm \sqrt{(4 - 13)}.$$

We therefore have that

$$P(E)y = (E + 1)(E - 2 - 3i)(E - 2 + 3i)y = 0.$$

The general solution is therefore

$$y = A(-1)^x + B(2 + 3i)^x + C(2 - 3i)^x$$

which we may rewrite as

$$y = A(-1)^x + B13^{x/2} e^{i\theta x} + C13^{x/2} e^{-i\theta x}$$

or, if only real solutions are of interest, as

$$y = A(-1)^x + b13^{x/2} \cos(\theta x) + c13^{x/2} \sin(\theta x)$$

where $\theta = \arctan(\frac{3}{2})$.

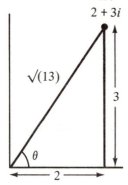

n	3	2	1	0	-1
p_n	1	-3	9	13	0
q_n	0	1	-4	13	0

5. In each case, the complementary function is the general solution of

$$P(D)y = (3D^3 + 5D^2 - 2D)y = 0$$

i.e.

$$D(3D - 1)(D + 2)y = 0.$$

The complementary function is therefore

$$y = A + Be^{x/3} + Ce^{-2x}.$$

(i) $D(3D-1)(D+2)y = e^x.$

Since $q = e^x$ is a solution of $(D-1)q = 0$, a particular solution η of our equation is to be found among the solutions of

$$(D-1)D(3D-1)(D+2)y = 0$$

which has general solution

$$y = A + Be^{x/3} + Ce^{-2x} + Ke^x.$$

From the form of the complementary function, we know that A, B and C may be chosen freely and the simplest choice is $A = B = C = 0$. We therefore seek a particular solution of the form

$$\eta = Ke^x.$$

To evaluate K we substitute in $(3D^3 + 5D^2 - 2D)y = e^x$. This gives

$$3Ke^x + 5Ke^x - 2Ke^x = e^x$$

$$6K = 1$$

$$K = \tfrac{1}{6}.$$

The required general solution is therefore

$$y = \tfrac{1}{6}e^x + A + Be^{x/3} + Ce^{-2x}.$$

(ii) $D(3D-1)(D+2)y = \cos x = \tfrac{1}{2}(e^{ix} + e^{-ix}).$

Since $q = \cos x$ is a solution of $(D+i)(D-i)q = 0$, a particular solution η of our equation is to be found among the solutions of

$$(D+i)(D-i)D(3D-1)(D+2)y = 0$$

which has general solution

$$y = A + Be^{x/3} + Ce^{-2x} + Je^{-ix} + Ke^{ix}$$

which we rewrite as

$$y = A + Be^{x/3} + Ce^{-2x} + j\cos x + k\sin x.$$

As in (i) we take $A = B = C = 0$ and seek a particular solution of the form

$$\eta = j\cos x + k\sin x.$$

Substituting in $(3D^3 + 5D^2 - 2D)y = \cos x$, we obtain

$$j(3\sin x - 5\cos x + 2\sin x) + k(-3\cos x - 5\sin x - 2\cos x) = \cos x$$

$$(5j - 5k)\sin x + (-5j - 5k - 1)\cos x = 0.$$

Thus $5j - 5k = 0$ and $-5j - 5k - 1 = 0$ – i.e.

$$\left. \begin{array}{l} j = -\dfrac{1}{10} \\[2mm] k = -\dfrac{1}{10} \end{array} \right\} .$$

The required general solution is therefore

$$y = -\frac{1}{10}\cos x - \frac{1}{10}\sin x + A + Be^{x/3} + Ce^{-2x}.$$

(iii) $D(3D - 1)(D + 2)y = x.$

Since $q = x$ is a solution of $D^2 q = 0$, a particular solution η of our equation is to be found among the solutions of

$$D^3(3D - 1)(D + 2)y = 0$$

which has general solution

$$y = A_0 + A_1 x + A_2 x^2 + Be^{x/3} + Ce^{-2x}.$$

As in (i) we take $A_0 = B = C = 0$ and seek a particular solution of the form

$$\eta = A_1 x + A_2 x^2.$$

Substituting in $(3D^3 + 5D^2 - 2D)y = x$, we obtain

$$-2A_1 + A_2(5 \cdot 2 - 2 \cdot 2x) = x$$

$$(-2A_1 + 10A_2) + x(-4A_2 - 1) = 0.$$

Thus $A_1 = 5A_2$ and $4A_2 + 1 = 0$ – i.e.

$$\left. \begin{array}{l} A_1 = -\tfrac{5}{4} \\[2mm] A_2 = -\tfrac{1}{4}. \end{array} \right\}$$

The required general solution is therefore

$$y = -\tfrac{5}{4}x - \tfrac{1}{4}x^2 + A + Be^{x/3} + Ce^{-2x}.$$

(iv) $D(3D - 1)(D + 2)y = e^{-2x}.$

Since $q = e^{-2x}$ is a solution of $(D + 2)q = 0$, a particular solution η of our equation is to be found among the solutions of

$$D(3D - 1)(D + 2)^2 y = 0$$

which has general solution

$$y = A + Be^{x/3} + C_0 e^{-2x} + C_1 xe^{-2x}.$$

As in (i) we take $A = B = C_0 = 0$ and seek a particular solution of the form

$$\eta = C_1 xe^{-2x}.$$

To evaluate C_1 we substitute in $(3D^3 + 5D^2 - 2D)y = e^{-2x}$. Since

$$D(xe^{-2x}) = e^{-2x} - 2xe^{-2x}$$

$$D^2(xe^{-2x}) = -2e^{-2x} - 2e^{-2x} + 4xe^{-2x} = -4e^{-2x} + 4xe^{-2x}$$

$$D^3(xe^{-2x}) = 8e^{-2x} + 4e^{-2x} - 8xe^{-2x} = 12e^{-2x} - 8xe^{-2x}$$

we obtain that

$$3C_1(12e^{-2x} - 8xe^{-2x}) + 5C_1(-4e^{-2x} + 4xe^{-2x}) - 2C_1(e^{-2x} - 2xe^{-2x})$$

$$= e^{-2x}$$

$$e^{-2x}\{C_1(36 - 20 - 2) - 1\} + xe^{-2x}\{-24 + 20 + 4\} = 0$$

and hence

$$C_1 = \frac{1}{14}.$$

The required general solution is therefore

$$y = \frac{1}{14} xe^{-2x} + A + Be^{x/3} + Ce^{-2x}.$$

7. $P(E)y = (E^3 - 5E^2 + 8E - 4)y = 2^x + x.$

We begin by finding the complimentary sequence. This is the general solution of

$$P(E)y = (E^3 - 5E^2 + 8E - 4)y = 0.$$

Since $z = 1$ is a root of $P(z)$, we obtain that $P(z) = (z - 1)Q(z) = (z - 1)(z^2 - 4z + 4) = (z - 1)(z - 2)^2.$

n	3	2	1	0	-1
p_n	1	-5	8	-4	0
q_n	0	1	-4	4	0

Thus
$$P(E)y = (E-1)(E-2)^2 y = 0$$

which has general solution

$$y = A + B2^x + Cx2^x.$$

This is the required complimentary sequence.

We now seek a particular solution η. Since $q = 2^x + x$ is a solution of $(E-2)(E-1)^2 q = 0$, η is to be found among the solutions of

$$(E-1)^3 (E-2)^3 y = 0$$

which has general solution

$$y = A_0 + A_1 x + A_2 x^2 + B_0 2^x + B_1 x 2^x + B_2 x^2 2^x.$$

From the form of the complimentary sequence, we know that A_0, B_0 and B_1 may be chosen freely and it is simplest to choose $A_0 = B_0 = B_1 = 0$. We then seek a particular solution of the form

$$\eta = A_1 x + A_2 x^2 + B_2 x^2 2^x.$$

Substituting in $(E^3 - 5E^2 + 8E - 4)y = 2^x + x$, we obtain

$$A_1 \{(x+3) - 5(x+2) + 8(x+1) - 4x\}$$
$$+ A_2 \{(x+3)^2 - 5(x+2)^2 + 8(x+1)^2 - 4x^2\}$$
$$+ B_2 \{(x+3)^2 2^{x+3} - 5(x+2)^2 2^{x+2} + 8(x+1)^2 2^{x+1}$$
$$- 4x^2 2^x\} = 2^x + x.$$

i.e.
$$(A_1 - 3A_2) + (2A_2 - 1)x + (8B_2 - 1)2^x = 0.$$

Thus $A_1 = 3A_2$, $2A_2 = 1$ and $8B_2 = 1$ – i.e.

$$\left. \begin{array}{l} A_1 = \tfrac{3}{2} \\ A_2 = \tfrac{1}{2} \\ B_2 = \tfrac{1}{8}. \end{array} \right\}$$

The required general solution is therefore

$$y = \tfrac{3}{2}x + \tfrac{1}{2}x^2 + \tfrac{1}{8}x^2 2^x + A + B_0 2^x + B_1 x 2^x.$$

9. The general solution of the differential equation of question 1(i) is

$$y = A + Be^{x/3} + Ce^{-2x}.$$

We must find values of A, B and C so as to satisfy the given boundary conditions. If $B > 0$, $y \to +\infty$ as $x \to \infty$. If $B < 0$, $y \to -\infty$ as $x \to +\infty$. Since

$y \to 1$ as $x \to +\infty$, we must have that $B = 0$. Then $y \to A$ as $x \to +\infty$ and hence $A = 1$. Finally we need that $y(0) = 0$. This means that

$$0 = 1 + Ce^0 = 1 + C$$

and so $C = -1$.

The required solution is therefore

$$y = 1 - e^{-2x}.$$

11. The equilibrium solution $\mathbf{y} = \mathbf{\eta}$ is found by solving

$$\begin{pmatrix} 1 & -1 \\ 1 & 1 \end{pmatrix} \begin{pmatrix} \eta_1 \\ \eta_2 \end{pmatrix} + \begin{pmatrix} -1 \\ -1 \end{pmatrix} = \begin{pmatrix} 0 \\ 0 \end{pmatrix}$$

which gives $\eta_1 = 1$ and $\eta_2 = 0$.

To discuss stability, we need the eigenvalues of the matrix. These are found by solving

$$\begin{vmatrix} 1 - \lambda & -1 \\ 1 & 1 - \lambda \end{vmatrix} = (1 - \lambda)^2 + 1 = 0$$

which gives $\lambda_1 = 1 + i$ and $\lambda_2 = 1 - i$. Since $\mathcal{R} \lambda_1 > 0$ (and $\mathcal{R} \lambda_2 > 0$), the system is *unstable*.

The general solution may be found by computing eigenvectors $\mathbf{u}$ and $\mathbf{v}$ corresponding to λ_1 and λ_2. We have that

$$\begin{pmatrix} 1 & -1 \\ 1 & 1 \end{pmatrix} \begin{pmatrix} u_1 \\ u_2 \end{pmatrix} = (1 + i) \begin{pmatrix} u_1 \\ u_2 \end{pmatrix}; \quad \begin{pmatrix} 1 & -1 \\ 1 & 1 \end{pmatrix} \begin{pmatrix} v_1 \\ v_2 \end{pmatrix} = (1 - i) \begin{pmatrix} v_1 \\ v_2 \end{pmatrix}$$

i.e.   i.e.

$$\left. \begin{array}{c} u_1 - u_2 = u_1 + iu_1 \\ u_1 + u_2 = u_2 + iu_2 \end{array} \right\} \qquad \left. \begin{array}{c} v_1 - v_2 = v_1 - iv_1 \\ v_1 + v_2 = v_2 - iv_2 \end{array} \right\}$$

and so   and so

$$u_1 = iu_2 \qquad\qquad\qquad v_1 = -iv_2.$$

We make the choice $u_2 = 1$ and $v_2 = 1$ which yields the eigenvectors

$$\mathbf{u} = \begin{pmatrix} i \\ 1 \end{pmatrix} \quad \text{and} \quad \mathbf{v} = \begin{pmatrix} -i \\ 1 \end{pmatrix}$$

The general solution is then

$$y_1 = 1 + \alpha i e^{(1+i)x} - \beta i e^{(1-i)x}$$
$$y_2 = 0 + \alpha e^{(1+i)x} + \beta e^{(1-i)x}$$

which we may express in the form

$$y_1 = 1 + (A \cos x - B \sin x)e^x$$
$$y_2 = 0 + (B \cos x + A \sin x)e^x.$$

We now take account of the boundary conditions $y_1(0) = 1$ and $y_2(0) = 1$. These give

$$\left. \begin{array}{l} 1 = 1 + A \\ 1 = 0 + B. \end{array} \right\}$$

The required solution is therefore

$$y_1 = 1 - e^x \sin x$$
$$y_2 = e^x \cos x.$$

Note that, far from converging to the equilibrium solution, these oscillate infinitely.

Index